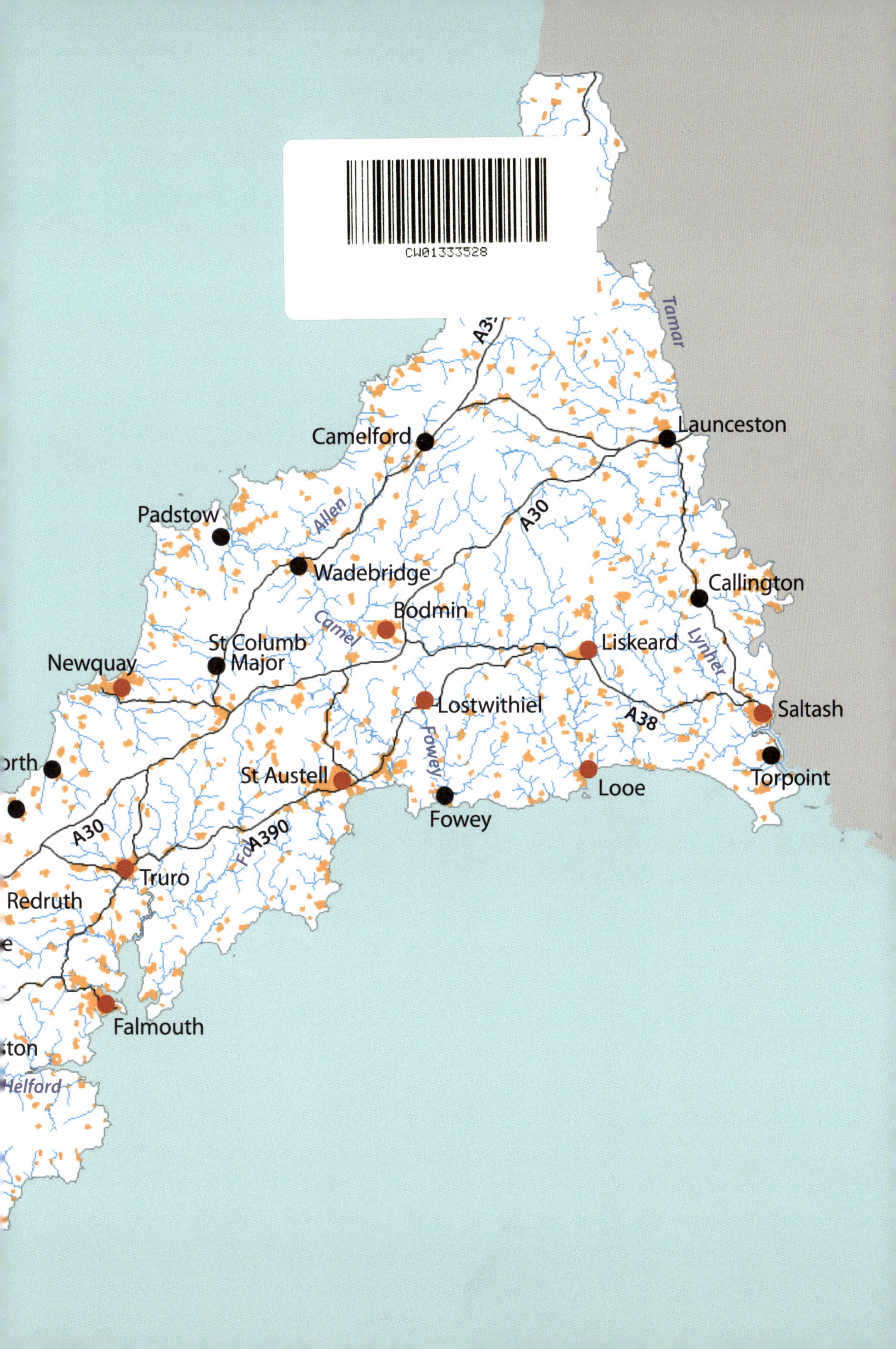
Tamar
Launceston
Camelford
Padstow
Allen
A30
Wadebridge
Callington
Bodmin
Camel
St Columb
Major
Liskeard
Lynher
Newquay
Lostwithiel
Saltash
A38
Fowey
St Austell
Torpoint
Looe
Fowey
A30
A390
Truro
Redruth
Falmouth

BUTTERFLIES of CORNWALL

ATLAS FOR THE TWENTY-FIRST CENTURY

Sarah Board, Tristram Besterman, Bob Dawson, Dick Goodere, Maggie Goodere and Cerin Poland

Cornwall Butterfly Conservation

Butterfly Conservation
Saving butterflies, moths and our environment
Cornwall Butterfly Conservation

pisces publications

Published 2021 by Pisces Publications for Cornwall Butterfly Conservation

Copyright © Cornwall Butterfly Conservation (2021)
Copyright © of the photographs remains with the photographers

All rights reserved. No part of this publication may be reproduced, stored in a retrieval system or transmitted, in any form or by any means electronic, mechanical, photocopying, recording or otherwise, without the prior permission of the publishers.

First published 2021.

British-Library-in-Publication Data
A catalogue record for this book is available from the British Library.

ISBN 978-1-913994-01-3

Designed and published by Pisces Publications

Visit our bookshop
www.naturebureau.co.uk/bookshop/

Pisces Publications is the imprint of NatureBureau,
2C The Votec Centre, Hambridge Lane, Newbury, Berkshire RG14 5TN
www.naturebureau.co.uk

Printed and bound in the UK by Gomer Press Ltd

MIX
Paper from responsible sources
FSC
www.fsc.org   FSC® C114687

All the paper used is FSC mixed source certified. All inks are colour-fast, vegetable derived (no petrochemicals involved). Each page is dried under UV in seconds, saving energy on hours of costly heating. All printing is digital, no metal plates involved. Production is by Gomer Press in South Wales, whose environmental credentials can be inspected at http://gomerprinting.co.uk/environment/ They are accredited to the Green Dragon Environmental Standard Level 2 https://www.greenbusinesscentre.org.uk/green-dragon-environmental-standard

Front cover: Silver-studded Blue *Plebejus argus* [Shaun Poland]
Phoenix United Mine, Bodmin Moor [Tristram Besterman]
Long Cove and Will's Rock, Porthcothan [Tristram Besterman]
Back cover: Botallack Mine, Botallack [Steve Batt]

Epigraph opposite: Rough Tor, Bodmin Moor [Tristram Besterman]

"The approach to Rough Tor started well enough. We crunched swiftly over King Arthur's Downs. The winter sun filled the moor and Rough Tor filled the sky ahead. Then we cut across a low valley, the ground softened and we were up to our calves in bog. We looked at each other, thirty metres of wet ground between us. 'Ah!' called Richard. 'Right!' I called back. The logic is always to go on – hopping boldly from tussock to tussock, even as the tussocks grow further apart, as they quiver at your footfall. It is a logic that should be firmly resisted. *Retrace your steps! Take the long way!*"

Philip Marsden, *Rising Ground: A Search for the Spirit of Place*

Dedicated to the unseen army of enthusiasts
who have submitted records of butterflies in Cornwall
over many years. This book was made possible
by their selfless devotion to a common cause.

Contents

vii   **Foreword by HRH Prince of Wales, Duke of Cornwall**

viii **A personal view by Kelly Uren, County Transect Co-ordinator**

ix   **Supporters**

x   **Photographic credits**

1   Chapter 1
    Introduction and acknowledgements

12 Chapter 2
    The Cornish environment

30 Chapter 3
    Recording and conserving butterflies

46 Chapter 4
    Butterfly species in Cornwall

50 HESPERIIDAE
50 Dingy Skipper
54 Grizzled Skipper
58 Small Skipper
62 Large Skipper

66 PIERIDAE
66 Orange-tip
70 Large White
74 Small White
78 Green-veined White
82 Clouded Yellow
86 Brimstone

90 NYMPHALIDAE
90 Wall
94 Speckled Wood
98 Small Heath
102 Ringlet
106 Meadow Brown
110 Gatekeeper

114 Marbled White
118 Grayling
122 Pearl-bordered Fritillary
128 Small Pearl-bordered Fritillary
132 Silver-washed Fritillary
136 Dark Green Fritillary
140 Red Admiral
144 Painted Lady
148 Peacock
152 Small Tortoiseshell
156 Comma
160 Marsh Fritillary
168 Heath Fritillary

172 LYCAENIDAE
172 Small Copper
176 Purple Hairstreak
180 Green Hairstreak
184 White-letter Hairstreak
188 Holly Blue
192 Silver-studded Blue
198 Brown Argus
202 Common Blue

206 Chapter 5
    Occasional visitors and rarities

214 Chapter 6
    Butterflies of the Isles of Scilly

222 Chapter 7
    Key sites for butterflies

245 Chapter 8
    Further reading

249 **References cited in the text**

254 **Locations mentioned in the text**

258 **Larval foodplants found in Cornwall**

261 **Index of butterflies**

CLARENCE HOUSE

It gives me great pleasure to introduce *Butterflies of Cornwall*, an important and timely publication, as we face mounting threats to our natural environment. Written and compiled in a county renowned for its rich, varied habitats and wild places, this volume draws on decades of observations and images made by the volunteers of Cornwall Butterfly Conservation. Through the systematic work of such environmental enthusiasts, our knowledge of the distribution of butterfly species, their preferred habitats and the state of their populations in Cornwall is beautifully presented.

I believe that it is through our appreciation and knowledge of these creatures that we will ultimately be successful in protecting and, I dearly hope, restoring their populations and, most importantly of course, their habitats. There are some encouraging examples of the early signs of recovery of national rarities, such as the Marsh Fritillary and the Silver-studded Blue, with descriptions of some well protected colonies. The rarest of all is the Heath Fritillary, now found in only one site in Cornwall, in whose conservation, I am proud to say, the Duchy of Cornwall plays a continuing role.

Indeed, each of us has a role to play in how we care for the environment if we are to ensure that future generations can enjoy its butterflies. Inevitably, there is much more work to be done in ensuring that other species continue to thrive. So, I can only hope and pray that this volume helps both those with lifelong expertise, and those with a new interest in butterflies, to extend their knowledge and enthusiasm for this most important marker of the health of our environment. The future of all species, including our own, depends on it.

A personal view

My love of nature began in childhood when I transformed the family garden into a wildlife haven. Armed with my trusty Reader's Digest field guides, I kept a careful log of the birds and other animals I could identify in our local area. So, the urge to record started early!

I feel very fortunate to have grown up in a county with some of the most stunning landscapes in Britain. Being out in the field, immersed in the environment, is one of the great pleasures of volunteering with Cornwall Butterfly Conservation. As transect recorder, I am building on decades of meticulous recording by so many people before me. The transect database is the most important and reliable source of data for tracking the fortunes of Cornwall's butterflies over time across the County. A weighty responsibility certainly, but also deeply satisfying. Every year, I look forward to the start of the transect recording season, which lasts from April to September. During these six months, my world becomes consumed by all things Lepidoptera, and I really would not want it any other way.

I feel very strongly that nature needs us now more than ever. Without stepping in to protect the natural world, it – and we – simply will not survive. Collectively and individually, we are responsible for the planet. So, I feel empowered to be one small part of a vast network of environmental activists, connecting across Cornwall, throughout Britain and beyond. Countless thousands of largely unseen volunteers, doing their bit for biodiversity, not out of a sense of duty but for the sheer love of it. What a privilege to be part of that!

Kelly Uren
Co-ordinator for Cornwall transect recording
and Wider Countryside Butterfly Survey
Cornwall Butterfly Conservation

Supporters

Cornwall Butterfly Conservation (CBC) is a county branch of the national charity, Butterfly Conservation. The publication of *Butterflies of Cornwall: Atlas for the Twenty-first Century* is a CBC initiative, whose delivery relied on a sustained collaborative effort by its members and supporters.

To publish the book, CBC had to raise £10,000, a target that was exceeded in 2021, during the coronavirus pandemic. That response to CBC's fundraising appeal was both humbling and heart-warming, and a generous award from the Postcode Local Trust got our project off to a good start.

Many personal contributions are celebrated as species sponsors, whose names are recorded at the head of the first page of the species that they chose to sponsor, in chapter 4.

In addition, we place on record our deep gratitude and appreciation to the following for their generous support:

Premium sponsors
- Sheila Ashby
- Philip and Faith Hambly
- In memory of Alison Norris

Major donors
- Sue Allen and family
- Anonymous
- Marcus Bede Colville
- Cooperesource
- Sue and Steve Cotton
- Mr Paul Drayson
- James Fowler
- L. Fowler and J. Hilary
- A local business in Gwithian
- Roger Hooper
- Kew Devon Cattle
- Morcom Construction
- Jo Poland
- Shaun Poland
- In memory of Ian Pryor
- Sophie Shaw
- The Smile Centre Local Giving Fund
- Dave Thomas
- Three Mile Beach
- Grace Twiston-Davies and David Rymer
- David White
- Marion Williams

Supported by players of PEOPLE'S POSTCODE LOTTERY

Awarded funds from POSTCODE LOCAL TRUST

Photographic credits

Mary Atkinson 184
Mark Bailey 32
Steve Batt 18 upper, 20, 21, 23, 26, 28 upper, 59 left, 61, 83, 91, 98, 104, 137 all, 174 upper, 179, 182, 204, 233, 236, 241, 242
Ian Bennallick 17, 117
Tristram Besterman 2, 4, 6 all, 9 all, 14, 15 all, 18 lower, 37, 45, 49 all, 75, 76, 77, 89, 97, 113, 116, 126, 135, 163, 187, 188, 225, 226, 228, 229, 232, 234, 238
Sarah Board 24 upper, 42 lower, 65, 121, 131, 162, 230, 235, 240, 244
Bob Dawson 12 lower, 28 lower, 29, 214, 215, 216, 217 all, 218, 221
Jerry Dennis 12 upper, 25, 93, 175, 227, 243
Sally Foster 38
Maggie Goodere 34 upper
Martin Goodey 219
Philip Hambly 82
Beth Harper 10, 40
Chris Jackson (Getty Images) vii
Adam Jones 51 right
Steve Jones 59 right, 63, 64 left, 69, 86, 102, 106, 112, 114, 118, 122, 136, 141, 150 right, 157 left, 169, 176, 181, 192, 198, 206, 210
Malcolm Pinch 22, 43 upper, 84
Cerin Poland 19, 24 lower, 42 upper, 43 lower, 44, 50, 51 left, 52 all, 53, 54, 55 all, 57, 62, 64 right, 66, 78, 79, 80, 87, 88, 90, 92, 94, 95, 99, 100, 103, 107, 110, 119, 125 all, 128, 132, 134 left, 142 left, 149, 155, 156, 160, 161, 165 all, 166, 168, 172, 174 lower, 177, 180, 183, 189, 191, 194, 196, 199, 200, 201, 202, 203 all, 231, 237, 239, 260
Jo Poland 34 lower
Shaun Poland 1, 30, 31, 58, 67 all, 70, 71, 72, 73 all, 74, 105, 108 all, 115, 124, 133, 139, 140, 142 right, 143, 144, 145, 146, 148, 150 left, 152, 153, 154, 157 right, 158, 173, 190, 195, 208
Dave Thomas 33, 111, 129, 134 right
Steve Townsend 27, 222
Charlotte Uren viii, 36
John Walters 68

CHAPTER 1
Introduction and acknowledgements

A Brimstone nectars on Red Campion

Inspiration

It is spring, and a Brimstone butterfly flies fleetingly into view. A little boy watches in fascination as the pale yellow visitor explores the garden they share for a few, magical minutes. Both will be famous one day. The boy will grow up to win a Nobel Prize in 2001 and be knighted for his work on genetics. And the butterfly that inspired him will be celebrated in the pages of his book *What Is Life?* In it, Sir Paul Nurse reflects on why the butterfly came into his childhood garden:

> "Was it hungry, looking for somewhere to lay its eggs, or perhaps being chased by a bird?... Of course I do not know why that butterfly was behaving as it did, but what I can say is that it was interacting with its world and then taking action. And to do that, it had to manage information. Information is at the centre of the butterfly's existence and indeed at the centre of all life… When it was flying about, its senses were building up a detailed picture of my garden. Its eyes were detecting light; its antennae were sampling molecules of the different chemical substances in its vicinity; and its hairs were monitoring vibrations in the air… The butterfly was combining many different sources of information and using them to make decisions with meaningful consequences for its future." (Nurse, 2020: 118–20)

And information, in the form of records, is at the centre of our book, also gathered from diverse sources. That information is interpreted by conservationists to have 'meaningful consequences' for the future of butterflies and all the species that share the Cornish peninsula, including, of course, us.

Us? Why us? Butterflies may have a certain allure, to be sure, but what has their future to do with ours? The short answer is, of course, everything.

We're all in this together

One thing that the coronavirus, or more precisely SARS-CoV-2, has taught us is that we forget our interconnectedness with the natural world at our peril. It is tragic that it takes a pandemic to bring the point home, but this particular virus is only the latest, and certainly will not be the last – or most lethal – in a long history of pathogens that began in animals and 'jumped' to humans. Margaret Wild, an expert on wildlife diseases, points out,

"These diseases don't transmit to humans often, and when they do, it's typically when we push natural systems by destroying animal habitat or crowding different species together with people in a marketplace… When we keep ecosystems healthy, we really reduce the risk of disease." (Bodin, 2020).

When, in 2018, I accepted the invitation to help with *Butterflies of Cornwall*, like most people on the planet I had never heard of a coronavirus. 2020 changed that, for all of us. Lockdown during a glorious English spring was a strange but remarkably compelling time to be thinking about wildlife and our relationship with the natural world. Along uncannily quiet roads, beneath skies free from vapour trails, nature took a long, deep breath and rewarded us richly. With our right to roam curtailed by lockdown rules, walks in local lanes were a necessary, soul-soothing, sensory feast of flowers, insects and birdsong. And in our restricted ambit we really paid attention, as we

Orange-tip nectars on Bluebell, roadside bank, Liskeard, April 2020

opened our eyes and ears to the intricate details of what had always been there. Along Cornish hedges, nodding bluebells fed Orange-tip butterflies, whose blue-green larvae a few weeks later could be found feeding on wayside Garlic Mustard. In May, nature's jaunty tricolour of Red Campion, Bluebell and Greater Stitchwort felt personal and universal. People lucky enough to live in the countryside reconnected with nature and rediscovered its value in their lives.

When we engage with nature, we feel better. The therapeutic value of walking along a country lane, in woods, around meadows rich in wild flowers, along a shore pounded by the sea or across open moorland is self-evident to anyone who has experienced it; the effect is even recognised in national policy. "Spending time in the natural environment – as a resident or a visitor – improves our mental health and… wellbeing. It can reduce stress, fatigue, anxiety and depression. It can help boost immune systems…" (HM Government, 2018: 71).

Cornwall provides all of this, and more. Visiting any of the best sites to see butterflies in the County, listed in chapter 7, can take you to parts of Cornwall that might be new to many. And for those who wish to venture offshore, the Isles of Scilly and their butterflies are described in chapter 6. An excursion with butterflies in mind adds something important to the experience. In his autobiographical account, *In Pursuit of Butterflies*, Matthew Oates, that doyen of butterflying, describes his book as, "Above everything else… a eulogy and a paean… on the beauty and wonder of British butterflies" (Oates, 2016: 17).

"All the different life forms that share our planet are our relatives" (Nurse, 2020: 73). The DNA that we share with birds, bluebells and butterflies speaks also of a shared evolutionary inheritance that stretches back billions of years. But you don't need to be a geneticist to understand our deep relationship with all other living things. Butterflies are telling us, silently and powerfully, about the state of the environment on which they, and we, depend.

The butterfly's tale

"Butterflies and moths have been recognised by the Government as indicators of biodiversity. Their fragility makes them quick to react to change so their struggle to survive is a serious warning about our environment" (Butterfly Conservation b). The UK Butterfly Monitoring Scheme (UKBMS) explains further:

"Butterflies have short life cycles and thus react quickly to environmental changes. Their limited dispersal ability, larval foodplant specialisation and close-reliance on the weather and climate make many butterfly species sensitive to fine-scale changes. Recent research has shown that butterflies have declined more rapidly than birds and plants emphasising their potential role as indicators. Butterflies occur in all of the main terrestrial habitat types in the UK and so have the potential to act as indicators for a wide range of species and habitats. Unlike most other groups of insects, butterflies are well-documented, their taxonomy is understood, they are easy to recognise and we have a wealth of information on their ecology and life-histories." (UKBMS)

Globally, it is predicted that "up to one million plant and animal species face extinction, many within decades, because of human activities" (Tollefson, 2019). In the UK, it is reckoned that 41% of UK species studied have declined since 1970 (Hayhow *et al.*, 2019). Data on butterfly populations from the UKBMS and the Wider Countryside Butterfly Survey (WCBS) tell a similar story. Habitat specialists, that is, butterflies with a highly specialised ecology, such as the Pearl-bordered Fritillary and Silver-studded Blue, declined by 59% between 1976 and 2019. Whereas the much less choosy wider countryside species, like the Small Tortoiseshell and Gatekeeper, have not fared so badly, declining by 20% over the same period (Defra, 2020a). A nationwide trigger for these declines is thought to have been the severe drought of 1976, which caused populations to crash. Habitat specialists were unable to recover to pre-1976 levels because of their greater vulnerability to habitat loss, whereas the wider countryside species have proved more resilient.

Butterflies – and many other pollinators and insects – depend on plants for their survival, both as a nectar source for the adult and for larval foodplants, so it is unsurprising that they have been particularly badly hit by the loss of 97% of wild flower meadows since the 1930s (Hayhow et al., 2019). Because of their importance to the butterfly's life cycle, larval foodplants are listed by name with the associated butterfly species near the end of the book.

The rate at which the biosphere is being degraded by human activity is unprecedented, not just in terms of human history, but in geological timescales. The causes are known and are felt everywhere, from Hayle to Honolulu. Burning fossil fuels heats the planet and stresses ecosystems; chemical waste and plastics pollute our waterways, oceans, the air we breathe and the soils on which all terrestrial life depends; intensive agriculture clears forests, destroys meadows and hedgerows, kills plants and animals with pesticides and trashes the soil; development displaces the biosphere with tarmac, concrete, glass and steel.

For most butterflies, there is, literally, nowhere else to go, and the data tell us that, starkly.

Putting Cornwall's butterflies on the map

The picture of Cornwall's butterflies that emerges from the data presented in this atlas is cause for concern. Just over three-quarters of the County's 37 resident or regular visiting butterflies have declined in abundance in South-west England between 1976 and 2018, a slightly steeper decline than the overall English trend. Most at risk are Cornwall's endangered species, designated originally between 1995 and 1999 in the UK Biodiversity Action Plan (BAP). Eleven of Cornwall's butterflies are UK BAP species listed in Section 41 of *The Natural Environment and Rural Communities Act* 2006 as 'species of principal importance', sometimes known simply as 'Section 41 species'. These are the Dingy Skipper, Grizzled Skipper, Pearl-bordered Fritillary, Small Pearl-bordered Fritillary, White-letter Hairstreak, Grayling, Wall, Small Heath, Silver-studded Blue, Marsh Fritillary and Heath Fritillary.

Small Pearl-bordered Fritillary on Germander Speedwell, South West Coast Path, West Penwith

A Cornish atlas

Atlas, in ancient Greek mythology, was condemned to bear the weight of the heavens for all eternity. He is usually depicted as an old, bearded guy, bent under the weight of his burden. Some might recognise the description. The load on his shoulders is represented by a globe decorated with celestial constellations. His name became synonymous with a book of maps because of the custom of placing an illustration of Atlas at the front, a tradition that began in the sixteenth century.

Atlas connects with Cornwall today. The Atlantic Ocean is literally the 'Sea of Atlas', a dynamic force that impacts the peninsula, its environment and its butterflies.

The primary subject of this book, which is subtitled *Atlas for the Twenty-first Century* is the ecology, distribution and population trends in Cornwall of the 37 species of butterfly that reside in or are regular visitors to the County, explored in depth in chapter 4. It is not intended as a guide to identification: a list of such books is given in chapter 8. Each species account in chapter 4 is supported by distribution maps and the best places to see the butterfly in Cornwall. Chapter 2 describes the County's environment and habitats, structured according to Natural England's eight National Character Areas mapped for Cornwall and the Isles of Scilly. In addition, a description with directions to the 20 best sites to see butterflies in Cornwall is provided in chapter 7, and, for completeness, a gazetteer lists in alphabetical order more than 150 locations mentioned in the text, with their national grid reference. Specially commissioned maps of the County provide an additional aid to navigation.

Featured on the front cover of this book is the Silver-studded Blue, one of the most iconic of Cornwall's legally protected butterflies. It is by no means the rarest: that title goes to the Heath Fritillary, whose sole Cornish location in the Tamar Valley is one of only a handful of sites in the UK where the species is found. The nationally rare Silver-studded Blue seems to be doing well in Cornwall: over the 10 years 2009 to 2018 it has become more widely distributed and has been seen in greater numbers. A challenge for conservation is to find out why. Understanding the drivers for success is just as important as knowing what drives numbers down. We know a great deal about the habitat on which the Silver-studded Blue depends, and in several of its strongholds careful habitat management meets its needs. As far as we know, in Cornwall a key part of its ecology involves an intimate and indispensable relationship with two species of ant. This is discussed further in chapter 4.

In chapter 3, the central importance of recording is described in detail, and how this provides the raw data that can be analysed to discern population trends over time for each species, and to map the changes in their distribution across the County as shown in chapter 4. Also in chapter 4, the local trends for each species in Cornwall are compared with regional and national trends for the butterfly. Just how that analysis, combined with research on species' ecology, then informs conservation priorities and methods is explored in both chapters 3 and 4. Some of the difficulty in interpreting the data lies in human choices, which can skew the evidence on trends. For example, a recurring theme in the accounts of individual species in chapter 4 is the fact that more people have submitted records in recent years. This probably results from raised public awareness of biodiversity loss and the success of national schemes that involve the wider public in recording, such as the Big Butterfly Count, described in chapter 3. More people recording produces more records, which can in turn generate an apparent upward trend, as exemplified by the Silver-studded Blue. Whatever the explanation – and there is always more to find out – the Silver-studded Blue seems to be one of Cornwall's success stories. Similarly, the Marsh Fritillary has been found on many more sites on Bodmin Moor, as a result of both Government-funded stewardship schemes and project funding targeted to look for it intensively over a three-year period, 2017–20 (Butterfly Conservation a).

For some species, such as the Clouded Yellow and Small Pearl-bordered Fritillary, sightings cluster along the coast. Closely cropped clifftop sward certainly favours the Small Pearl-bordered Fritillary, where its larval foodplant, Common Dog-violet is abundant amongst patches of dead Bracken and scrub where the adult can bask; just as the dune systems of the north coast provide an ecology well suited to the Brown Argus and Silver-studded Blue. But we also know that the South West Coast Path attracts more walkers than inland areas, raising the question of whether some of the distribution maps might result as much from human footfall as from butterfly wingbeats. These subjects are explored in further detail in chapter 4.

People, politics and power

Farming accounts for 78% of land use in Cornwall (Cornwall Council, 2019a: 17). All farmers value their land, and most landowners do not wish to harm nature. But the expectation over the last half-century for farmers to maximise food production, propelled by Government incentives, led to the high-input, high-yield model of farming that is now recognised as so environmentally damaging and economically unsustainable (Hayhow *et al.*, 2019: *passim*). Widespread pesticide use has reduced biodiversity, sending into freefall populations of the very insects and birds that are the farmer's natural allies. This kind of farming is one of many causes for the decline in bees, butterflies and birds.

As described in chapter 3, some of Butterfly Conservation's most successful conservation projects in Cornwall have resulted from partnerships with Cornish farmers and landowners. The Marsh Fritillary is the beneficiary of just such alliances on Bodmin Moor, where carefully managed grazing maintains the specialised habitat this protected species needs. When this is on land that is

Marsh Fritillary rests on Marsh Thistle, Bodmin Moor

designated for its wildlife importance, Natural England can support a low-input, low-yield, wildlife-friendly form of farming.

But we still have a very long way to go. In the twenty-first century, industrial-scale farming has brought commercial contract producers into a county unused and unsuited to their practices. The toll on the land is destructive and long lasting. Deep ploughing and heavy machinery trash the soil, pesticides destroy biodiversity, and the contract producer, who has no long-term investment in the land, moves on when it is no longer productive. The soil loses its structure, and with it its capacity to retain carbon and water. A resource that has taken many generations to create and on which we all depend is washed into roads and watercourses, ending as great slicks carried out to sea after heavy rain. The UK Government admits that "soil erosion and compaction from agriculture was estimated to impose an external cost in England and Wales of £305 million in 2010" (Defra, 2020a: 5), a production cost calculation rather than a cost to the environment. Cornish hedges, important havens of biodiversity discussed further in chapter 2, are stripped of vegetation by the wheels of heavy tractors and trailers too wide for the County's network of small lanes, and are torn apart to create or widen gateways for heavy machinery.

The four years following the 2016 EU referendum have seen a slew of Government policy documents putting the environment at the centre of agricultural policy and healthcare (HM Government, 2018; Defra, 2020b and c).

An old Cornish hedge degraded by farm machinery, near Liskeard

The much trumpeted 'public money for public goods' principle is enshrined in the Environment Land Management (ELM) scheme that will replace EU grants to farmers from 2024 (Defra, 2020b)). To mitigate climate change, the Government sets targets (in a politically expedient future) for reducing carbon emissions, eventually to 'net zero'. Meanwhile, the evidence for human-induced global heating is unassailable, as NASA reported that "Earth's global average surface temperature in 2020 tied with 2016 as the warmest year on record" (NASA, 2021).

In Cornwall, as elsewhere, the environment intersects with the economy and social justice. The need to earn a living and to provide housing for local people can often compete with environmental concerns. The key to success is to harness them together. Cornwall Council is committed to "Responsible use of the natural environment as a key economic asset" (Cornwall Council, 2013: 6). Indeed, people choose Cornwall as a holiday destination or as a place to live and work precisely because of its natural assets. Looking after these assets makes sound financial sense. Putting a monetary value on the environment might appear perverse or even distasteful, but environmental and economic sustainability are best served when they have common goals. Put another way, neither is sustainable without the other. Cornwall's Environmental Growth Strategy 2015–2065 (Cornwall Council, 2016) was developed jointly by Cornwall Council and the Cornwall and Isles of Scilly Nature Partnership as an attempt to reconcile these competing pressures over the next 50 years.

Cornwall has committed to developing "the commercial potential of cutting edge renewable energy and environmental technologies" (Cornwall Council, 2013: 15). In its unique maritime setting, Cornwall is well placed to be self-sufficient in renewable energy with wind both onshore and off, solar photovoltaic and technically elusive wave energy all in abundance and generating economic activity and jobs as well as electricity.

When Cornwall Council declared a climate emergency in 2018, it was one of the first local authorities to do so. Under the optimistic brand Carbon Neutral Cornwall, the Council has published its policy on how it intends to achieve that goal (Cornwall Council, 2019b). Improving biodiversity is now a key consideration in planning decisions, even if some of the science is wrong: planning documents insist incorrectly at the time of writing that the Large Blue butterfly occurs in Cornwall as a result of reintroduction (Cornwall Council, 2018). Getting the science right is crucially important to biodiversity recovery. Unsuccessful attempts to reintroduce locally extinct species – the Large Blue in 2000 and the Marsh Fritillary on Goss Moor in 2018 – show just how challenging it is to assess and manage the complex web of interdependent ecological factors correctly, to support a butterfly through its complete life cycle and build a sustainable population.

Derelict mines with acres of brownfield sites might at first appear obvious areas for both housing and commercial development, but as we now know, such sites are important for wildlife, including rare species of butterfly such as the Dingy Skipper and Silver-studded Blue, evidence collected in Cornwall Butterfly Conservation's Mining for Butterflies project, described in chapter 3. These are challenging decisions for planners and politicians alike, when the needs of a homeless family are weighed against the value of a protected species. Challenging or not, the decisions must be properly informed by science, and the environmental impact of developments mitigated as far as possible.

Sometimes, well-intentioned environmental policy can have unintended consequences. It is common knowledge that trees remove the greenhouse gas carbon dioxide from the air and upcycle it into the fabric of leaves and wood, thereby mitigating climate change. Consequently, planting trees, or so the reasoning goes, must be good for the environment. That simple logic underpins both UK Government policy on climate change and the Forest for Cornwall strategy, the implementation of which is backed by grants. "We will develop a mass woodland tree planting programme, once fully developed a Forest for Cornwall covering approximately 8000 hectares… or about 2% of Cornwall's land mass" (Cornwall Council 2019b: 41). Thus motivated, Cornish farmers and other landowners eye up their less productive, rough pasture and decide to plant trees on it instead of grazing livestock. There are two serious objections to this strategy. The first is general: that unimproved pasture in a temperate climate like Cornwall's, with a long-established flora, is already acting as a more efficient carbon sink than a wood; moreover, the carbon is stored securely underground. The second is specific to local biodiversity. Some of that rough pasture in Cornwall is home to rare butterflies, like the Small Pearl-bordered Fritillary and the Marsh Fritillary, not always on land that is legally protected. This is not a theoretical concern: it is already happening, and at least one colony of a Section 41 species of butterfly in Cornwall is threatened by an ill-considered tree-planting proposal part-funded from the public purse.

Established woodland, when left to its own devices, can benefit some butterflies and be a death sentence for others. Species that depend on the freshly cleared glades that were once a feature of regular coppicing are now struggling. Chapter 4 reminds us that the Heath Fritillary and Pearl-bordered Fritillary were known respectively as the 'woodman's follower' and 'woodman's friend' for their opportunistic habit of colonising freshly cut clearings, where a combination of shelter, warmth and early succession vegetation created the habitat they need. By the time the herbage began to close in, a new clearing provided their next safe haven as a result of coppicing. Those traditional forms of woodland management are long gone and must be replaced by deliberate conservation strategies, again in partnership with landowners, to maintain populations. The Duchy of Cornwall helps to maintain the Heath Fritillary in the Tamar Valley, and De Lank Quarry Ltd helps to support the Pearl-bordered Fritillary on the edge of Bodmin Moor. By contrast with the habitat requirements of these two species, the Speckled Wood thrives just about anywhere where there is sun-dappled shade, and is doing comparatively well along woodland margins and hedgerows.

A year into the pandemic, it is becoming clear that nothing will ever be quite the same again. Because people have fallen in love with the great outdoors, footfall in wild places has increased many fold. Converts to the joys of the countryside are most welcome, but the enjoyment of nature comes at a cost, witnessed by the toll on footpaths, sensitive landscapes and disturbance to wildlife. Access and conservation have always been uneasy bedfellows and their competing demands must be managed imaginatively if we are not to destroy the thing we love. At the same time, conservation charities have suffered a catastrophic fall in income. Working from home is the new normal and the obvious attraction of living outside the main conurbations is putting pressure on already hard-pressed rural housing and services. The domestic holiday market expands as overseas travel feels increasingly risky. Cornwall is in the eye of the storm and will need to be fleet of foot in how it responds to these challenges. As we meet the pressing needs of people, the County's habitats and species have never seemed more vulnerable. Perhaps one of the most important lessons of the coronavirus pandemic for policy makers is that 'natural capital' has a value on which society's more traditional measures of wealth, productivity and health depend.

Making a difference

National and local government, environmental organisations and business all have a role to play in tackling the biodiversity crisis and climate change. And so does each one of us. In the words of polar explorer and environmental activist Robert Swan, "The greatest threat to our planet is the belief that someone else will save it" (Ecologic, 2014).

Our countless everyday choices have an environmental impact, from car use to what we consume. If you are fortunate enough to have a garden, there are many ways in which you can benefit biodiversity, and butterflies in particular. I leave the grass unmown in one area of my garden so that the undisturbed tussocks of Cock's-foot produce an annual crop of Ringlets. Flowering Atlantic Ivy attracts the Holly Blue, and I plant for insects. A favourite nectar source is Argentine Vervain, which provides a regular feast for at least nine species

Ringlet rewards the relaxed gardener in the long grass of the author's garden, Liskeard

of butterfly, the occasional darting Hummingbird Hawk-moth, and bees and hoverflies.

Lobby and enlighten your local parish or town council about mowing road verges later in the year and leaving unmown the base of Cornish hedges, to benefit biodiversity. Get involved with a conservation organisation such as Cornwall Butterfly Conservation (CBC) or Cornwall Wildlife Trust (CWT).

As well as helping nature conservation, you get outside with kindred spirits to enjoy and learn about wild Cornwall. Young people especially are taking up the challenge, using social media platforms and embracing various forms of activism to combat the climate and biodiversity crises they have inherited from the profligacy of preceding generations.

CBC's annual field visit to Penlee Point, a species-rich nature reserve in East Cornwall, with a thriving colony of Marbled White butterflies and views over Plymouth Sound

Acknowledgements

I am grateful to HRH The Prince of Wales, Duke of Cornwall for his foreword and to Chris Gregory, Land Steward for the Duchy of Cornwall (Western District and Isles of Scilly) for his good offices in that connection.

The seed for this book was sown in 2018 by our then Chairman, Philip Hambly, when he pointed out that *A Cornwall Butterfly Atlas* (Wacher *et al.*, 2003), compiled by CBC, was 15 years old and, like him, getting on a bit. So he asked, in that guileless way of his, if it might not be a good idea to produce a new atlas with up-to-date information. Three years later and after many hundreds of volunteer hours of work, this book is the answer. In all walks of life, we build on the achievements of people who came before us. Getting started, we stood on the shoulders of John Wacher, John Worth and Adrian Spalding, the principal authors of the 2003 atlas.

Philip Marsden and Granta Books generously authorised the quote from *Rising Ground: a Search for the Spirit of Place* at the beginning of this book, a passage that describes a landscape familiar to anyone who has ventured into the mires of Bodmin Moor in search of the Marsh Fritillary butterfly.

Julie Williams, CEO of Butterfly Conservation, and her team in Dorset have given this atlas their support, and Dan Hoare and Jenny Plackett have provided specialist advice. We benefited from the knowledge and advice of those giants of Lepidoptera, Professor Jeremy Thomas, Matthew Oates and Peter Eeles, whom we consulted on some esoteric points. I have been privileged to chair a small but perfectly formed working group of CBC members, which at various times included Sue Allen, Jim Barker, Helen Barlow, Sarah Board, Bob Dawson, Jerry Dennis, Sally Foster, Dick Goodere, Philip Hambly, Roger Hooper, Cerin Poland, Jo Poland, Shaun Poland and Kelly Uren.

Books, self-evidently, do not write themselves. For their skilled and knowledgeable authorship, I thank Sarah Board for chapters 2 and 3, with input into chapters 4, 5 and 7; Dick and Maggie Goodere, with Cerin Poland, for chapter 5 and for the monumental chapter 4, to which Jim Barker contributed transect data and for which species champions Jerry Dennis, Sally Foster and Cerin Poland were co-authors respectively for the Grayling, Silver-studded Blue and Grizzled Skipper accounts; Bob Dawson for chapter 6; Sue Allen, Sarah Board, Jerry Dennis and Steve Batt for chapter 7; and Kelly Uren, Jo Poland, Cerin Poland, Dick Goodere, Jerry Dennis, Sally Foster, Richard Symonds and Russell Williams, who each contributed to chapter 3.

For the images that enliven nearly every page, I pay tribute to the camera work of Mary Atkinson, Mark Bailey, Steve Batt, Ian Bennallick, Sarah Board, Bob Dawson, Jerry Dennis, Sally Foster, Maggie Goodere, Martin Goodey, Philip Hambly, Beth Harper, Adam Jones, Steve Jones, Malcolm Pinch, Cerin Poland, Jo Poland, Shaun Poland, Dave Thomas, Steve Townsend, Charlotte Uren and

Members of the Atlas Working Group. From left to right: Sue Allen (secretary), Tristram Besterman (chair) and Sally Foster.

John Walters, who fulfilled the brief that every photograph must be taken in Cornwall, save for those rare visitors, the Brown Hairstreak, Monarch and Swallowtail.

For a small, volunteer-run local organisation to deliver a project of this scale is a challenge at every level, not least financially. Undaunted, Jo Poland, supported by Jim Cooper, led the successful campaign to raise more than £10,000 towards the cost of publication. That depended on the extraordinary generosity – and vote of confidence – of many individuals and organisations, without whom this volume would never have seen the light of day; their names are listed on the **Supporters** page near the beginning of the book, and as **species sponsors** at the beginning of each species account in chapter 4. And keeping us financially on the straight and narrow throughout the endeavour was our treasurer, Helen Barlow.

Peter Creed at Pisces Publications provided advice on content, layout, design and promotion. I am indebted to him and to Peter Eeles for permission to reproduce the chart of each species' life cycle through the year in chapter 4 from Eeles' 2019 book, *Life Cycles of British & Irish Butterflies*. In the same chapter, the 1km square distribution maps for each species in Cornwall, together with adult flight periods, were generated by the ERICA database, which is hosted by the Cornish Biodiversity Network, and relies on the support and capabilities of the indefatigable Dr Colin French, who also supplied the other maps in the book, except for the map of the Isles of Scilly in chapter 6, kindly provided by Teän Roberts.

I am grateful to the many conservation organisations that helped to promote *Butterflies of Cornwall*, including Butterfly Conservation, Cornwall Bird Watching and Preservation Society, Cornwall Wildlife Trust, RSPB Cornwall, Isles of Scilly Bird Group and Isles of Scilly Wildlife Trust.

At the centre of all the activity I was supported and kept up to the mark by a core group of co-editors and writers. They have been indispensable to making this book. Shaun Poland sourced and dealt with hundreds of images, bringing a keen eye for design to the selection. Helen Banks, Cerin Poland, Jo Poland and Kathy Wood trained eagle eyes on the text to rid it of inaccuracies. And presiding over everything has been our inimitable, unflappable, endlessly patient and super-organised Scientific Editor, Sarah Board. To these five individuals go my heartfelt thanks.

At a personal level, my wife, Perry, has shared me with the atlas with patience and encouragement for two years, most especially during the coronavirus pandemic of 2020–21. Born and brought up in Cornwall, Perry's love of her native county and its wildlife long predates my own.

Finally, echoing the dedication at the beginning of this book, I pay tribute to the hundreds of people who have submitted records of the butterflies they saw and identified in Cornwall, over so many decades in the twentieth and twenty-first centuries. This book is a tribute to every single one of those citizen scientists: long may their valuable work continue.

Endpiece

As I write this, I wonder what place butterflies will have in the Cornish landscape in 2035. *Butterflies of Cornwall* is a letter to the future, an environmental message in a bottle, written with a mixture of deep concern and cautious optimism.

I began by quoting Sir Paul Nurse, so who better to round it off?

"Biology shows us that all the living organisms we know of are related and closely interacting. We are bound by a deep interconnectedness to all other life: to the crawling beetles, infecting bacteria, fermenting yeast, inquisitive mountain gorillas and flitting yellow butterflies…

"As far as we know, we humans are the only life-forms who can see this deep connectivity and reflect on what it might all mean. That gives us a special responsibility for life on this planet, made up as it is by our relatives, some close, some more distant. We need to care about it, we need to care for it. And to do that, we need to understand it." (Nurse, 2020: 212).

Tristram Besterman
Chair, Atlas Working Group
Liskeard, February 2021

CHAPTER 2
The Cornish environment

Kenijack, north Cornish coast in West Penwith

The occurrence and distribution of butterflies generally is influenced by a variety of environmental factors including geology and soils, topography, plant communities, habitats and the presence of larval foodplants, rainfall, air temperature and microclimate, as well as human activity on the landscape. Cornwall is no exception.

The County lies at the western-most tip of the south-west peninsula of Britain, with its north coast pounded by the waves of the Atlantic Ocean, and the English Channel surging along the shores of the relatively sheltered south coast. With almost 435 miles (700 kilometres) of coastline on three 'sides' it is not surprising that the maritime influence is felt everywhere in Cornwall and is reflected in the weather (with a mild oceanic climate and above-average rainfall) and to a certain extent the habitats present. The County is approximately 86 miles (138 kilometres) in length, from Land's End in the west to Marsland Mouth in the north-east. Nowhere in Cornwall is more than 15 miles (24 kilometres) from the sea, with the majority of land within 5 miles (8 kilometres) of the coast (French, 2020).

The Isles of Scilly lie further south-west again, some 28 miles (45 kilometres) from

Isles of Scilly

Sites designated for wildlife and cultural importance

Cornwall is rich in wildlife and history. This is reflected in the fact that around 15% of the County's land area lies within a site designated for its important habitats and species, with 28% of the land area of the Isles of Scilly covered by a designation. All seven National Character Areas (NCAs) for Cornwall have sites designated for their importance at international, European, national and local levels. The Isles of Scilly NCA has sites of European and national significance. The legal framework for English sites of European status is subject to review at the time of writing.

INTERNATIONAL STATUS

World Heritage Site

The Cornwall and West Devon Mining Landscape was designated a World Heritage Site (WHS) in 2006 by UNESCO. The designation states, "The landscapes of Cornwall… were radically reshaped during the eighteenth and nineteenth centuries by deep mining for predominantly copper and tin. The remains of mines, engine houses, smallholdings, ports, harbours, canals, railways, tramroads and industries allied to mining… reflect an extended period of industrial expansion and prolific innovation." The Cornwall WHS is protected under the *Planning (Listed Buildings and Conservation Areas) Act* 1990, the *Town and Country Planning Act* 1990 and the *Ancient Monuments and Archaeological Areas Act* 1979.

Although this is not strictly wildlife legislation, Cornwall's mining heritage has had a profound, widespread and lasting impact on the environment of the County, as recognised by Natural England in the NCA descriptions.

EUROPEAN STATUS

Special Area of Conservation

Special Areas of Conservation (SACs) are those which were given greater protection under the European *Habitats Directive* 1992. In England, a SAC has statutory protection under *The Conservation of Habitats and Species Regulations* 2017 (as amended), which require the UK Government to establish a network of important, high-quality conservation sites that will make a significant contribution to conserving the habitats and species (excluding birds – see Special Protection Areas (SPAs) below) identified as being most in need of conservation at a European level. To qualify, sites must have been designated as Sites of Special Scientific Interest (SSSIs). (See below.) Together with Special Protection Areas (SPAs), these now form the national site network in England (known as the Natura 2000 network of sites in Europe). There are 16 SACs in Cornwall (excluding marine SACs), as well as the Isles of Scilly Complex SAC.

Special Protection Area

Special Protection Areas (SPAs) are protected areas solely for birds, classified under the UK *Wildlife & Countryside Act* 1981 (as amended) and the *Conservation (Natural Habitats, &c.) Regulations* 2010 (as amended) in accordance with the European Birds Directive. There are two SPAs in Cornwall, as well as the Isles of Scilly SPA (not including marine SPAs).

NATIONAL STATUS

National Nature Reserve

National Nature Reserves (NNRs) were established to protect some of the most important natural and semi-natural terrestrial and coastal habitats in Great Britain and to provide outdoor 'laboratories' for scientific research. They encompass land that has been declared under either the *National Parks and Access to the Countryside Act* 1949 or the *Wildlife and Countryside Act* 1981 (as amended). There are three NNRs in Cornwall.

Site of Special Scientific Interest

Sites of Special Scientific Interest (SSSIs) are sites of national importance for wildlife and/or natural features, notified under the *Wildlife and Countryside Act* 1981 (as amended). Together they form a network of sites supporting many characteristic, rare and endangered species, habitats and natural features. There are 136 SSSIs in Cornwall and 26 on the Isles of Scilly.

LOCAL STATUS

Local Nature Reserve

Local Nature Reserves (LNRs) are sites of local importance for wildlife, geology, education and public enjoyment. They are designated and managed by local authorities in England under the *National Parks and Access to the Countryside Act* 1949. There are 13 LNRs in Cornwall, which are managed to conserve their natural features for the future of both people and wildlife. There are no LNRs on the Isles of Scilly.

County Wildlife Site

County Wildlife Sites (CWSs) provide a network of wildlife sites that are important locally and complement the nationally designated sites. They have no legal protection but are chosen to represent local habitats, character and distinctiveness, having high nature conservation value within Cornwall. There are 498 CWSs in Cornwall and none on the Isles of Scilly.

Land's End, the most westerly point of the Cornish mainland. They are made up of over 200 islands, of which only five are currently inhabited. The diversity of natural habitats is more limited, but, as would be expected, all are heavily influenced by the sea.

Setting the scene

Running beneath the entire south-west peninsula, from Dartmoor in Devon to the Isles of Scilly, lies a vast granite body, the Cornubian batholith. Formed about 280 million years ago when hot magma cooled and crystallised deep in the Earth's crust, the resulting irregular mass of hard, igneous rock today breaks the surface to form the high granite moors and tors of Cornwall. These comprise, from east to west, Kit Hill, Bodmin Moor, Hensbarrow, Carnmenellis and West Penwith, extending out to sea to form the low-lying Isles of Scilly.

Towards the final stages of cooling, superheated steam trapped in the magma chamber, carrying dissolved metals, was forced under pressure into the crystalline mass of the granite and into fissures in the surrounding sedimentary rocks. Today, both on land and extending far out beneath the sea, these form metalliferous lodes, extending from the Tamar Valley in the east, through Bodmin Moor and the Redruth and Camborne area to Cape Cornwall in the west, and from the north coast around St Agnes to the south near Falmouth. The lodes were rich in tin, copper, tungsten, lead, zinc and iron, and, in places, small but significant amounts of gold and silver.

Surrounding the moorland granite outcrops are older marine sediments, laid down in the Devonian Period around 350 million years ago. This is the country rock, or 'killas' (a Cornish mining term), subsequently deformed and hardened in the same tectonic mountain-building event that emplaced the granite 380–280 million years ago. As a result of this geological upheaval, the Devonian sediments were metamorphosed into hard, grey, slaty rock, often traversed by veins of white quartz. The unimaginable forces involved can be seen in the spectacular folds visible today in the sea cliffs along the north Cornish coast, perhaps most dramatically at Millook. In the far north-east of the County, these Devonian slates are overlain by younger Carboniferous rocks of the Culm Supergroup, consisting predominantly of mudstones interbedded with siltstones and sandstones.

This neat sequence is complicated by the geologically unique Lizard peninsula, where much older metamorphic rocks, ancient sea floor basalt and even part of the Earth's mantle is present as serpentinite, brought to the surface by an earlier period of mountain building 490–390 million years ago.

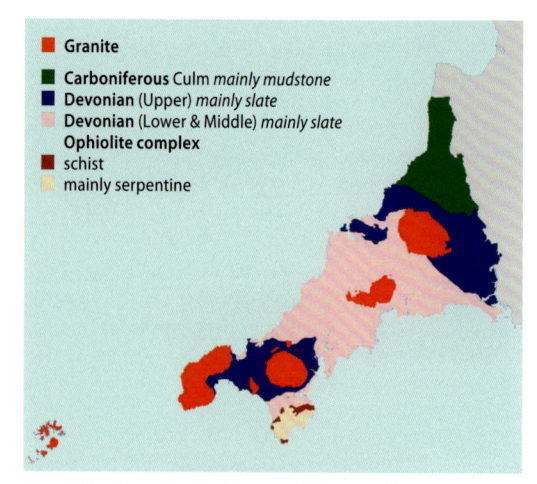

Simplified geological map of Cornwall
Based on Andy F at English Wikipedia

- Granite
- Carboniferous Culm *mainly mudstone*
- Devonian (Upper) *mainly slate*
- Devonian (Lower & Middle) *mainly slate*
- Ophiolite complex
- schist
- mainly serpentine

Folded rocks at Millook, south of Bude

The granitic, upland spine of Cornwall, resistant to weathering, makes an east–west watershed of high ground that broadly divides the County into two environmentally distinct catchment areas. To the north, a narrower, comparatively tree-less landscape has two main rivers, the Camel and the Hayle, with sandy estuaries, high, craggy cliffs of Devonian slate, and sand dunes. Draining to the south are the Tamar, Fowey and Fal rivers, with their many creeks and tributaries, characterised by tidal mudflats and valleys mantled with broadleaved woodland down to the high tide line.

The soils of Cornwall are produced by the interaction of weather and living organisms with the chemistry of the underlying rocks. Mainly along the north coast of the County, extensive sand dunes, reaching a height of 80 metres in places, have created a specialist habitat rich in calcium carbonate. Accumulated over millennia, 60% of the sand is composed of shell fragments of marine organisms, washed up onto the shore, ground down by the sea and blown by wind to form the dunes. Behind the more mobile seaward aspect of the dunes, the landward tracts of dune have stabilised to generate thin, well-drained, calcareous soils that support specialist plants, creating a habitat enjoyed by a diverse range of species.

As in most of Britain, Cornwall's landscape is the product of human activity. In prehistoric times, the forest that covered most of the peninsula was cleared, first by the settled farming communities of the Neolithic period, and then by the fuel-hungry smelters of the Bronze and Iron Ages. Traces of these settlements can still be seen scattered across the moors and headlands, or lie hidden beneath the lowland landscape. The most visible evidence of early, settled occupation of the landscape is the Cornish hedge. This is really a stone-faced wall enclosing a core of earth and rubble, forming 30,000 miles of interconnected, semi-natural habitat supporting around 500 native plant species and a range of butterflies, particularly the Wall *Lasiommata megera*. In west Cornwall, hedges were made by clearing granite boulders from the landscape to create fields during the Neolithic 6,000 to 4,000 years ago. On the moors above Zennor, hedges divide the landscape into small fields dating from the Bronze and Iron Ages, built between 4,000 and 2,000 years ago. Hedges continued to be constructed over subsequent centuries to enclose land for livestock and crops, particularly during the mining boom of the eighteenth and nineteenth centuries, to meet rising demand for food for the expanding population. Many of the beautifully constructed slate hedges of north Cornwall date from this period, recognisable from the fine 'Jack and Jill' herring-bone pattern of slate, stacked in rows.

Cornish hedge of granite boulders in west Cornwall

Cornish hedge of herring-bone slate in north Cornwall

People have long exploited the rocks of Cornwall to win metal, stone and clay. This began in the early Bronze Age around 4,000 years ago, when tin was 'streamed' from river gravel, and reached peak production of copper and tin from deep underground in the nineteenth century. As the industry dwindled and closed in the twentieth century, it left a scarred and rubble-strewn landscape over which the old Cornish engine houses stand silent witness, now protected as part of the Cornwall and West Devon Mining Landscape World Heritage Site. This altered landscape includes mine-dumps rich in arsenic and other toxic elements, creating a nutrient-poor, soil-impoverished micro-ecology today colonised by a range of species, including butterflies such as Silver-studded Blue *Plebejus argus*. These sites are the subject of a recent project by Cornwall Butterfly Conservation, Mining for Butterflies.

Above ground, there is hardly a square mile of Cornwall without its local quarry as a source of building stone. Major quarries produce aggregate and cut stone for the construction industry. De Lank Quarry, on Bodmin Moor, which supplied dressed granite for London's Royal Opera House, is still active and is a Site of Special Scientific Interest (SSSI). Its disused, sheltered, south- and south-west-facing slopes are home to the Pearl-bordered Fritillary *Boloria euphrosyne*. Kaolin, or china clay, was formed by the natural, hydrothermal alteration of Cornish granite. One hundred and twenty million tonnes of kaolin have been extracted since the eighteenth century, leaving a great chain of spoil heaps that dominate the skyline around St Austell, as well as deep pits, one of which is home to the Eden Project.

Cornwall's National Character Areas

Natural England, the Government's adviser for the natural environment, has divided England into 159 National Character Areas (NCAs). Each NCA is defined by a unique combination of landscape, biodiversity, geodiversity, history, and cultural and economic activity. Their boundaries follow natural features rather than administrative boundaries, so it makes sense to use them when considering the distribution of butterfly species, their habitats and larval foodplants. Cornwall is divided into seven NCAs, with an eighth for the Isles of Scilly:

149 The Culm
152 Cornish Killas
153 Bodmin Moor
154 Hensbarrow
155 Carnmenellis
156 West Penwith
157 The Lizard
158 Isles of Scilly

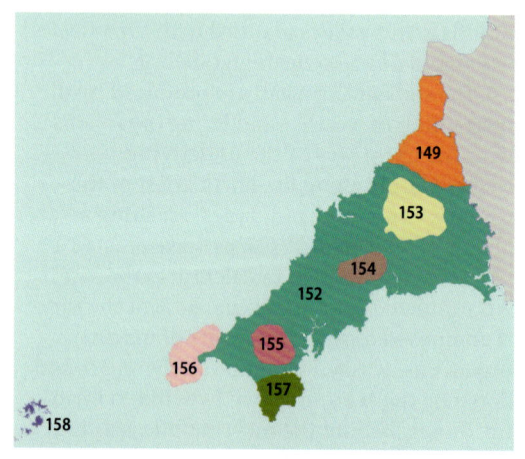

National Character Areas for Cornwall and Isles of Scilly
© Natural England copyright. Contains Ordnance Survey data © Crown copyright and database right 2013

Of the 37 resident/regular visiting butterflies found in Cornwall, 23 species are found in every Cornish NCA, with 20 of these being common butterflies with more general habitat requirements and larval foodplants that are common and widespread themselves. These include Small Skipper *Thymelicus sylvestris*, Large Skipper *Ochlodes sylvanus*, Large White *Pieris brassicae*, Small White *P. rapae*, Green-veined White *P. napi*, Speckled Wood *Pararge aegeria*, Ringlet *Aphantopus hyperantus*, Meadow Brown *Maniola jurtina*, Gatekeeper *Pyronia tithonus*, Peacock *Aglais io*, Small Tortoiseshell *A. urticae*, Comma *Polygonia c-album*, Holly Blue *Celastrina argiolus* and Common Blue *Polyommatus icarus*. Three species of conservation priority, namely the Wall, Small Heath *Coenonympha pamphilus* and Small Pearl-bordered Fritillary *Boloria selene*, are also found throughout the County in every NCA. These species and their larval foodplants favour both coastal and moorland heaths, as well as grassy areas with patches of bare ground and Cornish hedges. It is the coastal strip that is often richest in species, as semi-natural habitats remain with their diversity of nectar sources and larval foodplants, sandwiched between the sea and agricultural land.

The Culm National Character Area (NCA 149)

The Culm NCA, with its open, rolling plateaux and dramatic coastal cliffs, extends from Launceston northwards through Cornwall and into north-west Devon. It is a largely remote and sparsely populated landscape, with few main roads, a number of hamlets and isolated farmsteads, and no major settlements. The dominant influence on the Culm landscape is the clay that overlies the sandstones, shales and mudstones of the Upper Carboniferous Culm Supergroup. These rocks give rise to heavy, wet, clay soils, which are difficult to cultivate, and therefore farming is focused on animal grazing in pastures, many of which have retained a rich botanical diversity. Plant species of the Culm grasslands (also known as Rhôs pasture) include Meadowsweet *Filipendula ulmaria*, Ragged-Robin *Silene flos-cuculi*, Devil's-bit Scabious *Succisa pratensis*, Meadow Thistle *Cirsium dissectum*, Whorled Caraway *Trocdaris verticillata*, Wavy St John's-wort *Hypericum undulatum* and many types of orchid, including the Heath Spotted-orchid *Dactylorhiza maculata*. It is not just the grasslands, often dominated by Purple Moor-grass *Molinia caerulea* and rushes, that are floristically diverse, but also other habitats such as fen meadows, mires and wet heaths, with species such as Cross-leaved Heath *Erica tetralix* and Fen Bedstraw *Galium uliginosum* (French, 2020).

Most of the Culm grassland in Cornwall has been destroyed through agricultural activity, with only a small number of isolated sites remaining. It has been estimated that between 1984 and 1991 62% of Culm grassland sites in Cornwall and Devon were lost to drainage, re-seeding and agricultural chemicals; only 8% of the area of Culm grassland that existed in 1900 remains today (BRIG, 2011).

The Culm NCA has deciduous woodland present on the steeper valley sides of the River Tamar (and also the Rivers Taw and Torridge in Devon) and in a few coastal cliff locations, such as at Dizzard, north of Crackington Haven, with its unique dwarf oak *Quercus* spp. wood,

Grassland with Meadow Thistle, typical of The Culm NCA, at Greena Moor, near Bude

Marbled White thrives on the grassland of The Culm NCA

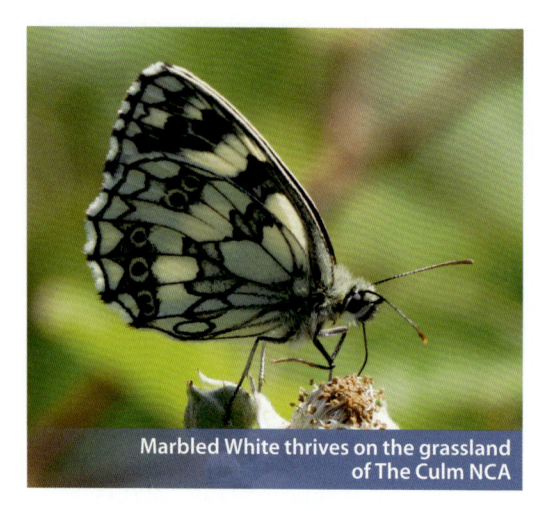

which has limited canopy height (between 1 and 8 metres) due to the exposed position it occupies, buffeted by the winds of the Atlantic. The Dizzard supports an internationally important lichen community.

Of the 37 resident/regular visiting butterfly species in Cornwall, 33 have been recorded within The Culm NCA. Marsh Fritillary *Euphydryas aurinia* favours the Culm grasslands, their wet Purple Moor-grass tussocks providing suitable habitat for the butterfly's larval webs alongside Devil's-bit Scabious, the butterfly's larval foodplant. The grasslands also allow Marbled White *Melanargia galathea* to thrive, whilst the species-rich hedge-banks lining the lanes and forming the field boundaries provide important habitats for Brimstone *Gonepteryx rhamni*, Speckled Wood and Ringlet. Green Hairstreak *Callophrys rubi*, Small Heath and Grayling *Hipparchia semele* favour the coastal grasslands.

The steep-sided valleys on the Cornwall/Devon border at Marsland with their oak woodland, traditional hay meadows, wild flower meadows and coastline attract a diversity of butterflies including Dingy Skipper *Erynnis tages*, Pearl-bordered Fritillary and Marsh Fritillary. As the summer progresses, Dark Green Fritillary *Speyeria aglaja* can be seen on the wing, along with the Purple Hairstreak *Favonius quercus*.

Cornish Killas National Character Area (NCA 152)

The Cornish Killas NCA is by far the largest in Cornwall. It essentially comprises the lower ground that abuts and excludes The Lizard and the granite outcrops of West Penwith, Carnmenellis, Hensbarrow and Bodmin Moor. It stretches from Penzance and St Ives in the west to the Devon border in the east. The landscape has an open character with views across the hills and out to sea. The gently rolling hills, sheltered coves and headlands of the south coast contrast with the rugged, rocky, high cliffs of the north coast, important for its geological features. The underlying geology of "sedimentary and metamorphic rocks… and a strong maritime influence are the unifying factors of the Cornish Killas NCA" (Natural England, 2014a: 7).

Coast of north Cornwall at Porthcothan, within the Cornish Killas NCA

The Cornish Killas NCA "contains 43,762ha (hectares) of the Cornwall Area of Outstanding Natural Beauty (AONB), covering 19 per cent of the NCA, mainly along the coasts. The NCA also includes 8,326ha of the Tamar Valley AONB and several stretches of Heritage Coast" (Natural England, 2014a: 3). It also contains Golitha Falls National Nature Reserve (NNR) and a small area of Goss Moor NNR.

Despite much of the area being farmed, the NCA has a diversity of habitats, including heathland (often forming mosaics with willow carr/woods and scrub), damp grasslands and wetlands, coastal grasslands and sand dunes, saltmarshes, broadleaved woodlands along the river valleys dominated by Sessile Oak *Quercus petraea* with occasional Ash *Fraxinus excelsior*, and the ubiquitous Cornish hedge dividing the fields. Parkland, mostly found in the southern part of the NCA, consists of mature trees, some veteran, and, along with dead wood, supports a rich diversity of mosses, liverworts, lichens and ferns. Former mining sites and quarries provide a range of habitats, from bare ground and sparsely vegetated areas to scrub and broadleaved woodland. These bare ground communities are particularly important for a diversity of invertebrates as well as specialist liverworts and mosses that tolerate heavy metals.

This is the only NCA in which all 37 of the Cornish resident/regular migrant butterfly species have been recorded, which is not surprising given its geographical spread and diversity of habitats. The stretches of coastal grassland, heathland and sand dunes are home to many species, including the Dingy Skipper, Small Heath, Grayling, Small Pearl-bordered Fritillary, Dark Green Fritillary, Green Hairstreak, Brown Argus *Aricia agestis*, Common Blue and Silver-studded Blue. Two key coastal sites for butterflies (as well as many other invertebrates and plant species) within this NCA are Penhale Dunes SSSI, which incorporates Gear Sands, near Perranporth, and the dune system encompassing SSSIs at Hayle Towans, Upton Towans and Gwithian Towans near Hayle. These two areas of sand dunes contain the County's best populations of Silver-studded Blue, with the species often being seen in its thousands at both locations.

Silver-studded Blue flourish on the dune systems of the Cornish Killas NCA

Penhale Sands also hosts the last Cornish stronghold of the Grizzled Skipper *Pyrgus malvae*, where the population appears to be stable. These calcareous dunes support communities of lime-loving plants in great diversity, including species providing rich nectar sources and larval foodplants, often being given shelter by sand hollows as well as scrub. Brown Argus has its stronghold in Cornwall in these two dune systems, with their areas of short turf and sheltered hollows providing warmth for the adults, as well as at Rock Dunes on the east side of the Camel estuary.

Many former mining sites have thin, highly mineralised soils that slow the rate of plant colonisation and keep areas in the early stages of succession. The resulting bare ground allows grassland and heathland to develop, supporting a range of butterfly species similar to those recorded within the coastal regions. Sites such as Poldice Mine/Valley have recorded Silver-studded Blue, Grayling, Small Heath, Common Blue and Small Copper *Lycaena phlaeas*. Small stands of oak support Purple Hairstreak.

The Cornish Killas NCA is also the only one in which Heath Fritillary *Melitaea athalia* and White-letter Hairstreak *Satyrium w-album* have been recorded in recent years, both of them in east Cornwall and mainly within the Tamar Valley, with Heath Fritillary now restricted to one site of woodland and heathland at Greenscoombe Wood SSSI. White-letter Hairstreak have always been scarce in Cornwall. The first specimen was authenticated in 1945, and it has been recorded in only a few sites along the south coast, to the east of Falmouth, as well as in the Bodmin area. It was thought to have become locally extinct after 1985, but it has been recorded close to the River Tamar nearly every year since 2013.

High summer sees Silver-washed Fritillary *Argynnis paphia* flying in mature woodlands towards the south coast. The unimproved meadows, predominantly in the south-east of the NCA, are home to a hitherto stable population of Marbled White, a butterfly that is also found on the north coast, east of the Camel River.

Bodmin Moor National Character Area (NCA 153)
True to its name, this NCA is defined by Bodmin Moor. Characterised by upland moor, it lies in the centre of Cornwall and includes Brown Willy, Cornwall's highest point at 420 metres above ordnance datum (AOD). It is the source of five of the County's rivers: the Fowey, Tiddy, Lynher and Inny flowing to the south, and the Camel to the north.

The geology of the NCA is granite, which crops out as tors, with their clitter slopes. It is also visible in disused quarries and their spoil heaps, such as the Cheesewring, and in the half-worked slabs of 'moorstone' nearby, often showing lines of feather-and-tare holes used to split the rock. Surface working of lodes for copper and tin has left long scars of hummocks and dells across the moor, and deeper workings are marked by derelict engine houses. The combination of geology and the high rainfall of the Atlantic coastal climate produces thin peaty soils, shallow, fast-flowing streams and wet grass heathland. The landscape is exposed and desolate, with large areas of common land (one third of the NCA). The edge of the moor is fringed with damp, deciduous-wooded valleys, including some ancient woodland in the steep-sided valleys of the River Fowey and the River Lynher. Here a range of Atlantic woodland lichens, mosses, liverworts and ferns flourish in the unpolluted air, mild climate and high levels of humidity and rainfall.

Small-scale settlements, commons and fields enclosed by Cornish hedges also characterise the NCA. The agriculture is dominated by grazing with some small areas of arable, creating "a more intimate landscape, reinforcing the transition between upland and lowland" (Natural England, 2013a: 7).

The habitats found within the Bodmin Moor NCA are varied and diverse and include heathland, acid and marshy grasslands, wet woodland and extensive peat bogs and mires, which support the County's most important network of Marsh Fritillary metapopulations. Of the 37 resident/regular visiting butterfly

Bodmin Moor NCA: King Arthur's Hall looking towards Rough Tor

species in Cornwall, 32 have been recorded within the Bodmin Moor NCA, reflecting the diversity of semi-natural habitats, particularly the agriculturally unimproved rough grassland, heathland and sheltered wooded river valleys.

Bodmin Moor is home to three nationally important fritillary butterflies, the Pearl-bordered Fritillary, Small Pearl-bordered Fritillary and Marsh Fritillary. Between 2017 and 2020, Butterfly Conservation's All the Moor Butterflies project resourced extensive fieldwork that confirmed 49 sites across Bodmin Moor for Marsh Fritillary, more than doubling the known sites from 20 recorded in 2015. This butterfly fares well on the open expanses of herb-rich wet grassland found across the moor, assisted by traditional farming techniques with appropriate grazing by cattle and ponies. Strategic partnership with sympathetic landowners to ensure that habitat is nurtured for the Marsh Fritillary will further contribute to the butterfly's breeding success on the moor. (Further details about this project are given in chapter 3.) One of the most important areas for Marsh Fritillary is Carkeet, located in the Fowey Valley. One of the best locations to see Pearl-bordered Fritillary and Small Pearl-bordered Fritillary within this NCA is Pendrift Downs; if the observer is lucky, they may also see the Marsh Fritillary. As the summer progresses, Silver-washed Fritillary can be seen flying around the woodland edges, along with Speckled Wood and Brimstone. Purple Hairstreak, a butterfly also associated with woodland on the eastern fringes of the moor, can also be seen flitting in the tops of oaks.

St Breward, on the western edge of Bodmin Moor, is the stronghold in Cornwall for Pearl-bordered Fritillary, which flies on warm and sunny days in the latter part of April and first half of May, gliding gracefully over bracken-covered slopes in search of violets *Viola* spp., on which the females lay their eggs. Small Pearl-bordered Fritillary is found on the same bracken-covered slopes, but is more widespread across the moor than the Pearl-bordered. It is also present in the wet meadows, where the caterpillar feeds on Marsh Violet *Viola palustris* subsp. *juressi* and Common Dog-violet *Viola riviniana*.

Bodmin Moor is one of the best places to see Small Heath, particularly in drier areas where the grass is short and sparse. The Wall and occasionally the Grayling make an appearance on the higher ground where the granite rock is exposed at the surface.

Marbled White has been recorded a handful of times on both the southern edge of the moor in the St Neot area and the north-eastern edge in the vicinity of Trewint.

Marsh Fritillary, whose strongholds are within Bodmin Moor NCA

Hensbarrow National Character Area (NCA 154)

This NCA is named after Hensbarrow Downs, the granite hills that are the focus of the china clay industry near its centre. The area can be divided roughly into three. On the western side the china clay workings and spoil tips dominate the landscape, to the north Mid Cornwall Moors SSSI forms an open and wild landscape, some of which is designated as a National Nature Reserve (Goss Moor NNR), and to the eastern side is an "area of contrast between the wild and open granite tors, the biodiverse heath and willow carr and an idyllic pattern of fields bounded by Cornish hedges and woodlands" (Natural England, 2013b: 3); the granite of Helman Tor and Roche Rock are notable features. The highest point is Hensbarrow Beacon at 312 metres AOD, and the lowest point is some 130 metres below sea level in the china clay pits. The River Fal rises in the heart of the NCA and cuts through a steep valley on its short journey to the sea.

There are two sites of international importance for their wildlife in the Hensbarrow NCA, both Special Areas of Conservation (SACs): Breney Common, Goss and Tregoss Moors SAC, and St Austell Clay Pits SAC, which cover 6% of the area. The dominant habitats within the NCA are the vast expanses of lowland heath and Purple Moor-grass and rush pasture, increasingly interspersed with wet woodland and scrub in recent years.

Of the many distinct wildlife sites within the area of national importance, the Mid Cornwall Moors SSSI comprises a network of 14 areas of closely connected semi-natural habitat covering an area of 1,653 hectares. This is characterised by wet and dry heathlands, fens, grasslands, broadleaved woodlands, carr, scrub and species-rich hedgerows, with ponds and waterways. These habitats support an assemblage of nationally rare and nationally scarce flowering plants and ferns such as Cornish Eyebright *Euphrasia vigursii*, Coral-necklace *Illecebrum verticillatum*, Chamomile *Chamaemelum nobile*, Lesser Butterfly-orchid *Platanthera bifolia*, Chaffweed *Lysimachia minima*, Lesser Water-plantain *Baldellia ranunculoides* and Allseed *Linum radiola*. The presence of the Marsh Fritillary butterfly is also a reason for the site's notification as a SSSI, and indeed as a site of international importance, as well as for its assemblage of invertebrates chiefly associated

Hensbarrow NCA: china clay spoil near St Austell

Dingy Skipper found in the Hensbarrow NCA

with scrub heath and moorland. In the Mid Cornwall Moors Marsh Fritillary breeds primarily in damp acidic grassland where the larval food plant, Devil's-bit Scabious, is abundant. Optimal breeding areas are typically a patchwork of short vegetation and long tussock grasses dominated by cattle-grazed Purple Moor-grass. Small areas of similar habitat are scattered through the NCA north of the clay working, and several projects have attempted to enhance the old spoil heaps to act as links between these important habitats. Woodland covers 14% of the area.

Although the Hensbarrow NCA is small in area, the diversity of habitats is reflected in the fact that of the 37 resident/regular visiting butterfly species in Cornwall, 31 have been recorded, with the complex network of semi-natural habitats in the Mid Cornwall Moors providing butterfly hotspots, particularly Goss Moor, Breney Common, Red Moor, Treskilling Downs and Caerloggas Downs. The agriculturally unimproved grassland, with its mix of short and long grasses, as well as the heathy areas, provide ideal habitats for Grayling, Dark Green Fritillary, Small Pearl-bordered Fritillary, Small Heath and Dingy Skipper, as well as the commoner Large Skipper and Small Skipper. Green Hairstreak are also found in areas receiving some shelter by adjacent scrub.

Marsh Fritillary have done well on Breney Common (part of the Helman Tor Cornwall Wildlife Trust reserve), but its disappearance from most other sites within the Mid Cornwall Moors makes monitoring of these metapopulations important. Goss Moor was once home to the Marsh Fritillary, but for a decade from 2011 there were no records of the species. In 2020, however, the sighting of four adults raises the possibility of a return of this rare butterfly to Goss Moor, although its future remains uncertain. With the support of major funding, Natural England plans to manage and improve habitats across Goss Moor for, amongst other species (and habitats), the Marsh Fritillary.

The Silver-studded Blue has been recorded in the area but not since 2010, where it was last recorded at Breney Common and on Goss Moor (on the disused railway line at St Dennis Junction). St Dennis Junction also contained one of Cornwall's most consistent populations of Grizzled Skipper. However, despite habitat management work being carried out on site in the years prior, the species was last seen there in 2011.

Carnmenellis National Character Area (NCA 155)

Carnmenellis is, like Hensbarrow, an inland NCA, although its proximity to the sea creates a mild climate. It lies between the towns of Camborne to the north, Helston to the south-west and Falmouth to the south-east. It has a predominantly agricultural landscape of "rolling hills divided by regular fields bounded by Cornish hedges… combined with the scattered areas of heath, woodland and moor" (Natural England, 2014b: 3). The NCA gets its name from its highest point, the hill of Carnmenellis, at 252 metres AOD, and has the popular bird-watching site of Stithians Reservoir at its centre.

The underlying geology of the Carnmenellis NCA is granite. The area had lodes exceptionally rich in copper, tin, lead, tungsten, silver and uranium, making it one of the most important tin- and copper- producing areas in the world during the eighteenth and nineteenth centuries. The landscape of this area has therefore been greatly shaped by mining, and 14% of the NCA forms part of the much bigger Cornwall and West Devon Mining Landscape World Heritage Site. The old mine dumps and disused buildings create a substrate rich in heavy metals, which has been colonised by a

Stithians Reservoir looking towards the hill of Carnmenellis

range of rare plant specialists, resulting in a network of sites within the NCA designated as the West Cornwall Bryophytes SSSI. Nationally rare liverworts found here include Greater Copperwort *Cephaloziella nicholsonii*, Lobed Threadwort *C. integerrima* and Lesser Copperwort *C. massalongi*, along with the nationally rare Gravel Thread-moss *Pohlia andalusica*.

Rough moorland once dominated this landscape, but now only fragments survive on the highest, most exposed hills, such as Crowan Beacon and Carnmenellis, where remnant heathland and wet moorland grassland remain. Most of the land is undulating and dissected by small streams radiating from the highest points of the granite. Fields are small and enclosed by Cornish hedges, their vegetation of gorse *Ulex* spp. and heathers often reflecting the former heathland upon which they were created. Woodland is generally scarce (only 6% of the total area), but there are areas of wet woodland and deciduous woodland in the valleys. These areas are all connected by a dense network of sunken lanes and Cornish hedges that provide important nectar sources for insects.

Of the 37 resident/regular visiting butterfly species in Cornwall, only 24 have been recorded within the Carnmenellis NCA, the lowest number of all the NCAs except for the Isles of Scilly. The more common species that fly along the hedged lanes and footpaths include Large and Small Skippers, Speckled Wood, Ringlet, Peacock and Small Tortoiseshell. The heathlands and the former mine sites, with their mosaic of habitats and bare ground, attract Wall, Small Heath, Small Copper and Dingy Skipper (last recorded at Carn Brea Castle, near Redruth, in 2013). Comma and Holly Blue have been recorded around the fringes of this NCA, with its deeper, more sheltered valleys. The Small Pearl-bordered Fritillary has been recorded on Porkellis Moor, where its larval foodplants, Common Dog-violet and Marsh Violet flourish.

Wall, found in the Carnmenellis NCA

Coast near Botallack, West Penwith NCA

West Penwith National Character Area (NCA 156)

West Penwith, which forms the Land's End peninsula, lies at the south-west tip of Cornwall, the toe of England thrust into the Celtic Sea. It includes the towns of St Ives and Penzance. West Penwith NCA has over 800 hectares of nationally and internationally important nature conservation sites, representing 4% of the land area, with a further 19% of the land being designated as local sites important for wildlife. The Cornwall AONB covers approximately 13,470 hectares, 67% of the total area, predominantly around the coast. The NCA also includes part of the Cornwall and West Devon Mining Landscape World Heritage Site (Natural England, 2013c).

The underlying granite and surrounding sea define the character of the peninsula. The higher ground is predominantly moorland and rough pasture with "wind-swept, rock-strewn heaths, ancient field systems and settlements, with frequent remains of a bygone mining industry clearly evident" (Natural England, 2013c: 7). The higher northern section of the plateau is dominated by a "north-facing heath-covered ridge sloping down through a farmed coastal plain and a narrow strip of coastal heath to steep cliffs that plunge into the Atlantic" (Natural England, 2013c: 7). Short streams channel the high rainfall of the NCA down through steep, wooded valleys to the sea via high waterfalls, sheltered coves, and narrow inlets known locally as 'zawns'.

The combination of high rainfall and granite gives rise to thin, peaty-topped soils, forming a mosaic of habitats with extensive dry and wet heathland, grassy marshes, gorse scrub, rocky outcrops, and cliff slopes with unimproved grasslands. The network of ancient Cornish hedges is a key feature of the landscape, described as one of "the world's oldest artefacts still in use" (Natural England, 2013c: 8). Due to the exposed nature of the landscape, woodland covers only 4% of the area, much of it being scrub woodland and willow carr that has developed in the valleys.

Of the 37 resident/regular visiting butterfly species in Cornwall, 31 have been recorded in West Penwith, although the Marsh Fritillary is believed to have disappeared from the area (with the last records of adults and webs, both in single figures, in 2010 at an inland site north-west of Penzance). The heathland and agriculturally unimproved grasslands, particularly along the coast, provide favourable habitats for Wall, Small Heath, Grayling, Small Pearl-bordered Fritillary, Green Hairstreak, Small Copper and Silver-studded Blue. A visit in July could well be rewarded with sightings of all these species on the wing.

Grayling, found in the West Penwith NCA

Both the Pearl-bordered Fritillary and Silver-studded Blue are mentioned in the citation sheet for the Aire Point to Carrick Du SSSI, which stretches almost the entire length of the north coast of the West Penwith peninsula. The Pearl-bordered Fritillary, recorded here in low numbers until 1999, has not been reported since. The Silver-studded Blue, by contrast, is still found in good numbers along this coastal stretch. This species is also mentioned in the citation sheet for Porthgwarra to Pordenack Point SSSI, south of Land's End, along with the Dark Green Fritillary. These two SSSI citation sheets highlight the importance of the unimproved, semi-natural habitats along the coast in supporting species diversity.

Dingy Skipper and Brown Argus have been recorded only sporadically within this NCA, with a handful of records of Brown Argus stretching from Morvah along the north coast to Gwennap Head on the south, and two records of Dingy Skipper, one at Pendeen in 1996 and the other at Levant in 2000.

As woodland here is scarce and restricted to scrub woodland and willow carr, species such as the Brimstone, Silver-washed Fritillary and Holly Blue are less common.

The Lizard National Character Area (NCA 157)

The Lizard peninsula forms the most southerly part of mainland Britain and is surrounded on three sides by sea, with the Helford River forming part of its northern boundary. It is dominated by a "gently undulating exposed heathland plateau cut by narrow river valleys" with a rugged coastline comprising "caves, enclosed bays and small rocky islands" (Natural England, 2013d: 3).

The Lizard has an exceptionally distinctive character, resulting from a combination of a mild maritime climate and its unique and complex geology and soils. It is recognised as being of outstanding conservation importance due to its range of semi-natural habitats, which support diverse flora and fauna, some of which are unique to this NCA or are nationally rare, such as Pigmy Rush *Juncus pygmaeus*, Wild Asparagus *Asparagus prostratus* and Cornish Heath *Erica vagans*. This has led to approximately 20% of the peninsula being designated a National Nature Reserve (managed by Natural England, the National Trust and Cornwall Wildlife Trust), with flower-rich coastal grasslands and heaths, as well as inland heathland. Forty-two per cent of the area of the NCA has either an international or national designation for wildlife and/or geological interest, and 98% of The Lizard is within the Cornwall AONB. More than 25% of the peninsula comprises important habitats ('priority habitats').

The plateau of The Lizard is an ancient erosion surface, planed flat beneath the sea about 4 million years ago, during the Pliocene epoch. Beneath it lie the geologically important rocks of The Lizard, some of which were formed in the Precambrian eon more than 600 million years ago. The NCA can be broadly divided into two geologically distinct areas, separated by a major fault. North of the fault are Palaeozoic sedimentary rocks, and to the south lies the more ancient complex of highly crystalline, metamorphic rocks, which include serpentinite, schist and gneiss. Serpentinite is a geological rarity, as it represents part of the Earth's mantle scraped up to the surface of the overlying crust.

The unique geology, particularly the serpentinite (due to its high magnesium content), gives rise to a similarly unique range of heathland and wetland habitats that support many rare plants and invertebrates. Sixteen per cent of The Lizard comprises heathland, with its mix of heathers. The heathlands are criss-crossed by ancient trackways, which, depending on the weather, vary from waterlogged to desiccated. These provide a habitat for specialist plants and animals that are more used to the seasonal pools of Mediterranean climates.

Across the peninsula are small, irregular, ancient fields bounded by Cornish hedges and winding, flower-rich lanes leading to the coast, with its hard-rock cliffs and valleys to the west and sandy beaches to the east. Woodland and trees are scarce on The Lizard and are largely confined to the north-east of the NCA, where small copses increase in size and abundance with proximity to the sheltered valley of the Helford River. As a consequence, species such as Silver-washed Fritillary, Purple Hairstreak and Speckled Wood are less common, although Speckled Wood can be seen flying along tree-lined paths and lanes.

A total of 141 Red Data Book species of plant (those that are nationally rare) have been recorded on The Lizard, representing 53% of the total for Cornwall (French, 2020). Many of these plants are found in the heathland, maritime grassland, mire and cliff habitats. The Lizard also supports nationally important communities of lower plants, with over 200 species of lichen recorded in the serpentinite area. Another rarity is the Chough *Pyrrhocorax pyrrhocorax*, which can be seen fossicking for grubs in short turf around sea cliffs. It arrived naturally in 2001 from Irish stock, and has been breeding successfully ever since.

Of the 37 resident/regular visiting butterfly species in Cornwall, 30 have been recorded on The Lizard, including recent though unconfirmed reports of Dingy Skipper around Kynance Cove. The peninsula is the stronghold in Cornwall for Small Pearl-bordered Fritillary and a key area for Marsh Fritillary, Grayling and Silver-studded Blue. Indeed, Marsh Fritillary is one of the reasons for the designation of the East Lizard Heathlands as a SSSI, with a breeding colony previously existing at Main Dale.

Heathland is the key butterfly habitat on The Lizard. The heaths start to come alive in late spring/early summer, with Small Pearl-bordered Fritillary and Small Heath taking flight. Marsh Fritillary might be glimpsed at any of its 14 key sites on The Lizard, where its

Heathland on Goonhilly Downs, The Lizard NCA

Small Pearl-bordered Fritillary, found in The Lizard NCA

larval foodplant, Devil's-bit Scabious, grows. Its habitats range from damp grassland dominated by Purple Moor-grass to coastal grassland that retains its moisture and damp heaths. As the summer progresses, Dark Green Fritillary and Silver-studded Blue fly low over the heathers, with Grayling flying in late summer.

Despite the abundance of its larval foodplants, the Green Hairstreak is represented by only a few records since 1999. The Wall and Small Copper favour the bare ground and rock outcrops of The Lizard's coastal areas.

Isles of Scilly National Character Area (NCA 158)

The granite mass of the Cornubian batholith continues beyond Penwith out into the Celtic Sea, where its ruggedly uneven surface forms the Isles of Scilly, located about 30 miles west of Land's End. The archipelago comprises over 200 islands and rocky islets distributed over an area of approximately 200 square kilometres.

The Isles of Scilly are all that remains of a drowned landscape. Ancient causeways, dwellings and walls submerged between the islands provide evidence that the present shorelines date from the eleventh to fifteenth centuries (Thomas, 1985). Until the Middle Ages there were three main islands with low hills and sand dunes surrounding a shallow plain, which transformed into the present-day landscape with its numerous islands as sea level began to rise after the last ice age (Parslow, 2007). The islands today cover an area of just over 14 square kilometres, and range from small rocky outcrops, barely breaking the surface at high tide, to larger, vegetated islands reaching 45 metres in height, five of which are inhabited by people.

The Isles of Scilly have a number of international and national designations for

View across Great Pool, Bryher, Isles of Scilly NCA

their important wildlife and geology covering both land and sea, and form an AONB. With maritime heath and grassland predominating, compared with the Cornish mainland the range of habitats is limited, with an absence of rivers and wetland, and with minimal woodland cover. The restricted land area and lack of habitat diversity have led to a paucity of species in all groups of flora and fauna as compared with mainland Cornwall (Parslow, 2007). However, due to their mild oceanic climate, the islands are the only UK location for many Mediterranean plant species, including Dwarf Pansy *Viola kitaibeliana* and Least Adder's-tongue *Ophioglossum lusitanicum*, as well as being strongholds for many unusual lichen communities because of the purity of the air. Rare arable plants survive in the islands' small bulb strips, where species found include Smaller Tree-mallow *Malva multiflora*, Corn Marigold *Glebionis segetum*, Small-flowered Catchfly *Silene gallica*, Shepherd's-needle *Scandix pecten-veneris* and Western Ramping-fumitory *Fumaria occidentalis*.

The islands are also an important staging post for migrating birds and are often the first landfall for rare species blown off course. There are very few resident species of butterfly. Parslow states that "perhaps the most interesting aspect of the butterfly fauna of Scilly is those species that do not occur in the islands or only as very rare vagrants. Several very common British butterflies are still rarities in Scilly" (Parslow, 2007: 298). Such rarities include Wall and Gatekeeper, both vagrants. Skippers and hairstreaks are absent from the islands, along with fritillaries, with the exception of the rare migrant Queen of Spain Fritillary *Issoria lathonia*, at least two of which were recorded on St Mary's in 2006 and two single vagrant records of Silver-washed Fritillary and Dark Green Fritillary.

Of the 37 resident/regular visiting butterfly species in Cornwall, only 15 have been recorded on the Isles of Scilly as residents or common migrants, including Large White, Small White, Green-veined White, Clouded Yellow *Colias croceus*, Red Admiral *Vanessa atalanata*, Peacock, Small Tortoiseshell and Holly Blue. Ringlet and Comma are more recent additions to the butterfly list of the islands, having colonised since the 1990s. A further six species of resident/regular visiting species in Cornwall are migrants/vagrants on Scilly: Orange-tip *Anthocharis cardamines*, Brimstone, Wall, Gatekeeper, Silver-washed Fritillary and Dark Green Fritillary.

Notable on the islands is the Speckled Wood butterfly, which is the subspecies *insula* (also found on the Channel Islands, which may have been the source of the colonisation). This differs from the subspecies *tircis* found on the Cornish mainland in that it has deeper orange spots, closer to the normal (nominate) southern European subspecies.

The Meadow Brown butterflies on Scilly generally show more differences in both spotting and colour than those on the mainland, tending to be brighter with more conspicuous orange spots. They are thought to be the subspecies *cassiteridum*.

Further details concerning the butterflies of the Isles of Scilly are given in chapter 6.

Speckled Wood subspecies insula, found in the Isles of Scilly NCA

CHAPTER 3
Recording and conserving butterflies

In this chapter, we describe the recording and conservation of butterflies in Cornwall undertaken by a number of organisations, farmers and landowners, conservationists and other individuals, working alone or together, many in a voluntary capacity. Everyone has their own reason for recording. Some do it purely for pleasure, but most want to reverse the decline of butterflies (and moths) across the County, in gardens and on farms, on nature reserves, along roadside verges, on common land, coastal cliffs and sand dunes, in woodland, on moorland and on former mining sites. The key is to understand the life cycle and ecology of each species of butterfly so that the habitat on which they depend can be conserved and restored using appropriate land-management techniques. The data obtained from recording, monitoring and surveying species are indispensable when making informed choices about targeting scarce resources where they are most needed and have the greatest impact on conserving butterflies in Cornwall.

Cornwall Butterfly Conservation

Cornwall Butterfly Conservation (CBC) is the County's branch of Butterfly Conservation (BC), the national charity, which works towards "a world where butterflies and moths can thrive and can be enjoyed by everyone, forever" (Butterfly Conservation, 2019: 2). Since its launch in 1993 at Greenscoombe Wood, CBC has grown to over 840 members in 2020.

CBC runs events throughout the County (often in partnership with other conservation-minded organisations), including field trips, practical conservation days, learning workshops, species survey days, and fund-raising events. Information on recording, volunteering and events is available on the CBC website http://www.cornwall-butterfly-conservation.org.uk as well as in CBC's twice-yearly magazine, *Cornwall Butterfly Observer*. On the website is a video that the branch commissioned from the award-winning, Cornwall-based documentary film-maker Nina Constable. Entitled *Cornwall's Butterflies: Back*

A highlight of the CBC year is the open day at Lethytep's eco-friendly flower meadows, lakes and woodland

from the Brink, this hauntingly beautiful film, narrated by Nick Baker, captures the fragility of Cornwall's butterflies, and CBC's volunteer work to protect them. The branch has also developed a strong social media presence through Facebook, Twitter and Instagram.

As with other branches of Butterfly Conservation, the Cornish branch is run by a committee of active volunteers, each with a role to play. Uniquely, however, CBC funds a part-time Volunteer Co-ordinator to recruit, train and support volunteers to record butterflies and undertake practical habitat conservation. One of the Volunteer Co-ordinator's responsibilities is the health, welfare and safety of every CBC volunteer in the field. Beginning in 2014, when the Volunteer Co-ordinator recruited 80 volunteers, so successful was the role that by 2019, 98 individuals gave their time to help CBC with a range of activities, amounting to 425 person-days of volunteer time, equivalent to £18,530 of donated time (calculated at the rate of the National Living Wage).

Volunteers learn the importance and techniques of accurately recording Lepidoptera in Cornwall. The number of individuals walking butterfly transects has doubled since 2014. Many volunteers have learnt to identify the larval webs of a target species like Marsh Fritillary *Euphydryas aurinia*, greatly increasing the effectiveness of recorders qualified to undertake butterfly surveys of priority species on the ground. These surveys have taken place on land owned, managed or overseen by partner organisations such as Natural England and South West Lakes Trust, which funded work by CBC. Five years of the Volunteer Co-ordinator contract has also raised the profile of CBC and increased membership.

CBC encourages volunteers to become Species Champions, an initiative that increases individual expertise in a single butterfly species whose ecology is insufficiently understood or whose survival is at risk in the County. The studies undertaken by a Species Champion will help CBC to understand the current status of a particular species in Cornwall and to target conservation resources effectively.

Jo Poland, Volunteer Co-ordinator

In 2014, through a process of competitive tender and interview, I landed the contract for the new role of CBC Volunteer Co-ordinator. With many years of volunteer co-ordination experience behind me at the Citizens Advice Bureau, it was my dream job!

The part-time role involves 30 days a year, and by communicating through email and social media I make the hours stretch. At least a third of my time is devoted to pre-visit planning, communication and risk assessment, to ensure that fieldwork goes without a hitch (weather and pandemics permitting!). I now support nearly 400 volunteers, carefully matching individuals to the work. Everyone has something to offer, and the following list illustrates the extraordinary scope of CBC volunteers' work, as they

- Clear scrub from many acres of land to create or reinstate habitat suited to target butterfly species
- Become regular butterfly recorders, Transect Recorders and contributors to the Wider Countryside Butterfly Survey
- Become Species Champions
- Find new sites for rare target species of butterfly
- Write and publish articles about butterflies and volunteering activities
- Manage the branch, oversee its activities, manage its finances, develop and review its business plan
- Promote CBC through social media
- Represent CBC at partner events
- Donate goods to sell/raffle, as well as making financial donations
- Compose and perform songs about CBC at folk clubs and CBC events

There will inevitably be things that I have missed, but I would like to thank each and every one of our volunteers for their dedication and hard work. CBC could do nothing without them!

Jerry Dennis, Species Champion for the Grayling *Hipparchia semele*

An interest in butterflies and moths has been with me since childhood. When I retired in 2017 and moved to Cornwall, I was delighted with the freedom to be able to walk in so many different places and to indulge in nature rambling and, of course, butterfly spotting. I began to record and report the butterflies that I saw.

In 2018, I took on a vacant butterfly transect at Gwennap Head, near Porthgwarra. After three years and many visits to this wonderfully atmospheric place, I have recorded thousands of butterflies and learnt so much about their life cycles and habits. I particularly remember a hot early July day in that first summer when the transect was full of butterflies; delicate Green Hairstreak, Small Copper and Silver-studded Blue were on the wing and the summer browns were flying in big numbers. I saw over 300 butterflies and 19 species on that one visit.

I was surprised and pleased to find the Grayling butterfly in one of its strongholds at the transect, having seen only a handful of them in my lifetime before moving to Cornwall. I volunteered to become the Grayling Species Champion in 2018 and set about the challenge of gaining a better understanding of this elusive butterfly. This has proved immensely interesting and satisfying; and the role is important, particularly for butterflies that are threatened at a national level. In 2020, for the first time, I visited most of the key sites for the species and recorded well over 1,000 Graylings over that one summer. Clearly Grayling colonies in Cornwall are currently doing well. They are precious, and we must make sure that we monitor and safeguard them for future generations.

A priority for CBC is to have a Species Champion for each of the 11 'species of principal importance' listed under Section 41 of the NERC Act 2006 (and former Biodiversity Action Plan (BAP) priority species) that occur in Cornwall. Besides those 11 priority species, however, a Species Champion is welcome to choose any Cornish butterfly that interests them.

Recording butterflies in Cornwall

Cornwall has a long history of recording butterflies, with individuals such as W. Noye, W.P. Cocks and J.J. Reading compiling lists and comments on the Lepidoptera of the County in the 1800s (Spalding, 1992). However, R.D. Penhallurick observed that butterflies were often overlooked in favour of a group such as birds (Penhallurick, 1996). Societies were established throughout Britain in the nineteenth century in response to a burgeoning interest in natural history. In Cornwall, these included the Cornwall Literary and Philosophical Institution, established in 1818, which later, as the Royal Institution of Cornwall, founded the Royal Cornwall Museum in Truro, with important collections and records of butterflies; the Cornwall Polytechnic Society in 1833 and the Penzance Natural History and Antiquarian Society in 1839. Although the term 'scientist' had yet to be coined, the expertise of Georgian and Victorian naturalists should not be underestimated. The journals of such societies exemplify a scientific rigour and knowledge amongst a largely male membership, in which landowners, lawyers, physicians and the clergy typically played a leading role.

Today, numerous organisations and individuals are involved in recording butterflies within the County, with CBC playing an important role. The majority of records used to be made opportunistically by individuals out for a walk; this still occurs, but to these valuable records are now added more systematic records of butterflies from routes (transects) walked regularly through the flight season every year; records from 1km squares collected at least twice a year for the Wider Countryside Butterfly Survey and timed counts. All these records are sent to CBC's County Recorder, who collates, manages and assures

How to record a butterfly

1 In the field
When you spot a butterfly, note down the following:
- Species name, if you know it (or, if you're not sure, take a photo to help identify it later).
- Number of individuals seen and their sex, if you can tell (assumed to be adults unless stated otherwise).
- Location (a place/area name, e.g. Cardinham Woods).
- Grid reference (Ordnance Survey (OS) grid reference to six (or more) figures, e.g. SW835547).
- Other field observations that might be of interest, such as weather conditions – high winds and wet weather account for fewer sightings; if nectaring, the species of plant being used.

2 Back at home
- Download from the CBC website the CBC Casual Recording Form, which you can find under the 'Recording' tab.
- Enter the information prompted on the form (an Excel spreadsheet). This includes:
- Species name – the common name is enough, but the Latin name in addition is helpful
- Grid reference (OS grid reference to six (or more) figures, e.g. SW835547)
- Location – where the butterflies were seen (nearest place name taken from an OS map)
- Date of sighting
- Quantity seen
- Your name
- Any comments

3 Submit your record
Please send your completed record(s) for Cornwall to the County Recorder at records@cornwall-butterfly-conservation.org.uk

Records can also be submitted by using the iRecord Butterflies app or via a form on the Cornwall Butterfly Conservation website.

the quality of the butterfly records collected in the County and enters them into the ERICA database. This database, under the aegis of the Cornish Biodiversity Network, holds records of the flora and fauna of Cornwall, with some records dating back to the nineteenth century.

Each year, the County Recorder submits all butterfly records collected locally to Butterfly Conservation's national team, where they are collated as part of the Butterflies for the New Millennium Scheme (BNM). This scheme, which maps the distribution of butterfly species throughout the UK, was launched by Butterfly Conservation and the national Biological Records Centre in 1995 to coordinate the recording effort "in a more comprehensive and systematic way than ever before" (Asher *et al.*, 2001: ix).

Recording on a CBC field trip at Penhale Sands

The total number of records received from all sources rose from 9,491 in 2009 to 34,316 in 2018. The number of butterflies seen increased from 42,372 in 2009 to 136,270 in 2018. The increased data provide better detail of what is happening to butterfly populations in Cornwall, including identifying new sites where rarer species have been found for the first time but may have been overlooked in the past. The impact of increased recording activity on population trends is still a matter of speculation, an issue that is discussed in chapters 1 and 4.

Most people record the adult stage of the butterfly, but it can be equally rewarding to search for and identify the eggs, larvae and pupae of each species. Observations of immature stages can be useful for guiding conservation work, particularly for the rarer species. Gaining further information on their ecology, such as preferred breeding areas or larval foodplants, can help to inform habitat management and aid in understanding periodic fluctuations in populations. The increasing number of informative and well-illustrated books and websites on each stage in the life cycle of British butterfly species make identification possible. (See chapter 8.)

Dick Goodere, County Butterfly Recorder (2011–20)

My wife and I arrived in Cornwall a few days after the millennium celebrations, when I started a new job in child protection. Maggie had more of a background in the environment than I, but both of us were new to butterflies. By chance, Sally Foster, an ecologist and expert in this field, happened to live in our village and was running a training course on butterflies in the village hall a few metres from our house. We were persuaded to attend and found it so fascinating that we never looked back! While I was working, Maggie walked a transect and became Hon. Secretary of CBC for seven years. I retired in 2007 and gradually became more involved with butterfly matters, first by walking my own transect. Then, when John Worth retired as County Recorder in 2011, in the absence of anyone else coming forward I took on the role. With no idea what an immense task I was embarking on, I initially shared duties with another CBC committee member, before taking it on fully from 2012. It has been extremely time-consuming and absorbing work. I am not technologically gifted and ERICA, our marvellous database, at first presented a challenge. I have particularly enjoyed analysing the end-of-year statistics and assessing the trends that have informed this book. Records have increased over the last decade from under 10,000 to over 37,000 and I feel that it is now time to pass the baton on to someone younger, already very knowledgeable, and equally enthusiastic.

Cerin Poland, County Butterfly Recorder (2020 onwards)

I feel privileged to have taken over the role of County Butterfly Recorder from Dick. His shoes will certainly be hard to fill! As County Recorder, I will carry on verifying records to a high standard, ensuring we maintain partnership working using our solid datasets to inform vital conservation work. Continuing to support the ever-growing recorder network, making recording even more accessible and available to more diverse communities, is an important goal for me. I am always happy to answer questions or queries and to help with identification, so do not hesitate to get in touch!

Contact details can be found on the CBC website under either the 'Recording' or 'Contact' tabs. Records can also be submitted by using the iRecord Butterflies app or via a form on the Cornwall Butterfly Conservation website.

Richard Symonds, submitting butterfly records in Cornwall

I first became interested in butterflies when I was around five years old, collecting larvae of Small Tortoiseshell and Peacock butterflies and then watching them complete their life cycle into adults. My interest in Lepidoptera was reawakened around 20 years ago after buying an old copy of *Butterflies* by E.B. Ford. I began to look for species that I had not seen before, and then a few years later when I had internet access I started to report my sightings to the branch of Butterfly Conservation in Hampshire, where I used to live. Since the advent of digital photography I have also enjoyed photographing butterflies while searching for aberrations of some species.

On moving to Cornwall in 2001, my wife and I settled in the Land's End peninsula. The local landscape of my village of Pendeen comprises a mixture of open moorland and coastal cliff paths, the latter supporting a number of colonies of Silver-studded Blue and Grayling. What I enjoy most about recording is gaining local knowledge of where to find certain species, to understand their habitats and requirements. My favourite spot to record butterflies stretches along the coastline from Pendeen Watch to Botallack, where in June and July a number of colonies of Silver-studded Blue are found. Penrose Woods near Helston is another favourite location in high summer to seek out Purple Hairstreak in the oak trees.

My favourite butterfly, the Purple Emperor *Apatura iris*, is not found in Cornwall. My first sighting was of a male in a wood in Hampshire feeding on the ground for around 10 minutes. I even persuaded it to rest on my hand and have a photo to prove it! My most exciting moment happened in the same wood a few years later when I met Purple Emperor expert Matthew Oates, who had been observing a butterfly, which took flight from the ground just after my arrival. This was a seldom seen aberration known as *lugenda*, with little or no white markings on the upperside. A rare moment shared!

National butterfly recording schemes

Butterfly Conservation has several recording programmes running nationally in which anyone can participate, whether a member or not (this also applies to county-run schemes). These recording schemes, which include the Wider Countryside Butterfly Survey and the Big Butterfly Count, bring together important data on the health of our more common species and the wider countryside as a whole.

The **Wider Countryside Butterfly Survey** (WCBS) was launched in 2009 to gauge the changing abundance of widespread species in the general countryside, as a means of redressing the disproportionate number of records submitted from nature reserves and areas rich in butterfly species. This survey was designed to gauge the health of the British countryside as a whole, particularly as 70% of the UK's land is in agricultural use. The WCBS involves volunteers carrying out surveys in 1km grid squares, selected at random to sample the whole country. Participants are asked to record all the butterflies within their allocated square at least once in July and again in August, with optional visits in May, June and September. In 2019, about 650 volunteers completed 1,957 surveys in over 800 1km squares across the UK, submitting records of 129,866 butterflies of 46 species (Wider Countryside Butterfly Survey Team, 2019).

The **Big Butterfly Count** was launched by Butterfly Conservation in 2010, as a means of engaging the wider public in a nationwide citizen science project. Participants are asked to count the number of each of 17 common species of butterfly (and two day-flying moth species) seen during a 15-minute period. This should be on a bright and sunny day at a site of their choice, during a three-week period in summer. From fewer than 16,000 counts submitted in 2010, participation soared to 116,009 counts by 113,500 people in 2019. The Big Butterfly Count has resulted in more than 7,000 additional records for Cornwall submitted annually in recent years.

Long-term butterfly recording on sites in Cornwall

Whilst all butterfly records are of value, those made every year at the same site (and ideally by the same person) are of greater value for monitoring populations in detail. Targeted recording and monitoring is particularly important for the long-term survival of rare species. Many butterflies are being regularly monitored in Cornwall, either through targeted surveys or by means of general butterfly transects.

Transect recording

Transect records provide valuable data about butterfly populations at a particular site over time, and contribute to monitoring trends at county, regional and national level. Data gathered consistently each year enable changes in butterfly numbers to be monitored over time, both in abundance and in the variety of species present. The method, which was devised in 1973 by the UK Centre for Ecology & Hydrology (UKCEH), involves an individual walking the same route each week from April to September in suitable weather conditions, counting the number of individuals of each butterfly species seen. Whilst personally satisfying, this method of recording demands long-term commitment from the recorder.

Kelly Uren, County Transect Co-ordinator

My first involvement with CBC was in 2013. During the summer of that year I spotted details of a field trip scheduled for a Saturday afternoon in June at Chapel Porth, near St Agnes. The notice stated that the Green Hairstreak was expected to be encountered on the walk. I had never heard of this very exotic-sounding butterfly before, let alone seen one, so I was instantly intrigued and couldn't resist going along.

Despite its being the height of summer, when the day arrived the morning started out pretty grey and a little windy, not the conditions usually associated with butterfly activity, but that didn't put me off. On arrival, I spotted a small group of people who had gathered in a corner of the car park, so I nervously headed over and was instantly greeted by a very enthusiastic man, who explained that I was indeed in the correct location for the butterfly walk. I immediately realised that I might not have been wearing the 'correct' apparel: everyone around me was kitted out in muted greens and browns, all blending in well with the natural surroundings, whereas I had turned up in a very bright pink coat! I had a really lovely afternoon, the sky soon cleared and we saw a great number of Green Hairstreak; I never realised that such beautiful and small butterflies existed in Cornwall before. The butterflies were very well hidden amongst the vegetation growing upon the cliffs, and I was in awe of the other people in the group, who were clearly very knowledgeable and had the most phenomenal eyesight. Numerous butterflies were being spotted all over the place, but all I could see was the greenness of the gorse. As a new face in the group, I was made to feel incredibly welcome and my initial nervousness of attending the field trip on my own quickly disappeared. Everyone was happy to help my 'new eyes' to adjust to see the beautiful butterflies around us.

I attended many more of the fabulous butterfly field trips that year, and to my amazement I discovered that Cornwall is home to more than just the small handful of butterflies I had believed there to be. And the lovely man who greeted me at the beginning of the walk? He turned out to be Jim Barker, CBC Transect Co-ordinator at the time and very knowledgeable about butterflies.

I was soon invited to attend a CBC committee meeting and then later to become a committee member. Jim has kindly taken me under his wing, sharing his vast butterfly knowledge with me, while I do my best to remember all that he tells me. I have since joined Jim in the position of Transect Co-ordinator for Cornwall, jointly overseeing the co-ordination of transects in the County. Becoming actively involved with CBC has without a doubt changed my life. It has opened up a whole new world to me. I have got to know a really lovely group of people who are incredibly passionate about conserving butterflies and indeed other wildlife in Cornwall, not only for us to appreciate and enjoy now, but also, hopefully, for many generations to come. I feel immensely honoured to be able to play a part in this.

Data from transect monitoring are particularly valuable when the same transect is walked for many years, with the methodology and recorder remaining constant. In the analysis of transect data, weekly counts for each species (and each brood) are summed (including estimates for missing weeks where appropriate) to calculate site annual indices of abundance. Although not an absolute measure of population size, these counts have been shown to be good indicators of population size and change. This information can then inform site conservation plans and priorities.

Richard Vulliamy, transect recorder at Greenscoombe Wood

Butterfly transect recording in Cornwall began in 1980 at Greenscoombe Wood in the Tamar Valley and Erisey Barton on The Lizard. Both of these sites are still actively recorded, although Erisey Barton was not walked during the 1990s or 2000s and was started again in 2013. Greenscoombe Wood, important for the Heath Fritillary *Melitaea athalia*, is unique in that data have been gathered from this site over a period of 40 years, only missing the two years 2009 and 2011. The number of transects walked in Cornwall has risen steadily from the year 2000, with a three-fold increase since 2010 to 50 sites monitored by transects, generating about a third of the total annual butterfly records gathered in the County by 2019.

Gwithian Green: an example of a Cornish transect (Foster and Dennis, 2020a)

Occupying an area of approximately 7 hectares near Hayle, Gwithian Green is situated about 800 metres from the sea on the edge of the village of Gwithian. It was designated as a Local Nature Reserve in 2002 and is floristically the richest and most diverse part of the Hayle Towans to Godrevy Towans sand dune complex. This relatively small area encompasses a wide variety of plant communities in habitats that range from calcareous, fixed dune grassland and dune slack/fen/marshland, to stream and pond communities and acid grassland and woodland.

The Gwithian Green butterfly monitoring programme started in 1997 with a 937-metre transect route being devised and sub-divided into eight sections passing through several different habitats. The recording of all adult butterflies along this transect has two main functions: first, to be part of the UK Butterfly Monitoring Scheme (UKBMS), and second, to inform land management relating to butterflies by a local group.

It is one of the earliest transects set up in Cornwall, and has one of the most continuous datasets, having been walked every year since 1997, generating a wealth of valuable data for analysis. Because the transect has been walked almost entirely by the same recorders over more than two decades, the dataset generated is consistent and reliable, enabling a detailed analysis of how butterflies have fared every year on the Green, with yearly occurrence and abundance data as well as longer-term trends plotted over a 24-year period. Species have been analysed in relation to how numbers have changed over time throughout the whole transect, over the various sections of the transect and in the differing habitats. Performance of grassland species have been looked at separately from those that are habitat specialists and those more wide-ranging species, with correlations made with weather conditions. There has also been a recent phenological analysis and comparisons made with South-west England and national data.

In the 24-year period from 1997 to 2020, 22,805 individual butterflies were recorded for 25 species. The total number of butterfly

Gwithian Green from the air

species recorded on the Green has been maintained over the years, with the addition of Silver-studded Blue *Plebejus argus* from 2013 onwards. There has been a gentle upward trend over the 24 years in the total number of butterflies recorded.

Of the 11 butterfly 'species of principal importance' (Section 41 of NERC Act 2006) resident in Cornwall, four have been recorded on Gwithian Green: Wall *Lasiommata megera*, Small Heath *Coenonympha pamphilus*, Small Pearl-bordered Fritillary *Boloria selene* and Silver-studded Blue. They all occur in small numbers and this makes data analysis problematical. What can be ascertained from the data, as well as from anecdotal evidence, is that there is concern about their apparent decline, apart from the Silver-studded Blue, whose numbers have increased since a small colony established itself on the Green in 2013. However, numbers of this species also declined in 2020, so all four species merit the continued attention of the local group.

Grassland species overall (including the Large Skipper *Ochlodes sylvanus*, Small Skipper *Thymelicus sylvestris*, Meadow Brown *Maniola jurtina* and Gatekeeper *Pyronia tithonus*) have declined slightly on Gwithian Green. These data allow comparison with trends elsewhere in Cornwall, as well as nationally. The graph below shows species on the Green initially declining more rapidly, compared with the rest of England, until about 2010, since when the fortunes of these grassland species have reversed and have bounced back on the Green, while national figures have done no better than level out. Numbers in 2020, however, dropped back slightly.

Grassland species: relative abundance since 1997

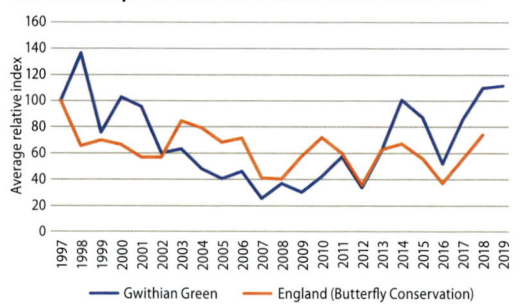

Initial work on the phenological analysis shows a progressively earlier emergence of individual butterfly populations through the 24 years, a trend attributable to yearly weather characteristics that exhibit progressive but subtle average warming of the local climate.

Timed counts

A timed count of individual butterflies on a site provides an estimate of the size of a population/colony of a particular species, often one of the rarer species, such as Marsh Fritillary, Pearl-bordered Fritillary *Boloria euphrosyne* or Silver-studded Blue. The number of adults seen flying within a defined area is counted, ideally at the peak flight period of the target species, with the start and end times of the count noted, along with the number of recorders. This method is useful to repeat within the same defined area a couple of times to ensure peak numbers are recorded. The method can also be applied to other life stages, for example when searching for the larval webs of Marsh Fritillary. CBC has trained members of partner organisations in Cornwall, including Natural England, the National Trust and Cornwall Wildlife Trust (CWT), to undertake such counts.

UK Butterfly Monitoring Scheme

Since 2006, a number of these national recording programmes have been brought under the umbrella of the UK Butterfly Monitoring Scheme (UKBMS), jointly overseen by the British Trust for Ornithology, UKCEH and the Joint Nature Conservation Committee. The UKBMS includes the long-running Butterfly Monitoring Scheme established by UKCEH in 1976, Butterfly Conservation's co-ordination of 'independent' transects (established in 1998) and, since 2009, the WCBS. It also includes the monitoring methods of timed counts of adult butterflies, single species transects, and egg and larval counts (known as 'reduced effort surveys of habitat specialists'). In 2019, the UKBMS co-ordinated the records of over 2.6 million butterflies on 3,003 sites across the UK (Brereton et al., 2020).

Conserving butterflies

Over-grazing, under-grazing and scrub encroachment (e.g. by Gorse *Ulex europaeus* and willow *Salix* spp.) have been identified as some of the main threats to butterflies such as Pearl-bordered Fritillary, Marsh Fritillary, Dingy Skipper *Erynnis tages*, Grizzled Skipper *Pyrgus malvae*, Small Pearl-bordered Fritillary and Grayling. Afforestation, especially with conifers, has been a problem in the past, particularly for Grizzled Skipper, Pearl-bordered Fritillary, High Brown Fritillary *Fabriciana adippe* and Heath Fritillary; this has the potential to become an issue again if ill-informed tree planting schemes proceed. (See chapter 1.) The absence of traditional broadleaved woodland management, such as coppicing, to create open glades is a further threat to several 'species of principal importance' in Cornwall. (See chapter 4.)

The agricultural improvement of wetlands has contributed to the decline of Marsh Fritillary and to a lesser extent Small Pearl-bordered Fritillary. The Grayling has declined inland due to the reclamation of post-industrial sites, particularly former mining sites, with the subsequent loss of bare ground habitat and associated short turf grassland.

The identification of these declines through ongoing recording and monitoring has led to a collaboration of organisations and individuals to pool knowledge and expertise for the conservation of butterfly species in Cornwall.

Several sites in Cornwall have been designated Sites of Special Scientific Interest (SSSIs) to protect Section 41 butterfly 'species of principal importance'. These include Aire Point to Carrick Du and Gwithian to Mexico Towans in the west, Greenscoombe Wood, Luckett in the east, and Penhale Dunes and Mid Cornwall Moors in central Cornwall (discussed in more detail in chapter 2).

Conservation organisations that manage part of their land holdings for butterflies in Cornwall include the National Trust, which owns more high-quality butterfly habitat than any other organisation in the County; Natural England, responsible for three National Nature Reserves (NNRs); CWT, which manages over 50 nature reserves, with Windmill Farm and Maer Lake jointly owned and managed by Cornwall Bird Watching and Preservation Society, and Marsland Nature Reserve in North Cornwall managed jointly with the Devon Wildlife Trust. Whilst these nature reserves are typically managed for a diversity of species and habitats, the conservation of butterflies becomes one of the main focuses of strategic site management on reserves that hold populations of butterfly 'species of principal importance'. The Gaia Trust manages its land on the Mid Cornwall Moors SSSI for the benefit of the Marsh Fritillary, in partnership with CBC and Natural England.

Organisations not primarily concerned with the natural environment also play an important part in butterfly conservation, in consultation with environmental bodies such as Natural England and Butterfly Conservation. The Ministry of Defence (MOD) owns extensive and enclosed land holdings at Penhale Sands (Penhale Dunes SSSI) and these are managed on their behalf by CWT for 'species of principal importance', including Silver-studded Blue, Dingy Skipper and Grizzled Skipper; and the Duchy of Cornwall manages the SSSI for the Heath Fritillary in its extensive woodlands at Greenscoombe Wood.

Farmers and landowners can access advice on habitat management for butterflies and wildlife in general from Natural England, the Farming and Wildlife Advisory Group, the Nature Friendly Farming Network, Plantlife and CBC. To qualify for grants under the Government's Environmental Land Management (ELM) scheme, designed to replace EU agri-environment schemes, farmers and landowners must comply with environment-friendly practices.

Practical conservation work

One of the main responsibilities of the CBC Volunteer Co-ordinator is to lead groups of volunteers outdoors to deliver practical conservation work, aiming to improve existing habitat and to create new habitat for target butterfly 'species of principal importance'. Complying with CBC committee's conservation priorities, and those of partner organisations, this work usually takes place during winter and early spring. Volunteers work in teams of around a dozen, and field days rely on careful planning. This includes risk assessment, transport to remote locations and the provision of properly maintained hand tools.

From chopping and sawing, to raking, dragging and piling, feeding a fire, dead hedging or distributing soup and cake, there are jobs and tasks to suit everyone, whatever their ability. Safety is the overriding priority, and all events are carefully risk assessed, with the use of safety equipment such as gloves, eye protection and hard hats mandatory where required. Some volunteers have been specifically trained in the use of brush-cutters, chainsaws and chemicals. These individuals have specific risk assessments for the work they carry out. At least one trained first aider is present at all events.

Scrub clearance by CBC volunteers for the Marsh Fritillary on Bodmin Moor, March 2019

CBC Practical Conservation Volunteer: Russell Williams

Russell began volunteering with CBC in 2018, inspired by conversations he had with Jo Poland, the Volunteer Co-ordinator. He had previously spent many years looking at plants and animals through the lens of a camera, creating wildlife documentaries. With CBC, he has been involved in recording butterflies as well as undertaking practical conservation work, providing him with opportunities to visit a variety of beautiful places around the County that he never knew existed. He has been particularly struck by the abundance of wildlife within former mining/post-industrial heritage sites while participating in the Mining for Butterflies project. Russell's favourite butterfly is the Small Tortoiseshell, which likes to over-winter in his hallway!

Russell is now a full-time carer, and he finds the time spent with CBC in the field boosts both his physical and mental wellbeing.

Russell Williams (right) with fellow CBC volunteers

Recent practical work has included scrub clearance on former mining sites, following the survey work undertaken as part of the Mining for Butterflies project, working in partnership with CWT at Penhale Sands to improve the habitat for the Grizzled Skipper (see **Conserving the Grizzled Skipper at Penhale Sands, Perranporth** below), and with the Gaia Trust and Natural England on the Mid Cornwall Moors.

Field conservation work by CBC volunteers has to be carefully prioritised because there is always more work than they can do, and it is never-ending, a bit like painting the Forth Bridge! CBC volunteers have earned a reputation for getting things done efficiently and safely. Many organisations now rely upon their services and willingly contribute to the cost of the work.

Projects to record and conserve butterflies in Cornwall

CBC has played a leading role in the design and delivery of time-limited recording projects focused on particular habitats and/or species, in the twenty-first century. These programmes have either included or led to strategic conservation to improve conditions in the field for habitat specialists such as Marsh Fritillary and Pearl-bordered Fritillary. Examples include Mining for Butterflies (2015 onwards) and All the Moor Butterflies (2017–20).

Mining for Butterflies

4,000 years of metal extraction have left their imprint on the Cornish landscape, from moorland hummocks and hollows to the rocky spoil alongside derelict mine buildings. These disused sites, scattered across the County, provide a niche environment for wildlife. Residual traces of metal in the scree-like mine dumps combine to create an ecology not found elsewhere in Cornwall. These specialist habitats, often with their own micro-climate, suit certain organisms very well, including species of butterfly.

Mining for Butterflies is a continuing project, which encourages volunteers to survey wildlife (particularly butterflies) and carry out conservation work on the former metalliferous mining sites that are part of the World Heritage Site designation.

CBC began to assess disused mining sites in 2015; in 2018, 37 mining sites were visited by 20 CBC volunteers, who recorded 26 butterfly species, including six 'species of principal importance' – Dingy Skipper, Silver-studded Blue, Small Pearl-bordered Fritillary, Wall, Grayling and Small Heath – as well as several day-flying moth species.

Mining for Butterflies could not have proceeded without the support of Cornish Mining World Heritage, Postcode Local Trust, Cornwall Council, Crofty Developments Limited, Butterfly Conservation and CBC volunteers. At the time of writing, a bid to the National Lottery Green Recovery Challenge Fund has been submitted to continue the project in partnership with Buglife.

The grants from our funding partners enabled CBC to engage a qualified field ecologist to assess 20 further sites, owned by Cornwall Council. Between July and September 2018, the butterflies and habitats at each site were recorded, and for the six 'species of principal importance' listed above noted the presence of larval foodplants and other factors on which these species depend. An assessment was also made of the suitability of each site for further work by volunteers.

The resulting report has helped CBC to decide its priorities for improving habitats for butterflies on the former mining sites and to identify where more fieldwork is required. And perhaps most important of all, the report provides independent scientific evidence for just how important these derelict mine sites are for wildlife and the importance of looking after them as part of Cornwall's World Heritage Site.

Silver-studded Blue

Former mining site, Wheal Peevor, near Redruth

Scrub clearance at De Lank Quarry for the Pearl-bordered Fritillary as part of the All the Moor Butterflies project

All the Moor Butterflies

Over the course of three years, 2017 to 2020, the All the Moor Butterflies project improved habitat conditions and raised awareness of six of our most threatened butterfly and moth species on Exmoor, Dartmoor and Bodmin Moor. Initiated and co-ordinated by Butterfly Conservation and funded by the National Lottery Heritage Fund, the project worked with 146 landowners across 201 sites to deliver gains for these special species. Over 5,000 people learned about the fascinating lives of these wonderful insects and were given opportunities to contribute to their conservation.

On Bodmin Moor the project focused on four 'species of principal importance': Marsh Fritillary, Pearl-bordered Fritillary, Small Pearl-bordered Fritillary and Narrow-bordered Bee Hawk-moth *Hemaris tityus*. One of the most important measures of success is that the number of known sites for Marsh Fritillary on Bodmin Moor increased by 152%, making this landscape one of the best in the UK for this species of butterfly.

CBC volunteers played a key role in delivering the project and will continue to do so over its 10-year legacy, focusing on De Lank Quarry, to ensure that its success is sustained.

Conserving the Grizzled Skipper at Penhale Sands, Perranporth (Poland, 2020)

Since 2014 the Grizzled Skipper has only been recorded at a single site in Cornwall: Penhale Sands, where its distribution spans over several 1km grid squares.

In 2018, CBC ran a two-day training course in partnership with CWT on the identification and ecology of the Grizzled Skipper. This was well attended by volunteers from both

Grizzled Skipper resting on an orchid, Penhale Sands

Scrub clearance at Penhale Sands, 2019

organisations, as well as individuals from other bodies such as the National Trust. The training included a field trip to Gear Sands and Penhale Sands MOD area, resulting in valuable sightings in locations where the species had not been accurately recorded before. In 2019, CWT funded a search day for CBC to target the butterfly in the Penhale Sands MOD area, with volunteers recording the species in further new areas. The annual CBC field trip into Penhale Sands MOD has also provided important records. In 2018 and 2019 the Species Champion for Grizzled Skipper developed pin-point geographical recording of the butterfly, so that its distribution across the Penhale site can be better understood. These efforts over the last two years, along with records from casual recorders, have resulted in the species being accurately recorded in at least 25 new 100m grid squares where there were no previous sightings other than those at 1km grid square resolution.

Between 1999 and 2019, 302 records of the Grizzled Skipper were submitted, consisting of 691 individuals spread across 18 1km grid squares. With Cerin Poland taking on the role of Species Champion for the Grizzled Skipper, survey intensity increased significantly from 2018, recording adults, larvae and eggs. In 2019 91 records of 130 individuals were submitted in six 1km grid squares.

There are plans to continue the current recording effort on Penhale to ensure that all areas of key breeding habitat and locations suitable to the Grizzled Skipper are surveyed to establish whether or not it is present. Further searches are planned in areas where there are historical records and in locations of suitable habitat surrounding the Penhale site. Continuing to raise awareness of the butterfly in Cornwall with a hope for more records and sightings in unknown locations is also a priority.

In partnership with CWT's Penhale ranger, two days of habitat improvement work for the Grizzled Skipper were undertaken by CBC volunteers within the Penhale Sands MOD area in December 2018 and 2019. These consisted of controlling privet *Ligustrum* spp., Bramble *Rubus fruticosus* agg. and other light scrub. All cuttings were raked and piled either in areas of unsuitable habitat or on top of scrub. The aim of this work was to create further suitable breeding habitat for the butterfly where larval foodplants can grow in a short sward on bare ground.

The work carried out in December 2018 has already proved to be successful with one of the larval foodplants, Wild Strawberry *Fragaria vesca*, appearing in abundance. A Grizzled Skipper larva was found on Wild Strawberry in this section in early August 2019, evidence that the butterfly is making use of the improved habitat. Further monitoring was completed in 2020 to ensure that the work carried out continues to improve habitat quality for the butterfly. This monitoring provided more positive results with eight Grizzled Skipper larvae and a single Dingy Skipper larva found where the work took place. CBC is grateful to CWT for funding this work, which should benefit the Grizzled Skipper in the longer term.

Cornwall Fritillary Action Group
The Cornwall Fritillary Action Group (CFAG) was reactivated in 2016 after some years of dormancy. As its name implies, the primary focus of CFAG is on the six species of fritillary that are native to the County, exchanging information, prioritising and co-ordinating conservation work and raising funds. The Pearl-bordered Fritillary has benefited from funding to undertake a programme of survey and research, with match funding provided by the costing of volunteers' time in lieu of a financial contribution. In time, CFAG's work might broaden out to include butterfly 'species of principal importance' in Cornwall. From 2020, the Group has been administered by the South West Regional Manager of Butterfly Conservation, and has representatives from CBC, Natural England, the National Trust, South West Lakes Trust, the Forestry Commission, CWT and volunteers from other local wildlife groups.

Everyone is welcome at Cornwall Butterfly Conservation

CHAPTER 4
Butterflies species in Cornwall

This chapter includes detailed accounts of every resident and regular migrant butterfly species in Cornwall. For each species, information is provided in a consistent format on its status and distribution, habitat and ecology, conservation, and the best places in the County to see it. (Butterflies that are occasional visitors or are rarities in the County are presented in chapter 5.) The species names (common and scientific) and the order of the species accounts follow the system of Agassiz *et al.* (2020). All accounts except three have been written in the main by Dick and Maggie Goodere. (Dick was County Butterfly Recorder from 2010 to 2019.) The Silver-studded Blue *Plebejus argus* account was written by Sally Foster, the Grayling *Hipparchia semele* by Jerry Dennis, and the Grizzled Skipper *Pyrgus malvae* by Cerin Poland, each as the respective Species Champion.

The following key sources – publications, databases and websites – have been drawn upon in the writing of this chapter:

- Data and maps from the Cornish Biodiversity Network's environmental recording database
- Butterfly Conservation website https://butterfly-conservation.org
- UK Butterflies website https://www.ukbutterflies.co.uk
- Asher, J., Warren, M., Fox, R., Harding, P., Jeffcoate, G. and Jeffcoate, S. 2001. *The Millennium Atlas of Butterflies in Britain and Ireland.* Oxford University Press, Oxford.
- Eeles, P. 2019. *Life Cycles of British & Irish Butterflies.* Pisces Publications, Newbury.
- Newland, D., Still, R., Swash, A. and Tomlinson, D. 2020. *Britain's Butterflies. A Field Guide to the Butterflies of Britain and Ireland.* WILDGuides. Princeton University Press, Princeton.
- Thomas, J. and Lewington, R. 2014. *The Butterflies of Britain & Ireland.* Bloomsbury Publishing, London.

The format for all the accounts is as follows.

Data summary
At the beginning of each species account, below an image of the butterfly, is a summary of key information about the species in Cornwall.

Conservation status refers to any legal protection afforded to the species as well as its current national conservation status, determined by any of the following:
- *European Commission Habitats Directive* 1992
- Schedule 5 of the *Wildlife and Countryside Act* 1981 (as amended)
- Section 41 of the *Natural Environment and Rural Communities Act* 2006 (NERC Act 2006)
- A new Red List of British butterflies (Fox *et al.*, 2010) listing those species that are threatened in Britain based on modified criteria devised by the International Union for Conservation of Nature and Natural Resources (IUCN)
- Butterfly Conservation priority (obtained from the Butterfly Conservation website, October 2020)

Trends and distribution presents two measures of the rate of change in abundance as percentages: first for the UK as a whole and second for South-west England, which covers Cornwall, Devon, Dorset, Somerset, Wiltshire and Gloucestershire. Both sets of data are derived primarily from the UK Butterfly Monitoring Scheme (UKBMS) (Brereton *et al.*, 2019), which largely draws on transect information. This has been supplemented by records from the Wider Countryside Butterfly Survey since 2013, along with other monitoring data such as timed counts. These recording schemes are discussed in more detail in chapter 3.

Both the UK and South-west England abundance data cover the period 1976–2018, with the trends for South-west England relating to the ten-year period 2009–2018 also presented. These data for the South-west have been produced for this book by Butterfly Conservation but have not been standardised and thus are not directly comparable in terms of significance with the UK trends. Monitoring of a few species only began a year or two after 1976, which is why, in some instances, the start date for both long term trends is later.

This is followed by the percentage of 1km and 10km Ordnance Survey grid squares in which the species is found in Cornwall in the two time periods, 1976 to 2018 and 2009 to 2018. These data have been obtained from the Cornish Biodiversity Network's environmental recording database, ERICA. Both the 1km and 10km grid squares are included because where a particular butterfly might be losing range overall, its colonies can be multiplying more locally.

Introductory paragraphs

The initial paragraphs describe key characteristics of the butterfly, focusing on significant features of the adult, which, along with the accompanying photographs, should aid identification, particularly when there is sexual or seasonal dimorphism or where there can be confusion with other species. That said, the purpose of this book is not to provide a guide to identifying butterflies: that is done far better in other publications. (See chapter 8.)

A photograph of the adult showing the upperside of the wings has been given for most species (except for those in which it is rarely seen, such as the Brimstone *Gonepteryx rhamni*, Small Heath *Coenonympha pamphilus* and Grayling); where relevant, an image of the underside has also been given, in some cases distinguishing between male and female. Other photographs have been included to help illustrate a point made within the text. All the photographs in this chapter have been taken in Cornwall, and photographic credits are given at the end of the book.

Distribution in Cornwall

Research for this book started in 2018. The decision was taken to compare data for the then most recent decade, 2009 to 2018 inclusive, with the period 1976 to 2008 inclusive. The latter period was chosen because it accords with national statistics, which started being calculated from 1976 onwards. The data for these two periods are presented in two separate distribution maps, showing the 1km grid squares occupied: red for 2009 to 2018, and blue for 1976 to 2008. These maps have been generated by ERICA and show changes in distribution over time. This is particularly important for the rarer butterflies. The 2009 to 2018 map also gives an idea of where the species is concentrated, which is useful for anyone seeking out a particular butterfly.

The data within ERICA have come from a variety of sources, with all records being verified and validated by the County Butterfly Recorder. The various methods of recording from which ERICA derives its data are discussed in chapter 3.

Because the two maps represent quite different time spans, any comparison should be treated with caution. Furthermore, in the more recent period (2009 to 2018) many more records have been submitted because of increased public involvement, for reasons discussed in chapters 1 and 3. The total number of records received from all sources has risen more than three-fold, from 9,491 in 2009 to 34,316 in 2018. The effect, in generating an increase in the number of butterflies recorded, from 42,372 in 2009 to 136,270 in 2018, could be presenting an unduly rosy picture. In addition, more records received from the public relate to commoner species, making these butterflies appear to be faring better compared with rarer habitat specialists.

Observer bias in Cornwall is also influenced by the draw of the coast and its accessibility via the South West Coast Path. We therefore need to question whether species whose distribution maps show greater occurrence along the coast are actually rarer inland. More recent surveying and studies of former mining sites, many of which are inland, have shown species assemblages similar to those on the coast, due to developing habitats on formerly bare ground. Further details are given in chapter 3, and in part in chapter 1.

Larval foodplants

The foodplants on which each species' caterpillars depend are described, including the most favoured foodplants in Cornwall and any alternatives. Knowing which foodplants are required for the development of the larvae is vital when species are sought or studied, particularly the rare species, or when collecting evidence of later broods. It is also important for habitat management, including vegetation-cutting regimes. The common and scientific names of plant species follow the terminology of the *New Flora of the British Isles* (Stace, 2019). A complete list of larval foodplants, ordered alphabetically by plant common name, used by each butterfly species in Cornwall, is given at the end of the book.

Habitat and ecology

This section contains information about the habitat requirements of each species, some more specialist than others, as well as details about their life cycle. This is especially useful in relation to some species, when finding eggs, larvae or larval webs can be as useful as recording the adult. Where a butterfly's life cycle in Cornwall diverges from the national picture, this is mentioned.

Each species' life cycle is illustrated by a chart taken from Eeles (2019) and is based on data for Britain and Ireland. Shown above the life cycle chart is a flight curve, illustrating when the adult is likely to be encountered in Cornwall. This flight curve is based on the cumulative total number of records per week from the 20-year period 1999 to 2018 inclusive drawn from the ERICA database (and not the total number of individuals counted).

The terms 'egg and 'ovum' are used interchangeably, as are 'caterpillar' and 'larva', 'pupa' and 'chrysalis' and 'caterpillar stage' and 'instar'.

Analysis of trends

Both occurrence and abundance trends are important for assessing how a particular species is faring.

Occurrence describes the spatial frequency of a species, measuring how much the species has expanded or contracted in terms of geographical range or distribution. Occurrence is used in *The State of the UK's Butterflies*, a report produced by Butterfly Conservation every five years or so, the most recent in 2015 (Fox *et al.*, 2015). Occurrence in Cornwall, particularly drawing attention to loss of sites, is mainly covered in the earlier **Distribution in Cornwall** section but is sometimes mentioned under this heading in relation to the national distribution picture, drawing upon information in Fox *et al.* (2015), which sometimes provides reasons for marked changes in a species' occurrence.

Abundance, on the other hand, considers the quantity, with the measurement derived from the total number of butterflies counted in a systematic way over a specified period of time. This indicates how well a species is doing in terms of population numbers in a given year or a longer time period. Abundance trends in South-west England also provide a useful comparison for Cornish data analysis.

Local abundance is hard to gauge accurately. Although transects generate the most reliable data on which to assess abundance, with their systematic weekly recording along the same route each year, not enough of them have been walked for long enough in Cornwall to draw dependable conclusions yet. To compensate for the paucity of transect data, records from the ERICA database that contain abundance information have been used in conjunction with transect data to inform some of the analysis.

Occurrence and abundance are equally important in assessing trends, with a **trend** showing a long-term direction in which a species is changing.

By contrast, short-term **fluctuations**, year on year, can sometimes be extreme and are often closely aligned to weather patterns. Disentangling the impact of short-term fluctuations from longer-term trends is a challenge and makes firm conclusions difficult to draw. The year 2018 was a good time to finish our analysis as it concluded on an optimistic note with a highly successful year for most butterfly species (as was the year that followed it). On the other hand, populations

Green Hairstreak along coast path near Mawgan Porth

Cornish hedge near Mawgan Porth

can crash when there is a particularly wet year (as in 2012), so there is never room for complacency. Even in good years with warm and sunny conditions, some species continued to decline. Populations can also crash in a drought year, and recent analysis has established that the drought of 1976 had a severe impact on long-term trends (Defra, 2020a).

Conservation

Under this heading, any conditions in which each species is known to flourish are outlined, as are the pitfalls caused by adverse land management or development, which affects each species differently. The information is based on published ecological research and local experience.

Best places to see

For the less common butterflies there is a list of suggestions for places where each species is likely to be encountered if visiting at the appropriate time of year and in the right weather conditions. This list is based on locations where relatively high numbers of the species have been recorded within the past decade. The grid reference mentioned for each location refers to the best place to see the species within the site. The life cycle chart and the flight period will help decide the best time to look for a species.

Upperside, male and female similar

Dingy Skipper
Erynnis tages

Sponsored by
*Steve Batt, Jasmine Besterman
and Jerry Board*

The Dingy Skipper is a species that is currently struggling in Cornwall and is a particular focus of recording effort. It is a small and active butterfly, aptly named for its drab, mottled, brownish-grey appearance. It is the most moth-like of our British butterflies with respect to wing pattern and positioning at rest in dull weather or at night, when its forewings are angled backwards in a manner similar to those of a noctuid moth. Consequently, it can be confused with day-flying moths such as Mother Shipton *Euclidia mi* and Burnet Companion *E. glyphica*, as well as with the Grizzled Skipper, which is the same size but has more distinctive black and white markings. All four species fly at similar times in the same habitat. In sunny weather, the Dingy Skipper can be seen basking on bare ground with wings spread wide. Freshly emerged, its patterning can be quite well defined, but it becomes increasingly drab as it loses its scales. Male and female are difficult to distinguish.

Resident in Cornwall
(not on the Isles of Scilly)

CONSERVATION STATUS
- Section 41 Species of Principal Importance (NERC Act 2006)
- Red List of British Butterflies; IUCN criteria: Vulnerable (VU)
- Butterfly Conservation priority: High

TRENDS AND DISTRIBUTION
UK rate of change in abundance
↓ 22% 1976–2018
South-west England rate of change in abundance
↓ 30% 2009–2018
↑ 17% 1976–2018
% of 1km squares occupied in Cornwall
1% 2009–2018 (22% 10km squares)
2% 1976–2008 (42% 10km squares)

Distribution in Cornwall

The Dingy Skipper is a species of great concern in Cornwall. Nationally, the butterfly has

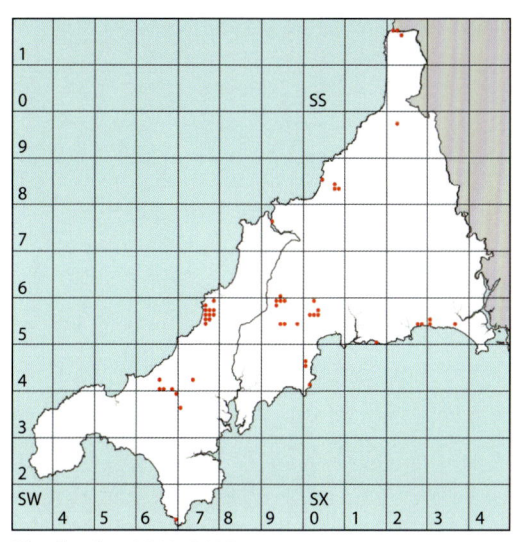

Distribution 2009–2018

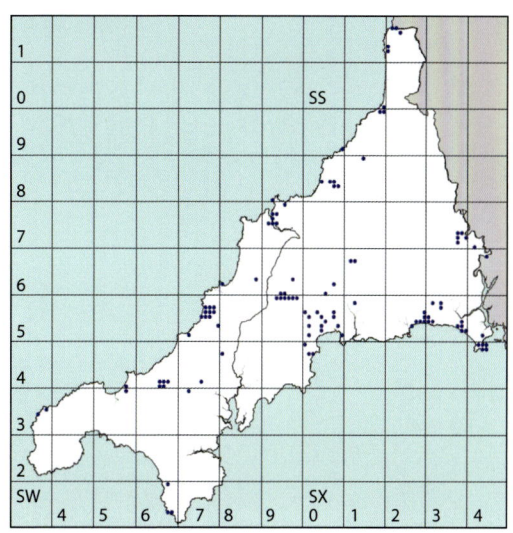

Distribution 1976–2008

declined in occurrence by 61% between 1976 and 2014 (Fox *et al.*, 2015). The number of both 10km and 1km squares in Cornwall where the Dingy Skipper has been recorded in the last 10 years has halved compared with the period 1976 to 2008. Former mining sites providing habitat favourable to this species have been lost to development, particularly in the Camborne/Redruth area.

There is also a decline in records from the Mid Cornwall Moors, especially the most prolific site at St Dennis Junction (Goss Moor), possibly due to many areas of breeding habitat being taken over or shaded out by the succession of scrub across the site. Fewer records have been received in recent times from south-east Cornwall, in particular from the Struddicks, Murrayton and Seaton areas, and there have been no sightings from Penlee Point and Rame Head in the last 10 years. Penhale Sands remains the best site at which to see this butterfly, with many hundreds flying in good years.

Occasionally there are credible reports from areas where the Dingy Skipper has not been seen for many years. A recent example is Millook, on the north Cornwall coast, where the last previous record was in 1924. It is possible that the apparent loss of sites in certain areas could be a consequence of under-recording, as the butterfly is easy to miss and only flies for a short period. Former sites need

Typical resting pose of the Dingy Skipper

Mother Shipton moth, sometimes confused with the Dingy Skipper

Butterflies of Cornwall   Hesperiidae

Egg on Common Bird's-foot-trefoil

Fifth instar larva on Wild Strawberry

further surveying to check more rigorously for the presence of this species.

Larval foodplants

The normal foodplant is Common Bird's-foot-trefoil *Lotus corniculatus*, but Greater Bird's-foot-trefoil *L. pedunculatus* can be used in damp areas.

Habitat and ecology

The adult butterfly favours well-drained, rough pasture in sunny places, with a mixture of taller vegetation for shelter. In Cornwall this encompasses dunes, coastal grassland, old railway lines, disused quarries, industrial and former mining sites, and woodland margins wherever larval foodplants grow in a sparse sward. Generally the butterfly prefers an intermediate state of grassland succession (Thomas and Lewington, 2014).

Dingy Skipper live in discrete colonies, many of which are small. Although a rather sedentary species, the butterfly is capable of flights of up to a few kilometres, which enable it to establish new colonies in favourable habitats. It is normally on the wing from late April to early June, with a rare second generation appearing in August, which has been observed in Cornwall.

The bright orange eggs are laid singly on the upperside of the base leaflets of its foodplant. The caterpillars are fully grown by August, when they create, and live within, a tent of leaves spun and twisted together, in which they hibernate for eight months (Eeles, 2019).

Analysis of trends

From 1976 to 2018 Dingy Skipper declined in abundance by 22% nationally. Over the same period, South-west England appears to have done significantly better; however, between 2009 and 2018 there was a reversal, with the south-west also showing a marked decline, a trend reflected in the Cornish data. Both the number

Flight curve based on number of records in Cornwall 1999–2018; and UK life cycle

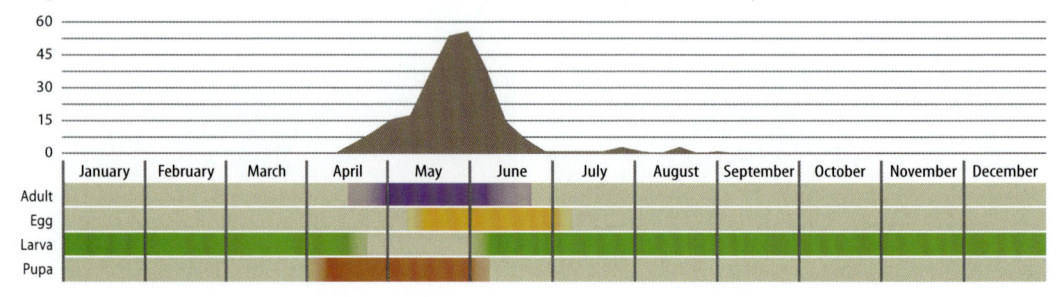

St Dennis Junction, one of the best places to see Dingy Skipper

of records and the number of individuals of this butterfly fluctuate, the data largely dependent on how closely and how frequently its Penhale Sands stronghold is monitored. This seems to be the only area where significant numbers have been recorded recently.

The decline in Cornwall can be partly attributed to the destruction of habitat, losses due to the reclamation of wasteland and former mining sites, and conversion of land to agriculture or development. Shading out of woodland clearings, overgrazing and scrub regeneration have also played a role.

Conservation

The colonies at Penhale Sands are probably secure so long as the Ministry of Defence remains in possession of some areas and continues to work with Cornwall Wildlife Trust to manage them.

All populations of this species need to be monitored carefully so that suitable habitats are provided and maintained by occasional scrub control. Elsewhere, tidying up former industrial and mining sites should be resisted, although it is essential to control the spread of scrub and Bracken *Pteridium aquilinum* (Spalding with Bourn, 2000).

Best places to see
- Gear Sands (Penhale Sands) SW773557 and Mount SW782570
- Seaton Sea Wall SX307541
- St Dennis Junction (Goss Moor) SW936597

Upperside, male and female similar

Grizzled Skipper
Pyrgus malvae

Sponsored by
Gill Broad, James Fowler
and Cerin Poland

The Grizzled Skipper is easy to overlook: it is one of Cornwall's smallest butterflies and can resemble a moth. When freshly emerged, the upperside of the wings is a charcoal brown, chequered with a pattern of white spots. The underside, by contrast, is russet-brown, dashed with white chequering. A black and white chequered strip runs along the outer margin of the upper- and underside of both forewing and hindwing. As the butterfly ages, the base colour fades and the species can resemble the Dingy Skipper and day-flying moths such as Burnet Companion *Euclidia glyphica* and Mother Shipton *E. mi*, which often share the same habitat. (Mother Shipton is illustrated on the Dingy Skipper pages.) Male and female are hard to tell apart. In the aberration *taras*, reported frequently at Penhale Sands in Cornwall, the white spots on the forewing merge to form radial white bars, which cover a larger area than the normal (nominate) form.

Resident in Cornwall
(not on the Isles of Scilly)

CONSERVATION STATUS
- Section 41 Species of Principal Importance (NERC Act 2006)
- Red List of British Butterflies; IUCN criteria: Vulnerable (VU)
- Butterfly Conservation priority: High

TRENDS AND DISTRIBUTION
UK rate of change in abundance
↓ 55% 1976–2018
South-west England rate of change in abundance
↓ 52% 2009–2018
↓ 67% 1976–2018
% of 1km squares occupied in Cornwall
<1% 2009–2018 (5% 10km squares)
<1% 1976–2008 (15% 10km squares)

Distribution in Cornwall
The Grizzled Skipper has never been common in Cornwall, but there are historical records

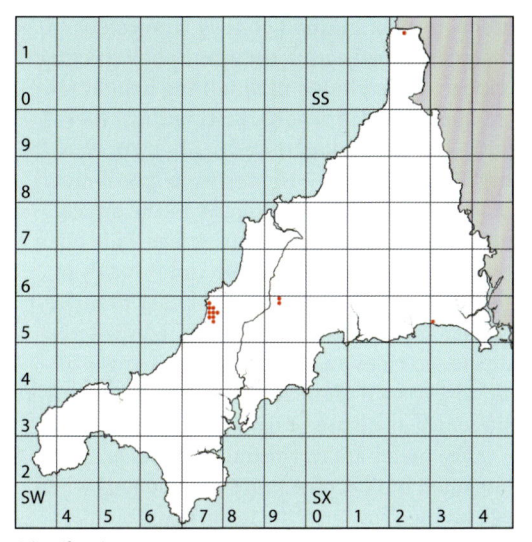

Grizzled Skipper

Distribution 2009–2018

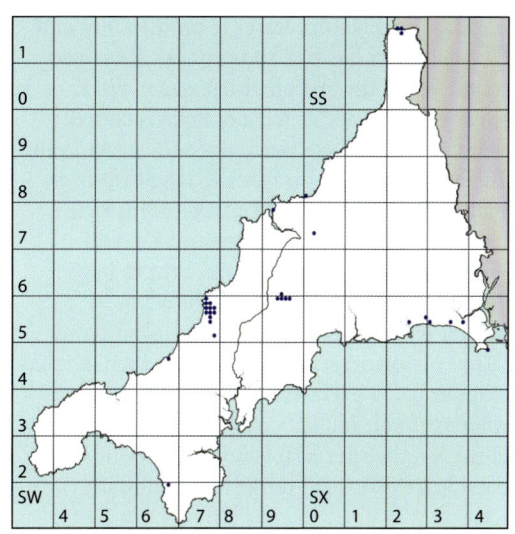

Distribution 1976–2008

from many parts of the County. In recent years it has declined dramatically and is now found on only one site, on the north coast, making it one of the County's rarest butterflies. This residual population resides in an area of dune scrub and grassland that extends from Gear Sands to Penhale Sands, near Perranporth. Adults have been found there annually since at least the 1980s, but seldom in higher numbers than a hundred or so. The Ministry of Defence land adjoining the Penhale site contains other colonies, which are now being accurately mapped so that the butterfly's distribution across the site can be better understood.

A population once existed at St Dennis Junction on Goss Moor. However, the species has not been seen at this location since 2011, although visits have been made in the flight period most years since, evidenced by records of the Dingy Skipper, which is still present on the site. Despite conservation work being carried out prior to and one year after 2011, the Grizzled Skipper seems to have been lost, and what was once prime habitat is now dominated by scrub. There are some small patches of suitable habitat left at St Dennis Junction and in other areas on Goss Moor where the butterfly may still exist at very low density; future searches are planned.

Underside, male and female similar

Aberration *taras*

Marsland Nature Reserve on the Cornwall–Devon border has had sporadic records over recent years, the last of which was in 2013. However, the species has not been recorded there in consecutive years since the early 1990s. The butterfly was also recorded at Seaton Sea Wall in 2009, and there are several historical records for the Seaton Valley and Keveral Wood, where the colony is thought to have become extinct in the late 1980s when its habitat was shaded out by mature woodland. Other historical records show that the Grizzled Skipper had a preference for coastal areas, where suitable habitats may still remain. However, the species has not been found outside its historical range for quite some time.

Larval foodplants

The Grizzled Skipper uses a wide range of species in the rose family Rosaceae as larval foodplants. On the Penhale/Gear Sands site the butterfly has been observed using Agrimony *Agrimonia eupatoria*, Barren Strawberry *Potentilla sterilis*, Silverweed *P. anserina*, Creeping Cinquefoil *P. reptans*, Salad Burnet *Poterium sanguisorba* subsp. *sanguisorba*, Bramble *Rubus fruticosus* agg. and Wild Strawberry *Fragaria vesca*. Wild Strawberry was thought to be the primary foodplant for the Goss Moor colony. Other foodplants, including Dog-rose *Rosa canina* agg., Tormentil *Potentilla erecta* and Wood Avens *Geum urbanum*, are also used.

Habitat and ecology

The preferred habitat is a relatively low, sunny sward, with occasional low scrub and some undulations in the terrain and is similar to that required by the Dingy Skipper. Thus, on the dunes at Penhale, where the foodplants are widespread and plentiful, the Grizzled Skipper seems to exist in small, loosely connected colonies spread across the site. Rabbit grazing appears to be vital to maintaining habitat for the species in Cornwall, creating short turf with bare ground on the edge of scrub, ideal for egg laying. The butterfly seems to spend much of its time in sheltered hollows in the dunes, nectaring or basking with its wings wide open on bare ground. It is an agile and rapid flyer that is readily disturbed, so it is easy to miss. Its former sites in the County included sheltered, unshaded rides and glades, sunny fringes of woods, coastal grassland, and former industrial sites such as disused railway lines.

The males are territorial. Females are seen less often but become more conspicuous when egg laying. The butterfly is normally single-brooded in Britain; in Cornwall it usually appears on the wing in early May, with a peak later in the month, and it can continue flying until late June. In warm springs it has been seen as early as April. Rare sightings in July may be attributed to an uncommon second brood, but this has not been recorded in the County for many years.

Small, light green eggs are usually laid on the underside of leaves of the larval foodplant. Once the caterpillar has hatched, it spends the next two to three months in a tent held together by silk on leaves of the host plant. The pupa is then formed in vegetation near to the ground, where it will spend the next nine months until the butterfly emerges the following spring.

Analysis of trends

Long-term trends between 1976 and 2014 show that the Grizzled Skipper has declined in both occurrence and abundance across its range in England and Wales; however, the more recent trend (2005 to 2014) shows that both are

Flight curve based on number of records in Cornwall 1999–2018; and UK life cycle

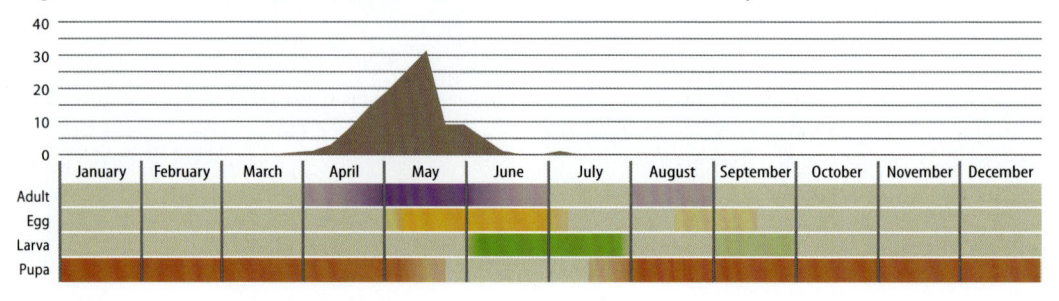

Penhale/Gear Sands, the only place in Cornwall to see the Grizzled Skipper

relatively stable (Fox *et al.*, 2015). Nonetheless, in South-west England the species still appears to be declining, with the recent trend (2009 to 2018) showing a 52% decline in abundance. In Cornwall, the distribution of Grizzled Skipper has shrunk considerably, with only a single site remaining. It is important to note that many of the locations where the species formerly occurred were only represented by a couple or even single records in most cases. The distribution at Penhale appears to be stable, and the increased recorder effort since 2017 has found the species breeding in at least seven different 1km squares. Abundance in the County is harder to monitor, relying solely on sightings from casual recorders, which can vary highly from year to year, depending on recorder activity. Annual fluctuations may be associated with weather patterns, with a previous very wet winter thought to deplete numbers significantly, although recovery can occur following warmer, drier seasons. The exceedingly wet summer of 2012 could well have contributed to the demise of the species at St Dennis Junction, as numbers of records and individuals for the two previous years appeared to be stable.

The outlook for the Grizzled Skipper in Cornwall is uncertain. Sustained and consistent monitoring at Penhale, involving timed counts, will generate data that can be statistically analysed. This will be valuable for understanding the environmental pressures on the butterfly, so that it can be better conserved in the County.

Conservation

At the time of writing, Penhale/Gear Sands has had extra effort put in by Cornwall Butterfly Conservation in partnership with Cornwall Wildlife Trust (CWT). This was done to increase knowledge on the distribution of the species across the site through butterfly search days undertaken by volunteers and a two-day training event held in 2018. There has also been an annual practical conservation day (funded by CWT) to clear scrub and maintain suitable habitat for Grizzled Skipper. This work has already proved successful, with evidence of the butterfly's larval stages using the area where habitat management had taken place the previous year.

A longer-term, county-wide strategy is needed to conserve Grizzled Skipper in Cornwall. The butterfly does not disperse far, with just over 1.5km the maximum travel distance recorded in a previous mark-release-recapture study (Brereton, 1997). A focus on the management of suitable habitat that is within this distance from Penhale would therefore be a first step towards expanding the Grizzled Skipper's territory in Cornwall. Further searches are needed in areas with historical records to determine whether the butterfly is still present in other locations. If it is found elsewhere, conservation efforts should be targeted appropriately to ensure that the species' future in Cornwall is secured.

Best places to see
- Penhale: Gear Sands SW773557, Mount/Pennans Field SW782570

Upperside, male

Small Skipper
Thymelicus sylvestris

Sponsored by
Jim Barker, Muriel Hilder
and Luna Mae Miller

The Small Skipper is one of the two 'golden' skippers that are widespread in Cornwall (the other one being the Large Skipper *Ochlodes sylvanus*). It is a vivid orange, rather moth-like little butterfly and can be seen regularly 'skipping' amongst tall grasses and flowers, usually at low level, in high summer. When it eventually rests, it often adopts a characteristic pose with forewings angled upwards in a V shape above the flattened hindwings. The triangular tip to the underside of the forewing is an olive buff colour. The Small Skipper is slightly smaller than the Large Skipper but is best distinguished from it by the lack of light chequered patterning on the upper- and underside of its wings. The male, which in all other respects is similar to the female, has on the forewing a sex brand, which is slightly curved and thinner than that of the male Large Skipper.

It is far more difficult to tell the Small Skipper apart from the Essex Skipper *Thymelicus lineola* (see chapter 5), which it closely resembles. Whether the Essex Skipper is actually resident in Cornwall remains a subject of debate, as it is possible the butterfly's eggs may occasionally be imported amongst hay (although the County is unlikely to be a big importer of hay, as it can produce its own).

Resident in Cornwall
(not on the Isles of Scilly)

CONSERVATION STATUS
■ Butterfly Conservation priority: Low

TRENDS AND DISTRIBUTION
UK rate of change in abundance
↓ 76% 1976–2018
South-west England rate of change in abundance
↑ 49% 2009–2018
↓ 80% 1976–2018
% of 1km squares occupied in Cornwall
11% 2009–2018 (76% 10km squares)
16% 1976–2008 (73% 10km squares)

Small Skipper

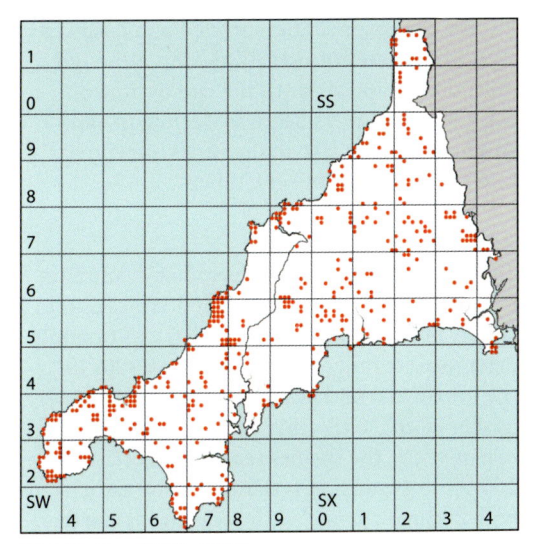

Distribution 2009–2018

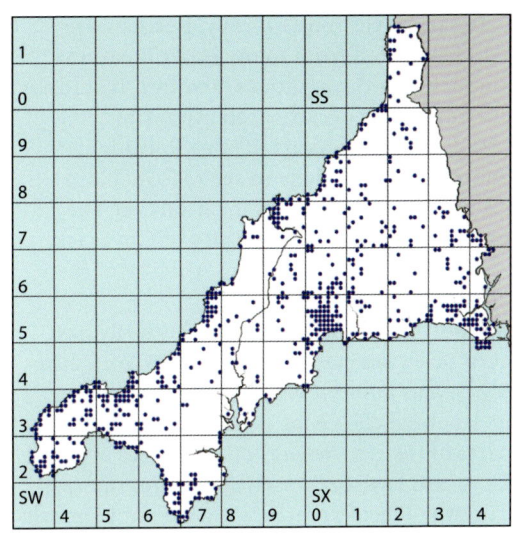

Distribution 1976–2008

One way to distinguish the two species is by looking at their antennae. The Essex Skipper has black tips to the underside of its antennae, whilst the Small Skipper's are brown. However, several instances of Small Skipper with black antenna tips have been observed, shedding doubt on the reliability of this method. If in doubt, supporting photographic evidence for independent confirmation of the species is valuable.

Distribution in Cornwall

The Small Skipper is widespread but patchy in its distribution in Cornwall, and is probably under-recorded due to its small size and habit of flying and basking among tall grasses. For the novice, it may even be confused with other flying insects, particularly moths.

Comparing the maps and the numbers of squares where this butterfly has been seen over time gives cause for concern. Although the number of records submitted in the last 10 years is slightly higher than in the period 1976–2008, the range of places it inhabits in Cornwall seems to have declined since 2008.

The butterfly lives in fairly compact colonies, although larger numbers inhabit the sand dunes of Penhale and the Towans

Male at rest, forewing angled upwards with characteristic V shape

Underside

between Hayle and Godrevy. Records are also submitted from many other locations and habitats throughout Cornwall, including clifftops, quarries, disused mine sites, meadows and gardens. Penlee Point is another reliable place to see reasonable numbers, as are various locations on The Lizard, such as in the vicinity of Coverack.

Larval foodplants

The only known foodplant the butterfly has been observed using in Cornwall is Yorkshire-fog *Holcus lanatus*, which is thought to be its favoured choice for egg laying in other parts of the UK. Several other grasses are known to have been used elsewhere, such as Timothy *Phleum pratense*, Cock's-foot *Dactylis glomerata*, Meadow Foxtail *Alopecurus pratensis* and Creeping Soft-grass *Holcus mollis*.

Habitat and ecology

The Small Skipper is generally found on rough, unimproved grassland, where it favours longer grass. Roadside verges and field margins, when left untouched, are sufficient to support colonies. It is a robust little butterfly and a strong flyer, and it tolerates moderately windy places.

The female lays her eggs in small batches inside the leaf sheath of the foodplant, where they hatch about two to three weeks later, in July/August. The larva eats its eggshell and spins a silk cocoon around itself inside the grass sheath, where it remains in hibernation until the following April, when it feeds up on grasses, going through four moults until it is fully grown. It starts pupating in early June at the bottom of grass tussocks and emerges in around two to three weeks.

The species is single-brooded and has been seen in the County as early as May, although most do not appear until the end of June or early July. Numbers begin to decline in early August, with the last adults rarely surviving into the first week of September.

Analysis of trends

The Small Skipper is widespread and common throughout most of England and Wales but is largely absent from the Lake District. It rapidly expanded its range throughout the 1990s in the north-east and north-west of England (Asher *et al.*, 2001). This has continued in recent years, the species reaching Scotland in 2007. The expansion is thought to be associated with climate change, with areas in the north becoming warmer, allowing the species to survive (Eeles, 2019).

During the late nineteenth and early twentieth centuries, the Small Skipper was often described as common in Cornwall. Its range has declined in recent years, to some extent reflecting national trends, although in the UK a loss of some former sites has been compensated by its spread northwards. Abundance fluctuates a great deal, and local statistics can be heavily influenced by a few transects where it is seen on a weekly basis in large numbers.

There has been some recovery in abundance in South-west England over the last 10 years, reflecting the national trend. The evidence suggests that numbers increase significantly when the weather before and after the butterfly's emergence is warm and dry. The Small Skipper is also less affected by drought than most other butterfly species because its early hibernation enables any parched foodplant to recover, so the following year's emergence can still be a healthy one.

Flight curve based on number of records in Cornwall 1999–2018; and UK life cycle

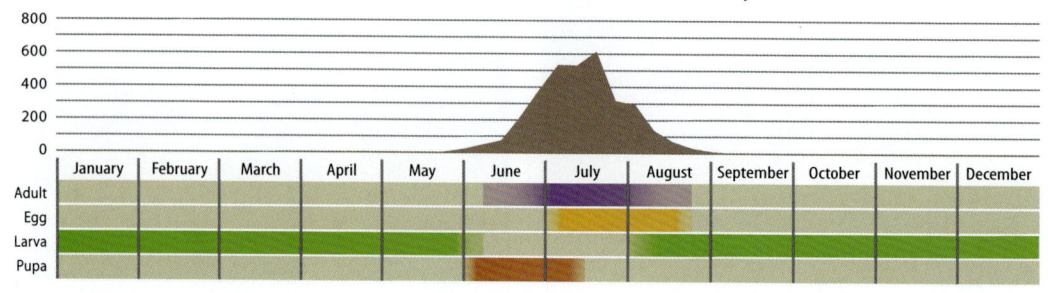

Seaton Valley Countryside Park, one of the best places to see Small Skipper

Conservation

Overgrazing or changes to cutting regimes, together with the elimination of grasses on agricultural land and roadside verges, results in the loss of eggs and larvae. Similarly, the butterfly's habitat is being destroyed by ever-increasing housing and tourist developments and the tidying up of public spaces, including disused mine sites. The largest colonies are either on nature reserves or on dunes and consequently are reasonably safe, since these areas are usually managed to contain a varied composition of habitats. Fortunately, rabbits usually graze short turf, leaving untouched the longer grasses, which the Small Skipper favours.

Best places to see

- Mexico Towans SW562385 / Upton Towans SW576397 / Gwithian Towans SW578407
- Penlee Point SX438489
- Seaton Valley SX302555
- Kynance SW692130
- Gear Sands (Penhale) SW769562
- Porth Joke SW773602

Upperside, male, showing prominent dark sex brands on forewing

Large Skipper
Ochlodes sylvanus

Sponsored by
Sarah Hutchings, Leon Truscott
and Charlotte Uren

The Large Skipper is a small butterfly whose brown-flecked, amber wings catch the eye. When it alights, the wing configuration identifies this unmistakably as one of the UK's 'golden' skippers, with forewings typically angled upwards and hindwings kept flat. In Cornwall, the only butterfly with which it might be confused is the similarly coloured but plainer and more diminutive Small Skipper, which, however, lacks the Large Skipper's brown chequering and has much less prominent black sex brands on the forewings of the male. The Large Skipper often has a greenish tinge to the body and the underside of the hindwings. These distinguishing marks are important, because the two species share the same habitat and have overlapping flight periods, although the Large Skipper emerges a couple of weeks earlier, usually in late May.

Males are often found perching in a prominent, sunny position awaiting passing females, which are generally less conspicuous. Although the males also exhibit patrolling behaviour (mainly in mid- to late morning), they tend to be more commonly seen at rest than the more active Small Skipper.

Resident in Cornwall
(not on the Isles of Scilly)

CONSERVATION STATUS
■ Butterfly Conservation priority: Low

TRENDS AND DISTRIBUTION
UK rate of change in abundance
↓ 27% 1976–2018
South-west England rate of change in abundance
↓ 5% 2009–2018
↓ 59% 1976–2018
% of 1km squares occupied in Cornwall
23% 2009–2018 (77% 10km squares)
17% 1976–2008 (78% 10km squares)

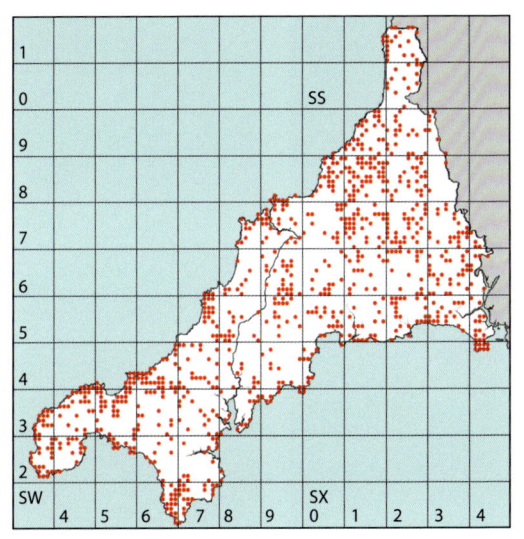

Distribution 2009–2018

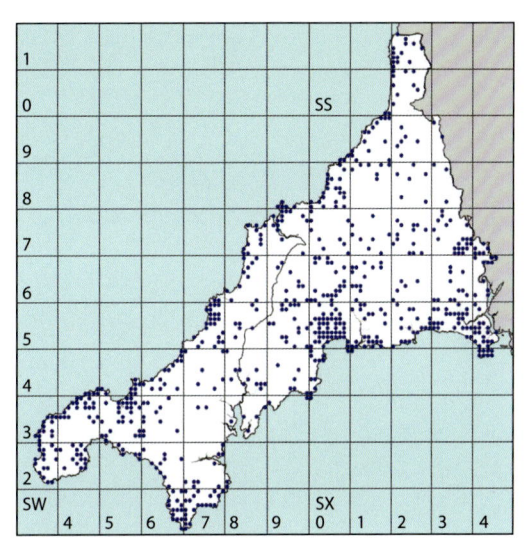

Distribution 1976–2008

Distribution in Cornwall

The Large Skipper can be found throughout most of the County. Colonies tend to comprise fewer than 30 individuals and it is unusual for more than one or two to be seen at any single spot. The colonies are particularly numerous in the Penhale and Hayle dune systems and in a few other areas along the north coast. There are also several locations along the south-east coast, including areas adjacent to Rame Head, where larger numbers of this species have been recorded. Some woodland areas also yield significant numbers, including Cabilla Wood and Greenscoombe Wood. Of the transects, Ruan Lanihorne, near Truro, produces particularly large end-of-year totals. However, at the right time of year the Large Skipper can be seen in almost any location where their larval foodplant is plentiful. There has been a slight increase in range in the last 10 years, which may be due to the increase in regular recorders, as well as the Large Skipper's inclusion in the target list of the Big Butterfly Count for a few years (although it has since been removed because of possible confusion with other skippers). This resulted in the submission of records from a wider range of locations during this period.

Larval foodplants

Cock's-foot *Dactylis glomerata* is the only foodplant recorded in Cornwall. Purple Moor-grass *Molinia caerulea* and False Brome *Brachypodium sylvaticum* have both been reported as occasional foodplants elsewhere.

Habitat and ecology

The favoured habitat of the Large Skipper is sheltered, often damp, areas with long, uncut grass; even very small patches of long grass are sufficient to support a colony. Coastal footpaths seem to be the most favoured habitat

Underside

Male with female on right

Larva

in Cornwall, perhaps because there is little grazing and grass cutting is minimal, although they are also likely to be routes favoured by recorders! Large Skipper have also been seen in good numbers on dunes and old industrial sites, and in meadows, woodland rides, parks, graveyards, roadside verges and rural gardens.

The butterfly is single-brooded and commonly flies from the end of May, although in Cornwall it occasionally appears in early May. It has a relatively long flight period, peaking in June, which in the County can occasionally extend into early September. The male is indefatigable in his pursuit of females; the mated female, by contrast, spends much of the day basking or resting, interspersed with periods of egg laying. Eggs are laid singly on the underside of the blades of the foodplant. The caterpillar hatches after just over two weeks and immediately constructs a grass tube by binding the two edges of the grass blade with silken threads. It continues feeding until the fourth moult, when it goes into hibernation. In the spring it moults twice more; the caterpillar is not hard to find at this stage, being a blue-green colour with a brown and cream head. The chrysalis stage lasts about three weeks in May–June.

Analysis of trends

Many colonies have been lost because of ploughing and reseeding of meadows with grasses such as Perennial Rye-grass *Lolium perenne*, which is unsuited to butterflies. Nonetheless, the ability of Large Skipper to thrive in small colonies in restricted areas means that the butterfly is holding its own in Cornwall, in terms of both range and abundance, although this has fluctuated over the last 10 years, showing notable increases during warm summers, particularly in 2014. UK and South-west England trends over the

Flight curve based on number of records in Cornwall 1999–2018; and UK life cycle

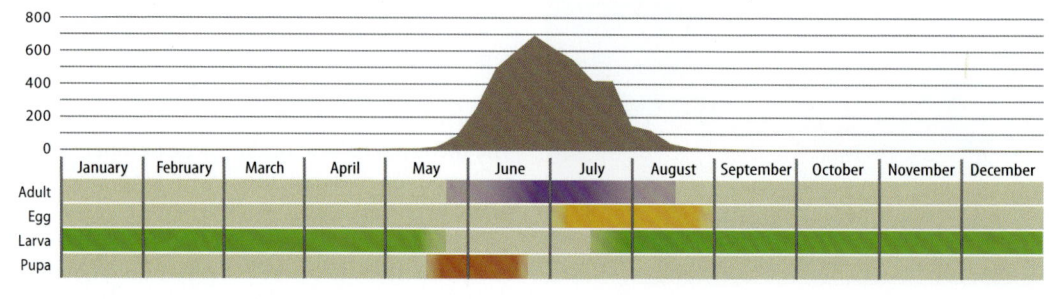

Penlee Point, one of the best places to see Large Skipper

past 40 years show a significant decline in abundance, although there has been a slight recovery in the last 10 years.

Conservation

In the short term, no management at all is probably best for this species; cutting long grass can destroy the immature stages, since all these live at 10–40cm above ground in the stems or leaves of grasses. The practice of leaving grass verges uncut along lanes and bridleways can only benefit the Large Skipper. In the longer term, scrub management to prevent grassland being shaded out will also be beneficial. The creation of areas containing long grass in an agricultural landscape subject to environmental management incentive schemes can provide further habitat for the species, albeit a temporary measure in some situations (Asher *et al.*, 2001).

Best places to see

- Gear Sands SW769562
- Cabilla Wood SX134653
- Upton Towans SW575400
- Penlee Point SX442487
- Ruan Lanihorne SW894419
- Godolphin Warren SW596313

Upperside, male

Orange-tip
Anthocharis cardamines

Sponsored by
Louis Philip Besterman, John O'Sullivan
and Marion Williams

The Orange-tip is a medium-sized butterfly and one of the first non-hibernating species to emerge in early spring. The male is unmistakable, with bright orange tips to the upperside of his forewings. Lacking this colour, the less conspicuous female has sooty grey wing tips and can at first glance be confused with one of the 'whites'. The underside of the hindwings of both sexes dispels any such confusion, however, with its distinctive, mottled pattern of yellow and black (which together give a mossy green appearance) on a white background, providing near-perfect camouflage, with wings closed, in sun-dappled shade. Males are more active as they search for mates, and it is thought that the orange wing tips warn predators that they are unpalatable. The female is often found in the vicinity of the larval foodplants and is able to test their suitability for egg laying by first 'tasting' with her feet.

Resident in Cornwall
(vagrant on the Isles of Scilly)

CONSERVATION STATUS
■ Butterfly Conservation priority: Low

TRENDS AND DISTRIBUTION
UK rate of change in abundance
↑ 17% 1976–2018
South-west England rate of change in abundance
↑ 4% 2009–2018
↑ 97% 1976–2018
% of 1km squares occupied in Cornwall
18% 2009–2018 (76% 10km squares)
19% 1976–2008 (75% 10km squares)

Distribution in Cornwall

The Orange-tip is widely distributed in the County. The two distribution maps show little change in the number of 1km and 10km squares where the butterfly has been seen, although the pattern of locations has shifted slightly, probably as a result of a different set of

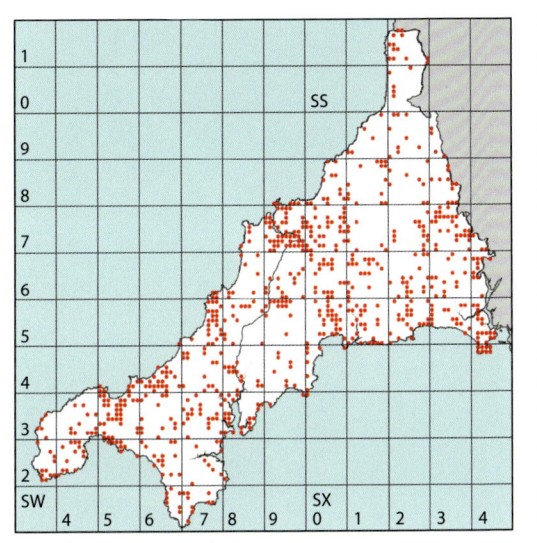

Orange-tip																				Butterflies of Cornwall

Distribution 2009–2018

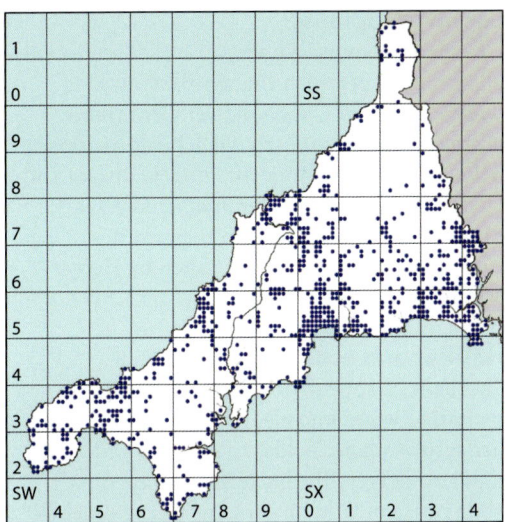

Distribution 1976–2008

recorders visiting new locations. It has become more common in north Cornwall in recent years, but it is still hardly ever seen either in the higher areas of Bodmin Moor or on the hills of West Penwith, suggesting that it prefers lower-lying ground. Nevertheless, there are also some smaller gaps within the distribution maps where it has never been recorded. These gaps are difficult to explain: the absence of suitable larval foodplants might be a factor, although these are widely distributed across the County. Another reason could just be under-recording, as the species flies for only a few months early in the year.

Upperside, female

Underside, female

67

Larval foodplants

Orange-tip can use a wide variety of foodplants, all crucifers, of which the most favoured is Lady's Smock or Cuckooflower *Cardamine pratensis*, Jack-by-the-Hedge (Garlic Mustard) *Alliaria petiolata*, Honesty *Lunaria annua* and Hedge Mustard *Sisymbrium officinale*. Garlic Mustard is less common on the granites of west Cornwall, where its place is taken by Hedge Mustard.

Habitat and ecology

This species appears in a variety of habitats, including woodland clearings, gardens, hedgerows, damp fields, riverbanks and roadside verges. The butterfly nectars on a wide range of wild flowers, including Bluebell *Hyacinthoides non-scripta*, Greater Stitchwort *Stellaria holostea*, Ragged-Robin *Silene flos-cuculi*, Cow Parsley *Anthriscus sylvestris* and Dandelion *Taraxacum officinale* agg., as well as the flowers of its larval foodplants. It also uses a variety of garden flowers.

The Orange-tip is normally single-brooded; adults have been seen as early as the beginning of March in Cornwall, although mid- to late April is more normal; Penhallurick (1996: 41) states that no other butterfly "so characterises the advent of spring" as the Orange-tip. Nonetheless, the butterfly's main flight period is from May to June, extending sometimes into July. Individuals have occasionally been recorded in both August and September, possibly belonging to a rare second brood. Other theories suggest that these late individuals may be the result of a delayed emergence for reasons unknown, as instances of the species remaining in the pupal stage for up to three years have been observed in captivity (Eeles, 2019).

Pupa

The male butterfly usually emerges earlier than the female, which when first appearing is inclined to seek shelter in undergrowth, where she is eagerly pursued by the male. After mating, the female flies in search of suitable larval foodplants. Later in the season the male is observed less frequently than the female, which can nevertheless be overlooked amongst the other white butterfly species unless seen at rest with the underside exposed. Neither males nor females are ever seen in great numbers in one place, and the species is something of a wanderer, not forming tight colonies.

Eggs are laid singly on a flower stalk, initially greenish-white and then gradually turning orange, making them easier to find. They hatch relatively quickly, in approximately seven days. After hatching, the larva devours its eggshell and, as it is cannibalistic, will eat

Flight curve based on number of records in Cornwall 1999–2018; and UK life cycle

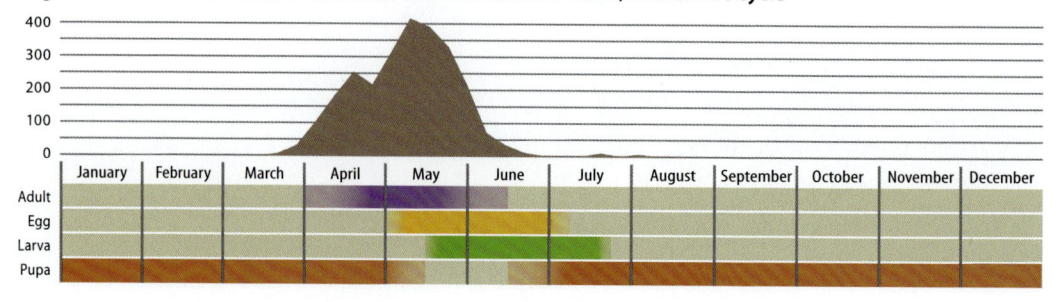

Windmill Farm, one of the best places to see Orange-tip

any other eggs it can find, as well as other immature larvae of the same species. It feeds mainly on crucifer seed pods, where the slim-line caterpillar, in its later, green instars, is superbly camouflaged. Only a minority of caterpillars reach the pupation stage, because of predation by a multitude of insects and birds. The larva pupates after three or four weeks and is attached by a silk girdle to a suitable location on a plant, the pupa resembling a seed pod. Because the annual larval foodplants collapse and perish over winter, the pupa generally attaches to a more resilient plant nearby, where it remains for up to 10 months.

Analysis of trends

The Orange-tip is subject to periodic fluctuations, probably caused at least in part by the abundance of the tachinid fly *Phryxe vulgaris*, which is largely confined to the coastal regions of Cornwall and parasitises the caterpillars (Thomas and Lewington, 2014). Cannibalism among young larvae may increase as a result of a period of bad weather during the egg laying season, causing the female to be less careful in her choice of plants and resulting in over-crowded, hungry caterpillars sharing inferior foodplants. Yet populations of this butterfly soon recover in the right conditions after a poor year, and in the last decade there has been a slight upward trend in both records and numbers. In the past 10 years the Orange-tip appears to be faring slightly better in Cornwall than in South-west England and the rest of the UK.

Conservation

Environmentally sensitive grassland management is crucially important to this species: mowing in spring or early summer destroys populations, as does cutting grass verges in June. Cutting wild crucifers before they set seed also removes the butterfly's larval foodplant from its annual cycle. Agricultural 'improvement' eliminates foodplants from damp meadows.

Best places to see

Although this butterfly is widely dispersed, good numbers can be seen at the following locations:
- Cotehele Woods SX418682
- Prussia Cove SW552278
- Breney Common SX055610
- Windmill Farm SW690149
- St Dennis Junction (Goss Moor) SW936597

Upperside, male

Large White
Pieris brassicae

Sponsored by
*Alasdair Beth Garnett, Sophie Shaw
and in memory of Christine Hawke (née Lawrence)*

The Large White is the biggest of our white butterflies and, together with the Small White *Pieris rapae*, is often referred to as a 'cabbage white', because the caterpillars of both species voraciously consume cultivated brassicas, an appetite that does not exactly endear them to gardeners and farmers. Even butterfly enthusiasts have been known to squash those rampant caterpillars and the distinctive clusters of yellow eggs when spotted on precious crops! The butterfly is predominantly white, although the underside has a pale yellow tinge. The upperside of the forewing of both sexes has black tips, and there are two black spots on the underside. The female has two black spots on the upperside of the forewing, with a dash of black where it meets the hindwing. Both male and female have a small black mark on the leading edge of the hindwing. Markings on the spring brood tend to be less pronounced than on the summer brood, with the wing tips more grey than black.

Resident and migrant in Cornwall
(resident and migrant on the Isles of Scilly)

CONSERVATION STATUS
■ Butterfly Conservation priority: Low

TRENDS AND DISTRIBUTION
UK rate of change in abundance
↓ 33% 1976–2018
South-west England rate of change in abundance
↓ 29% 2009–2018
↓ 51% 976–2018
% of 1km squares occupied in Cornwall
46% 2009–2018 (84% 10km squares)
36% 1976–2008 (81% 10km squares)

The most reliable way to distinguish Large and Small White is the presence of much more pronounced and extensive black tips to the forewings of Large White. Despite the implication of their names, size is not a reliable determinant, as the dimensions of both species can vary: a large Small White may not differ greatly

Large White

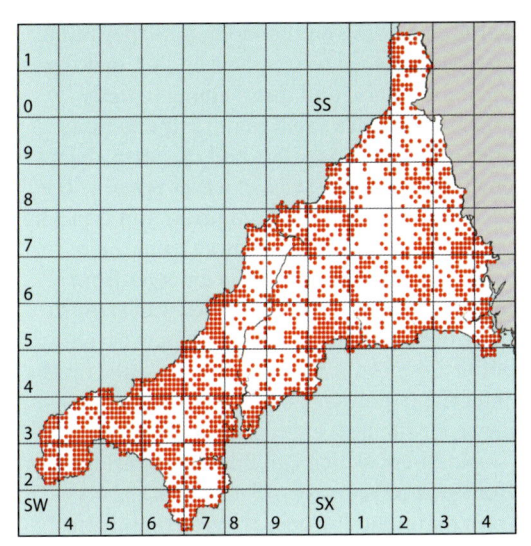

Distribution 2009–2018

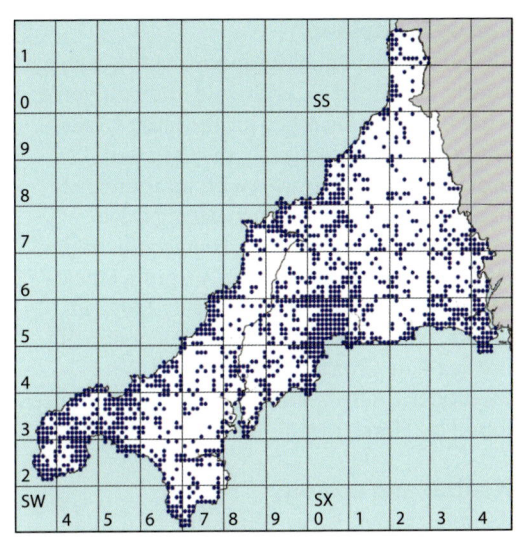

Distribution 1976–2008

from a small Large White, despite there being a measurable size difference on average. The Small White male has one spot on the forewing, whereas the Large White male has none.

Distribution in Cornwall

This highly mobile butterfly is a powerful flier. It can be seen anywhere in Cornwall, even occasionally on the higher moors, but is most frequently found around fields of broccoli, cauliflower, cabbage and Oil-seed Rape *Brassica napus* subsp. *oleifera*, crops grown widely in the County. It is also a common visitor to gardens and urban areas. The number of 1km squares it now occupies has increased slightly from earlier times. The concentration of records has changed on the two distribution maps; this probably reflects the increase in recorders and sightings from the Big Butterfly Count rather than any significant changes to favoured areas.

Upperside, female

Larval foodplants

The larvae feed on many species of the cabbage family, Brassicaceae, both wild and cultivated, with a strong preference for the latter (Asher et al., 2001). Consequently they are well provided for in Cornwall with its abundance of cultivated brassicas, including garden plants such as Nasturtium *Tropaeolum majus*, Wallflower *Erysimum cheiri*, Virginia Stock *Malcolmia maritima* and Honesty *Lunaria annua*. Near the coast, they will occasionally use Sea-kale *Crambe maritima* and Wild Mignonette *Reseda lutea*. Larvae have also been found on Horse-radish *Armoracia rusticana*.

Habitat and ecology

As noted above, Large White can be found almost anywhere, especially where larval foodplants are abundant and where there are good nectar sources. The adult butterfly is attracted to a wide range of flowers, including many members of the daisy family, Asteraceae, such as Common Knapweed *Centaurea nigra* and most species of thistle.

The butterfly has two main broods a year, with a partial third brood in Cornwall. Since it overwinters as a chrysalis, the first brood is often seen flying as early as March in the County, with occasional individuals being recorded in February. Normally, however, significant numbers do not appear until the second week of April. Each female can lay up to 600 yellow, skittle-shaped eggs in several separate smaller batches, which usually hatch within two weeks. Other closely related butterflies such as Small White are deterred from laying in the same area, because the female of the Large White leaves a repellent chemical signal (Eeles, 2019). The caterpillars are a distinctive mix of black, grey and yellow. Through feeding on the foodplants they build up toxic mustard oils in their bodies, making them distasteful and thus helping to deter predators such as small mammals and birds. They are voracious eaters and can strip a plant down until only the ribs of leaves remain. They become solitary in the final instar and typically pupate in a sheltered spot away from their foodplant. The second brood emerges from July into August and September, and is much more numerous than the first. A few of these butterflies produce a third brood, whilst most enter the pupal stage and overwinter in this phase for the next eight months.

Numbers, which can fluctuate considerably from year to year, are inflated by migrants from mainland Europe, arriving mainly in the high summer. These start their own breeding cycle, which may not be in phase with indigenous populations; as a result, it is often possible to find all four stages of the life cycle at the same time. Vast immigrations of Large White have been recorded in Cornwall, most recently in 1986 and 1988 (Penhallurick, 1996). Butterflies seen in October and even

Mating pair, showing undersides

Flight curve based on number of records in Cornwall 1999–2018; and UK life cycle

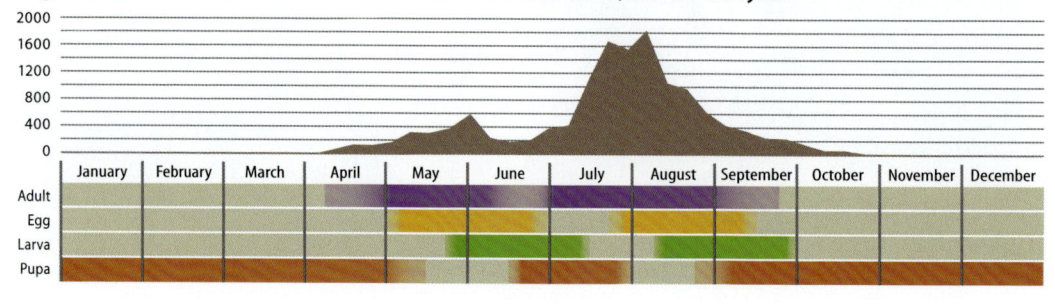

Caterpillars

Parasitic wasps recently hatched from their yellow cocoons. By this stage the skin of the host Large White caterpillar has fallen away.

November may be the progeny of late migrants or of a partial third brood of residents. There is also some evidence to show that a reverse migration from the UK to continental Europe occurs in late summer and early autumn (Asher *et al.*, 2001).

Despite the butterfly's ability, in its various stages, to deploy odour and taste to repel most predators, it nevertheless has many enemies that are undeterred by such defences, particularly a range of parasitic wasps and flies that are capable of destroying both larvae and eggs. The third brood may help the species to withstand such assaults if the larvae can survive the cold, as there are fewer parasites around to attack it in the late autumn months.

Analysis of trends

Population size can vary considerably from year to year, because of changes in migration and parasitic activity. It is reckoned that parasitic infestation of caterpillars, by the wasp *Cotesia glomerata* for instance, can cause mortality rates of larvae to be as high as 95% (Asher *et al.*, 2001), an important factor in the natural control of the species, which can otherwise be a serious pest. The data for Cornwall reflect this fluctuating pattern, and indicate that abundance rates have dropped quite significantly in the last 10 years, compared with the previous 32 years. Average transect numbers in Cornwall over the last 10 years reflect this drop in abundance, which was also noted in the South-west England regional trends.

In the past, the butterfly often reached epidemic proportions (noted in a variety of historical documents), seriously damaging crops. However, the use of modern insecticides has substantially reduced the butterfly's impact on commercial crops. Nonetheless, broad-spectrum insecticides such as neonicotinoids have also devastated other insect populations, including bees and moths, which are indispensable to farmers for pollination, so their general use was banned in the UK in 2018. Alternative pesticides also pose environmental risks and could kill the very parasitic insects that naturally control Large White populations.

One prime example showing regional fluctuation of the species occurred in 2012 on the Isles of Scilly, when the transect walked on St Mary's reported by far its highest number, whereas records from the Cornish transects that year show the Large White, along with most other butterfly species, having its worst year since transect records began.

Conservation

No conservation of this species is necessary.

Best places to see

Anywhere in Cornwall.

Upperside, male

Small White
Pieris rapae

Sponsored by
*Rose Barlow, Ian Beavis
and Mary Crewes*

As with its near relative the Large White, the Small White is popularly known as a 'cabbage white' because of its preference for cultivated brassicas, although its caterpillars do not do quite as much damage to the foodplants. There are seasonal differences in appearance, the summer brood having stronger charcoal marking than the spring brood. There are also differences between the sexes: the female has two charcoal spots on the upperside of each forewing, whereas the male has just one; undersides are similar.

'Whites' are notoriously difficult to tell apart, and there are sometimes problems distinguishing Small White from Large White, Green-veined White *Pieris napi* and female Orange-tip. Although the Large White is on average quite a bit bigger than the Small White, the size of the two species varies and therefore is not a solely reliable means of identification. The Small White tends to have smaller and

Resident and migrant in Cornwall
(resident and migrant on the Isles of Scilly)

CONSERVATION STATUS
■ Butterfly Conservation priority: Low

TRENDS AND DISTRIBUTION
UK rate of change in abundance
 ↓ 25% 1976–2018
South-west England rate of change in abundance
 ↑ 7% 2009–2018
 ↓ 40% 1976–2018
% of 1km squares occupied in Cornwall
 43% 2009–2018 (84% 10km squares)
 28% 1976–2008 (78% 10km squares)

sometimes fainter charcoal markings, which do not extend far around the back edge of the tips of the forewings, compared to the Large White's more extensive markings. Both species have yellowish undersides to the hindwings and two spots on the underside of each forewing. Small White males usually have one

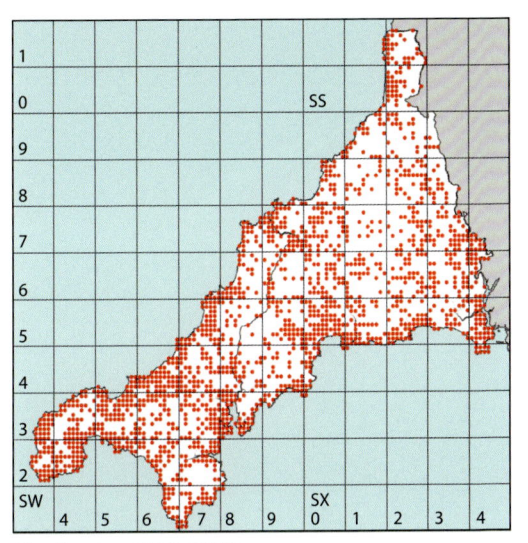

Distribution 2009–2018

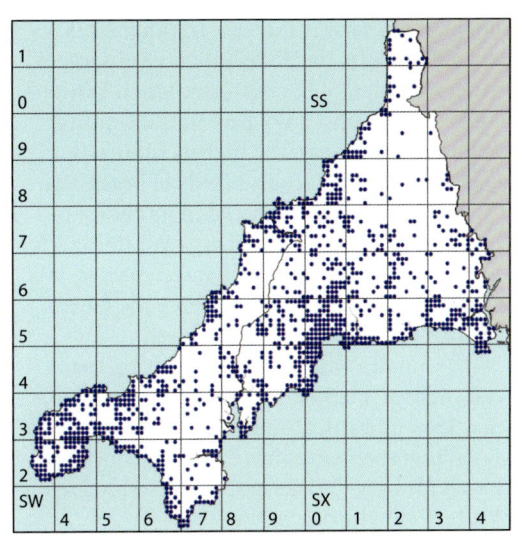

Distribution 1976–2008

spot on the upper forewings, whereas Large White males have none.

The easiest way to decide whether a butterfly is a Green-veined White or a Small White is to look at the underside of the wings. The distinctive pattern of 'veins' is unmistakable in the Green-veined White and absent in the Small White. However, the Small White is a strong flier and not always easy to catch in a stationary position, and the freshly emerged Green-veined White male is very active when looking for a mate, so behaviour can be similar. Even experienced observers can find it difficult to distinguish the three whites on the wing and rely on the butterfly at rest to make a certain identification.

Distribution in Cornwall

The distribution of Small White in Cornwall is similar to that of Large White, and although

Upperside, female

there has been some increase in range, this probably reflects the distribution of observers rather than that of the butterfly. Small White can be seen almost anywhere in the County, with the exception of the highest moors, but tend to congregate around fields of brassicas. The butterfly is less restricted in its choice of larval foodplants than the Large White, and is therefore less reliant on cultivated species. It is migratory, and migrants reinforce the native Cornish population in the summer.

The Small White has been described as 'common' or 'abundant' in Cornwall from at least 1846 onward, although some writers have stated that the species has been less common than both Large White and Green-veined White (Penhallurick, 1996).

Larval foodplants

As well as using all cultivated brassicas, the Small White finds Garlic Mustard or Jack-by-the-Hedge *Alliaria petiolata*, Hedge Mustard *Sisymbrium officinale*, Charlock *Sinapis arvensis* and Wild Cabbage *Brassica oleracea* var. *oleracea* equally attractive. In fact any member of the cabbage family (Brassicaceae) will suffice, including cultivars like Wallflower *Erysimum cheiri*. Nasturtium *Tropaeolum majus* is also used.

Habitat and ecology

Small White enjoy many of the same habitats as Large White, frequenting brassica fields, hedgerows, lanes, rough ground and gardens, wherever there are larval foodplants and plentiful nectar sources; the butterfly is particularly attracted by white or pale flowers.

In Cornwall, the Small White usually produces three generations a year. The adults begin to appear in numbers during April, although individuals have been seen in March and even in January, albeit that the one recorded in January had probably overwintered as a pupa in a greenhouse (Penhallurick, 1996). The female often mates several times after emerging from the pupal stage in spring. She receives a nourishing 'nuptial gift' from her partners along with their sperm, and this helps to prolong her life. Males attempt to prevent further mating by spraying an anti-aphrodisiac, but in so doing can attract a parasitic wasp, which parasitises the eggs. Unlike those of the Large White, the pale green bottle-shaped eggs are laid singly on the chosen plant, usually on the underside of a leaf, a few days after mating. After changing colour to yellow and then grey, the eggs usually hatch within a week. The small caterpillar eats its way in towards the heart of the plant, inflicting hidden damage. The larva is green and relies on camouflage to avoid predation, particularly as it returns to

Underside, female

Flight curve based on number of records in Cornwall 1999–2018; and UK life cycle

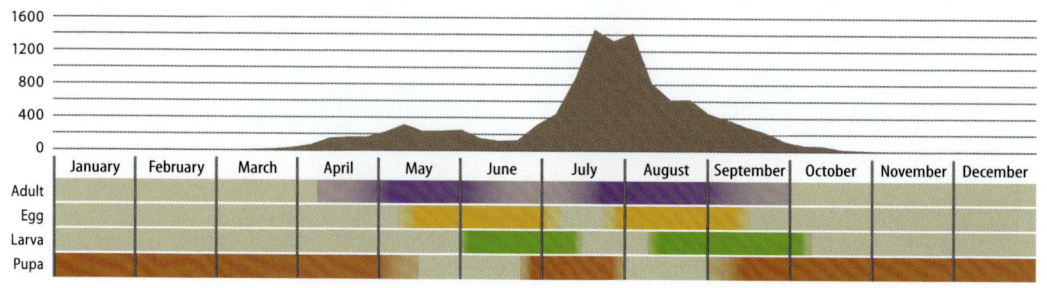

the outer leaves in its final stage, unlike the larvae of the Large White, which deter predators by their startling colours and off-putting taste.

Pupation usually takes place away from the foodplant. Some pupae will go straight into hibernation and these have waxy, sturdier shells, whilst those that go on to produce the second brood have casings that are thinner and more translucent.

The first brood is generally over by the end of June. The second brood and partial third brood occur in rapid succession, so that butterflies are seen continuously from July until October, during which period numbers are augmented by migrants. Research continues into the migratory habits of this butterfly. There is some evidence that the first brood instinctively flies north in the spring and the offspring fly back south in the summer. The second brood is far more abundant than the first, despite some of the first-brood caterpillars not developing that summer beyond the pupation stage.

Analysis of trends
Numbers of Small White fluctuate from year to year, at least in part due to migration patterns. The ups and downs seem largely in line with those of the Large White, although there has been less of a decline in the Small White population, and indeed trends in South-west England indicate a slight increase in abundance in recent years, despite control through crop-spraying. The House Sparrow *Passer domesticus* and the Garden Warbler *Sylvia borin* are known to eat the eggs of this species and so exert a measure of natural control (Emmet and Heath, 1990). A range of birds also predate the caterpillars, as do parasitic wasps and other invertebrates such as beetles and harvestmen. Nevertheless, this butterfly remains one of the most common British species.

Conservation
No conservation of this species is necessary.

Best places to see
Anywhere in Cornwall.

Cornish allotment with brassicas favoured by Large and Small White

Underside, male and female similar

Green-veined White
Pieris napi

Sponsored by
Ros Fryer, Jenny Wellsteed
and Kathy Wood

The Green-veined White is more complicated to describe than most other butterflies, not only because of sex differences, but also because there are seasonal variations as well as similarities with other medium-sized 'whites', particularly the Small White and the female Orange-tip (which flies in spring and early summer in the same habitat). As a result, under-recording is likely to occur. Inexperienced recorders can sometimes assume they are just seeing 'cabbage whites', and even experienced recorders can misidentify when recording butterflies seen in flight or with just the upperside of the wings visible.

The behaviour of the Green-veined White can help to distinguish it from the Small White. Green-veined White have a more fluttering and less vigorous flight, settling more frequently and for longer periods on the chosen nectar plant. Unfortunately this characteristic is shared with the female Orange-tip. The easiest

Resident in Cornwall
(resident on the Isles of Scilly)

CONSERVATION STATUS
■ Butterfly Conservation priority: Low

TRENDS AND DISTRIBUTION
UK rate of change in abundance
  ↓ 12% 1976–2018
South-west England rate of change in abundance
  ↓ 29% 2009–2018
  ↓ 11% 1976–2018
% of 1km squares occupied in Cornwall
  43% 2009–2018 (85% 10km squares)
  28% 1976–2008 (78% 10km squares)

way to identify the Green-veined White is when it rests with wings closed, exposing the undersides with their uniquely characteristic prominent veins, a powdery green-grey against the surrounding whitish-yellow. This is especially noticeable in the spring brood and distinguishes it from all other whites. The green

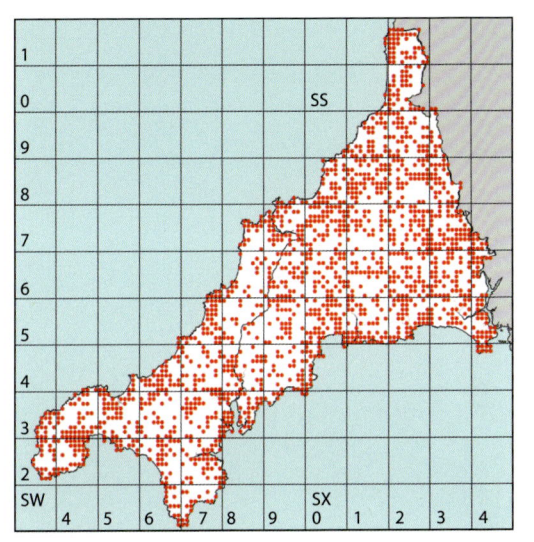

Green-veined White

Distribution 2009–2018

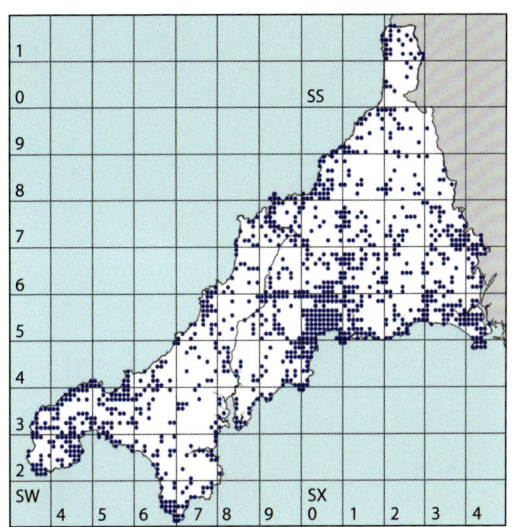

Distribution 1976–2008

colour is in fact an optical illusion created by black scales on the yellow background, and is usually duller in later broods. This patterning harmonises with blades of grass and rushes on which the butterfly roosts, providing a degree of camouflage.

The uppersides of the wings vary according to sex and to which brood the butterfly belongs. Both sexes have a greyish-black tinge to the tips of the forewing, although this is not always pronounced. The early-brood female tends to have a more prominent dusting of grey scales on the upperside of her wings. Females from all broods have two black spots in the centre of the forewing. The male, by contrast, has a single black spot on the upperside of the forewing, which can be absent or very faint in the first brood. The black markings on the upperside of the wings of both sexes from the summer brood are darker and more easily seen than those on the spring brood.

The normal (nominate) form does not occur in the British Isles, but there is considerable controversy over the naming of the British subspecies; the subspecies *sabellicae* is found throughout England and Wales, so without evidence to the contrary it is reasonable to assume that this is the form seen in Cornwall.

Distribution in Cornwall

The distribution maps and the records from which they are derived are hard to interpret. It would seem that the range over the last 10 years has expanded considerably, so that the butterfly can now be seen almost anywhere in Cornwall. This has happened despite the number of records submitted in these two time periods having increased only slightly. This is likely to be the result of more recorders sending in more sightings from inland locations, perhaps helped by the Big Butterfly Count. This in turn would seem to indicate that coastal records

Upperside, male

Upperside, female

have decreased proportionately. The 2009–2018 distribution map is perhaps a more accurate representation of where the butterfly is to be found, since its preference for sheltered, damp locations made it hard to explain the previous coastal bias.

Larval foodplants

Almost any of the brassica family seems to be attractive, with Cuckooflower *Cardamine pratensis* (also known as Lady's Smock), Garlic Mustard *Alliaria petiolata*, Hedge Mustard *Sisymbrium officinale*, Water-cress *Nasturtium officinale* and Charlock *Sinapis arvensis* mainly used. Hairy Rock-cress *Arabis hirsuta*, which occurs in Cornwall only on calcareous dunes at one or two isolated places on the north coast, can also be used.

Habitat and ecology

Green-veined White can be found in a wide variety of habitats, preferring moist, humid conditions where foodplants are plentiful. The butterfly therefore favours damp meadows and gardens, hedgerows and Cornish hedges bordering lanes and tracks, and the banks of ponds, lakes and waterways. It is also to be found in woodland glades, on wet heathland, and along road verges, and is even seen on the higher moors. It does not seem to be excessively attracted to cultivated crops, so is not seen as a pest, unlike Large White and Small White, although occasionally it lays its eggs on Nasturtium *Tropaeolum majus* in gardens and garden 'overspill'.

The butterfly has two main generations, and in warm years it produces a third. First-

Flight curve based on number of records in Cornwall 1999–2018; and UK life cycle

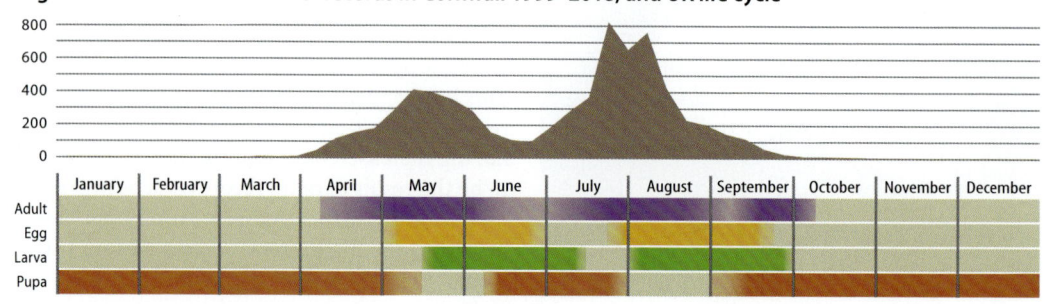

brood adults usually emerge in April, reaching a peak in May and tailing off in June. The flight period of the second generation peaks at the end of July into August. The mating process is quite fascinating and is described in detail in Thomas and Lewington (2014). Having located a female, the male will first shower her with lemon-scented 'love dust', which is so powerful that it can even be detected by humans. During mating, the female is smeared with anti-aphrodisiac to deter rival males. However, the effect does not last long, and females that are genetically programmed to be promiscuous will consent to be mated again. As far as the fertilisation of the eggs is concerned, this is unnecessary, but females which mate several times get a boost from the 'nuptial gift' injected along with the sperm. The 'gift' conveys 15% of the male's body weight in the form of proteins and other nutrients, enabling females that mate several times to live longer and lay more and larger eggs (Thomas and Lewington, 2014).

The female will then locate a suitable larval foodplant by 'tasting' with her feet, looking particularly for the presence of mustard oils, which are distasteful to predators. She lays her eggs singly, usually on the underside of a leaf. When the larvae emerge, they will not be in competition with any Orange-tip caterpillars, which share some of the same foodplants, because the larvae of Green-veined White eat only the leaves rather than the seed pods favoured by the Orange-tip. Caterpillars emerge in about four or five days and then pupate after only about two and a half weeks (and four moults), usually at a distance from their foodplant. A proportion remain in the pupal state to overwinter until next spring; these possess thick, waxy shells, unlike the thinner-shelled, semi-translucent cases housing the butterflies that will emerge 10 days later, usually towards the end of June (Eeles, 2019). Despite some of the pupae entering hibernation, the second brood of adults is usually larger than the first, and in favourable summers a third generation can be produced in late August and September. Occasionally adults are seen flying as late as October.

Analysis of trends

As noted above, Green-veined White now appear to be more widespread in Cornwall than previously indicated, probably because of a greater number of recorders submitting records, along with the increasing number of transects walked.

There is a long-term decline in this species both in the UK as a whole and in South-west England, and this seems to have worsened over the last 10 years; drainage of damp grasslands has probably contributed to the situation. Cornwall data are inconclusive. The average number of butterflies seen for each record on the Cornish database seems to have dropped in the more recent period. There have been considerable fluctuations during the last 10 years, although there seems to be a slightly upward trend in both records and numbers, helped by a bumper year in 2018. Transects are hard to turn to for evidence, because the first two years under consideration (2010 and 2011) produced very large numbers from just two transects that have not been walked since, thus distorting overall comparisons. If these early results are discounted, the average yearly number of butterflies per transect roughly corresponds with the slightly upward trend seen overall for this species.

More training would be beneficial to help recorders to identify these butterflies accurately and distinguish them from the other whites.

Conservation

Preservation of wetlands and damp meadows, together with Cornish hedges, is essential for this species.

Best places to see

Anywhere in Cornwall where there is suitable habitat.

Uppersides displayed, male above, female below

Clouded Yellow
Colias croceus

Sponsored by
*Rannoch Goldfinch, Johanna Snel
and Ecological Surveys Ltd*

Butterfly enthusiasts are always excited to see this medium-sized migrant butterfly, which appears in Cornwall in very variable numbers each year but seldom survives British winters. The upperside of the Clouded Yellow is a rich orange-yellow colour, sometimes described as saffron, with a single black dot on each forewing. Both sexes also have a broad black band edging both the forewings and the hindwings, the females differing slightly by having a number of yellow dots within those bands. The underside of the wings, which is the most common view when the butterfly is not in flight, is more of a pale greenish-yellow hue. There is a notable figure-of-eight marking (white within a brown outline) in the middle of each hindwing, and a row of brown dots parallel to the delicate pink-coloured border trim. The antennae and legs are also a delicate shade of pink.

At a distance, it is possible to confuse this species with the Brimstone, which is the only other yellow butterfly in the UK, although the Brimstone is larger and has a different wing shape. When the Clouded Yellow is at rest, it almost invariably has its wings closed and is easy to identify correctly. About 10% of females are of the *helice* form, which can be so lacking in colour that it may be confused with one of

Migrant in Cornwall
(migrant on the Isles of Scilly)

CONSERVATION STATUS
■ Butterfly Conservation priority: n/a

TRENDS AND DISTRIBUTION
UK rate of change in abundance
↑ 274% 1979–2018
South-west England rate of change in abundance
↑ 210% 2009–2018
↑ 194% 1989–2018
% of 1km squares occupied in Cornwall
10% 2009–2018 (75% 10km squares)
20% 1976–2008 (81% 10km squares)

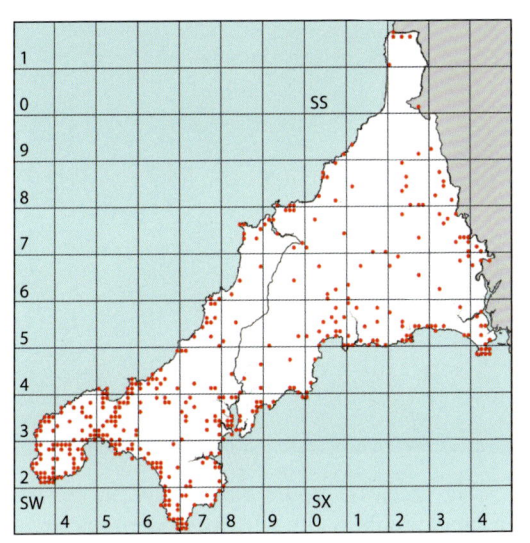

Clouded Yellow

Distribution 2009–2018

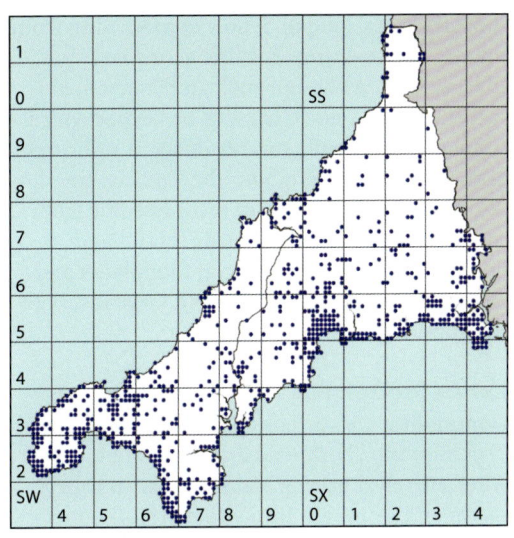

Distribution 1976–2008

the other white butterflies, although the black border on the upperside and the characteristic markings on the underside should distinguish it from other species. The *helice* form is probably responsible for over-optimistic reports of the rare Pale Clouded Yellow *Colias hyale* or the even rarer Berger's Clouded Yellow *C. alfacariensis*, as these butterflies are so difficult to determine unless examined closely. (See chapter 5.)

Distribution in Cornwall

As the Clouded Yellow is a migrant to the UK, it is unsurprising that numbers and distribution in Cornwall vary greatly from year to year, but it is possible that the great invasions of the past may never be repeated. There was a report, for instance, in 1868, quoted by Frohawk in his 1934 book, *The Complete Book of British Butterflies*, recounting clouds containing

Underside, the usual view when not in flight

thousands of Clouded Yellow approaching from the sea and covering the cliffs near Marazion (Eeles, 2019). Although such an occurrence is rare, in recent years Cornwall has experienced some very large influxes, resulting in sightings of the butterfly throughout the County. In poor years, the majority of records come from the coast; the butterfly spreads inland and northwards in search of larval foodplants and nectar-rich areas when there is a large influx in the good years.

Larval foodplants
Commonest are various species of clover *Trifolium* spp., both wild and cultivated. Alternatively, Lucerne *Medicago sativa* subsp. *sativa* and Common Bird's-foot-trefoil *Lotus corniculatus*, along with other trefoils and vetches, can be used.

Habitat and ecology
Clouded Yellow migrate from North Africa or southern Europe on a regular basis; in good years they travel in large groups. The first migrants usually arrive in May or June and appear to move in a north or north-easterly direction, reversed in the late summer or autumn to a south or south-westerly direction as they commence the return migration. The vulnerability of larva and pupa to cold and wet means they cannot overwinter in most areas of the UK.

The adult can be seen almost anywhere in fields, gardens or meadows or on waste ground. It delights especially in fields of clover or Lucerne, or other flower-rich areas with trefoils and vetches where it can also lay its eggs. It is a fast, strong flier, tolerant of open windy places, as demonstrated by its ability to survive the long flights across the sea in good condition.

Underside of the helice form

The species is reported to survive the winter in small numbers annually, as a caterpillar or pupa, along the south coast of England (Eeles, 2019), an occurrence that might become less exceptional as a result of climate change. In 2007 and 2017 Clouded Yellow were thought to have overwintered near Seaton Sea Wall in south-east Cornwall, with adult records in those years submitted from that location from early April. Although there is no certainty about the stage of development in which the overwintering occurred, it is thought likely that it would have been in the larval or chrysalis stage, as no adults were seen in that vicinity on warm winter days.

The early migrants will usually produce a first brood that emerges in late July to August. In good years this can be extremely large, swelling the numbers of later migrants, which sometimes continue to arrive until late autumn. As the species has a short life cycle, it is capable of producing three broods in favourable years. The peak for the final brood in Cornwall is late September, but butterflies have been seen as late as December. The female

Flight curve based on number of records in Cornwall 1999–2018; and UK life cycle

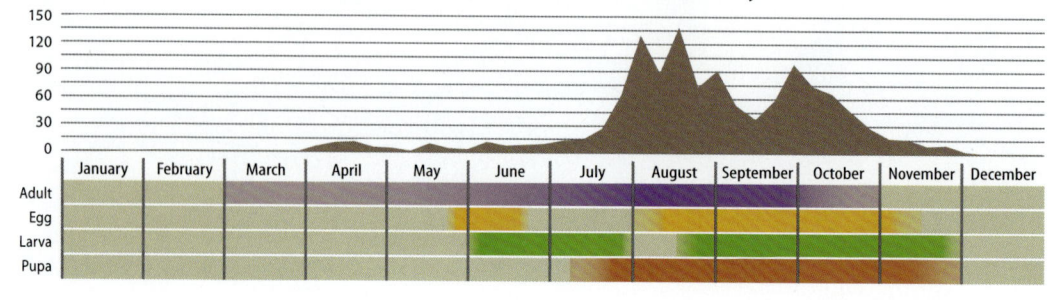

lays bottle-shaped eggs singly and in great numbers – "up to 600 [eggs] have been recorded from a single female" (Eeles, 2019: 108) – on the upperside of leaves of the larval foodplant and hatch in about a week. The caterpillars feed voraciously and mature within about a month, emerging as adults about two weeks later following pupation. It has been noted in Cornwall that immigrant Clouded Yellow almost always have a greater wingspan than UK ones, presumably because they have bred under more favourable conditions (Wacher et al., 2003).

Analysis of trends

The Clouded Yellow has had a series of good seasons, appearing in large numbers in several years during the 1940s. There followed a 33-year gap from 1950 until the next mass immigration in 1983 (Thomas and Lewington, 2014). The early part of this century saw good numbers of Clouded Yellow in Cornwall in 2000, 2002 and 2003, leading up to the best year in recent times, 2006, noted nationally as a 'Clouded Yellow year'. This species has been recorded every year since 1976, although numbers vary from just 18 adults recorded in 2009 to 881 recorded in 2014. Both 2013 and 2014 were the best years in the decade according to both transect data and *ad hoc* records. The relatively large numbers seen in those years are still much lower than the 2,726 butterflies recorded in 2006, despite the fact that at that time there were fewer recorders. The wide variation in the numbers reaching this country in different years is most likely caused by the weather systems in Europe and Africa. Given reasonably good weather, those arriving in the County relatively early usually produce at least one brood, but a poor, wet summer may reduce the numbers.

Cornwall is fortunate that, even when national abundance is low, it is one of the best counties in which to see this butterfly.

Conservation

Little can be done to support this largely immigrant species, but since the foodplant of the caterpillar is still cultivated by farmers, it is one of the few butterflies that can benefit from agriculture, provided that insecticides are not used. An abundance of wild flowers close by will also help to sustain the butterfly.

Best places to see
- Porthgwarra SW370215
- Cudden Point SW550278
- Predannack Head SW661163
- Windmill Farm SW692151
- St Keverne (Dean Quarries) SW800205
- Treluggan Cliffs SW887374
- Seaton Sea Wall SX307541

Underside, male

Brimstone
Gonepteryx rhamni

Sponsored by
Pam Cox, Faith Hambly
and Jenny Tagney

The Brimstone is a large and spectacular butterfly. The buttery-yellow male and the pale, greenish-white female are hard to miss when they fly in February/March – one of the first butterflies to take to the wing in Cornwall. Both have heavily veined wings, with a distinctively scalloped outline, resembling a leaf when at rest, so the butterfly is perfectly camouflaged on foliage when nectaring, roosting or hibernating. Both male and female have a brown dot on the underside of the fore- and hindwings, and an orange dot on the upperside, the markings being slightly less pronounced in the female. Despite these unique characteristics, misidentification is possible, especially by inexperienced observers. The male Brimstone can be confused with a Clouded Yellow, and the female Brimstone with a Large White. The absence of any black markings or borders on the Brimstone is a further distinguishing feature.

Resident in Cornwall
(vagrant on the Isles of Scilly)

CONSERVATION STATUS
- Butterfly Conservation priority: Low

TRENDS AND DISTRIBUTION
UK rate of change in abundance
↑ 15% 1976–2018
South-west England rate of change in abundance
↑ 63% 2009–2018
↑ 46% 1976–2018
% of 1km squares occupied in Cornwall
10% 2009–2018 (71% 10km squares)
11% 1976–2008 (67% 10km squares)

Distribution in Cornwall

The Brimstone has a widespread but patchy distribution across Cornwall. It is uncommon west of Truro and very scarce in West Penwith and on Bodmin Moor and The Lizard peninsula. This may be due to the restricted range of its primary larval

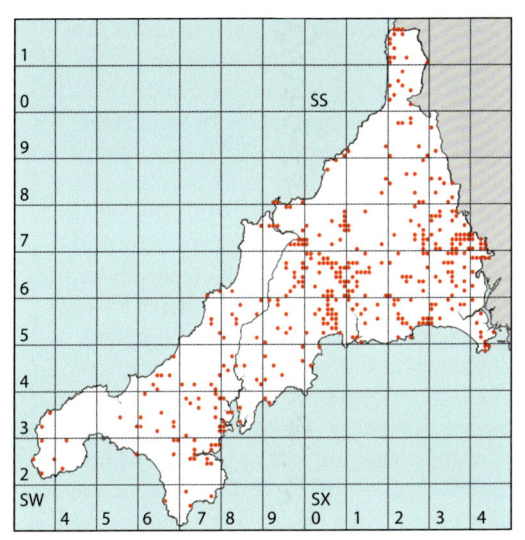

Distribution 2009–2018

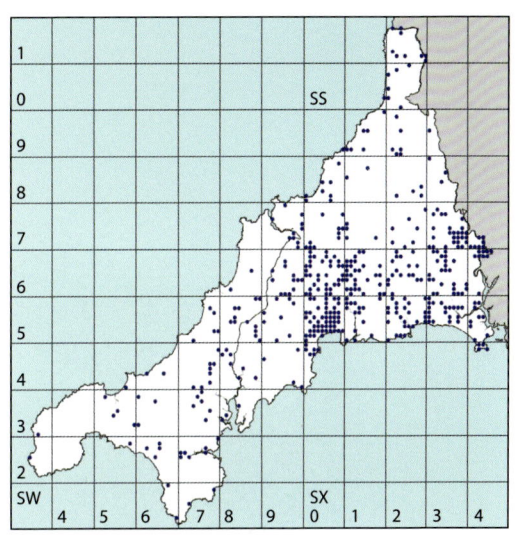

Distribution 1976–2008

foodplant, which is rarely found in these parts of the County. However, the female especially has a wandering habit, and both sexes have occasionally been spotted by credible recorders in these areas in recent years.

Larval foodplants

The principal foodplant is Alder Buckthorn *Frangula alnus*, which "has been much planted to encourage the population of Brimstone Butterflies" (French, 2020: 181). Buckthorn *Rhamnus cathartica* is used as an alternative primary foodplant elsewhere, but it is not native to the County and there are only records of planted specimens in several areas of east Cornwall (French, 2020).

Habitat and ecology

The Brimstone's strongholds within the County are primarily in woodland, where

Underside, female

Eggs attached to the underside of Alder Buckthorn leaves

the species is often seen flying up and down rides or taking advantage of available nectar. Suitable breeding habitat is found in woodland clearings, scrub and hedgerows, where the larval foodplant grows in sheltered but sunny locations. However, the butterfly's wandering habit means that it can be seen almost anywhere; gardens are visited often, where, as avid feeders, the Brimstone nectars on cultivated flowers, particularly the nectar-rich flowers of Buddleja *Buddleja davidii*. Other favoured nectar plants are Wild Teasel *Dipsacus fullonum*, thistles *Cirsium* spp. and Common Knapweed *Centaurea nigra*.

Males are usually the first Brimstone to be seen in spring, often resting with their wings angled to maximise the warmth of the sun. They later start patrolling woodland edges and hedgerows on the lookout for a mate. When a virgin female appears, the two butterflies spiral upwards together and then tumble down into a bush, where they remain coupled for a few hours or more (Eeles, 2019). The female waits a few weeks for her eggs to mature and is choosy when selecting which particular Alder Buckthorn she deems suitable for egg laying: some plants receive large numbers of eggs, whilst others close by receive none. She favours sheltered, sunny sides of young bushes, where the ribbed, bottle-shaped, pale green eggs are laid singly, usually on the underside of unfurling leaflets. The caterpillars take one or two weeks to hatch and are easily found, especially where feeding damage is detected. They emerge a mottled olive-brown colour, turning green and changing slightly in appearance throughout the instars. When fully grown, the caterpillar is bluish green with a white stripe running along each side. In the later stages of growth it exudes a bitter fluid that is thought to deter predators, but a large number still fail to reach maturity. Many are either predated by birds or attacked by parasites such as the tachinid fly *Phryxe vulgaris*, which in Cornwall is mainly found in coastal regions. Caterpillars that survive secure themselves to the underside of a leaf and pupate for about two weeks before emerging in July or August.

With only one generation a year, the Brimstone is one of the longest-lived British butterflies, surviving as an adult for as long as 11 months. It overwinters as an adult, usually within dense foliage of Holly *Ilex aquifolium*, ivy *Hedera* spp. or Bramble *Rubus fruticosus* agg. and is one of the earliest butterflies to appear in the spring: it is not uncommon

Flight curve based on number of records in Cornwall 1999–2018; and UK life cycle

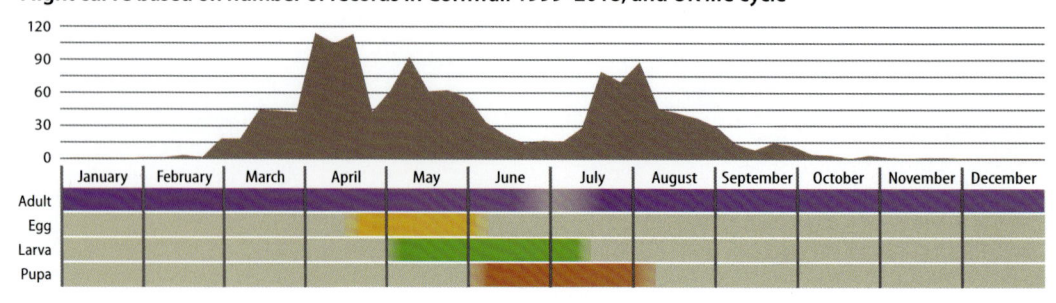

to see it as early as February in Cornwall. These adults often live long enough to overlap the next generation, although by then they may be somewhat battered in appearance. Consequently, there is no month of the year when a Brimstone butterfly might not be seen flying on a warm, sunny day.

Analysis of trends

The range of this butterfly does not seem to have changed much in recent years, although it appears to have been recorded slightly less in the wider St Austell area, a fall in records possibly attributable to the death of a particularly prolific recorder locally.

The yearly statistics show considerable fluctuations over the last decade, with a very slight upward trend in abundance. This lags behind recent South-west England trends, which show a 63% increase over the same time period.

There are generally more records in the spring than from the next generation in the summer. Logically this cannot reflect the true picture. Possible explanations are that the visibility of the butterfly declines when there is increased leaf cover and vegetation, that the butterfly flies more often in the spring (Penhallurick, 1996), and that numbers are boosted because of the enthusiasm of recorders excited to report one of the first harbingers of spring! Most records of Brimstone are of males.

Conservation

Efforts have been made in west Cornwall to plant Alder Buckthorn grown from locally sourced seed. There is no evidence that this has made any difference to Brimstone populations, although the infrequent sightings in the west may be a result of this initiative. Nevertheless, the planting of this foodplant in gardens is encouraged and it can be rewarding to observe each stage of the butterfly's life cycle close to home.

Ensuring that woods are coppiced and cleared may help to create the sunny sheltered conditions that this butterfly enjoys and enable vigorous new growth of Alder Buckthorn. Hedgerows are also important and should be retained and managed appropriately for biodiversity.

Best places to see

- Greenscoombe Wood SX392726
- Cabilla Wood SX134653
- Marsland Nature Reserve SS2316
- Breney Common SX055610
- Ponts Mill SX073563

Greenscoombe Wood, one of the best places to see Brimstone

Upserside, male

Wall
Lasiommata megera

Sponsored by
Marcus Bede Colville, Shaun Poland
and in memory of Stan Burgoyne

The Wall, or Wall Brown, is a medium-sized butterfly. Its glowing, golden-brown wings are heavily patterned on the upperside with black lines, and there are black eyespots with small white centres on the tips of the forewings and the margins of the hindwings. Though the sexes are similarly marked, the male is slightly smaller and is easily distinguished from the female by the broad black sex brand that runs across his forewings. The undersides of the forewings are a paler version of the upperside, whereas the hindwing underside is a buff colour with a border of seven eyespots, the outermost two fused in a figure of eight. The butterfly is named for its habit of alighting on walls, rocks and stony places, where it is clearly visible as it basks with its wings open. When threatened, it closes its wings, dropping the forewing behind the hindwing, the dull underside of which provides good camouflage. Occasionally, if seen from a distance and in flight, a Wall can be confused with a fritillary or a Comma *Polygonia c-album* because of its strong colour, but its eyespots and drab underside reveal its true identity.

Resident in Cornwall
(vagrant on the Isles of Scilly)

CONSERVATION STATUS
- Section 41 Species of Principal Importance (NERC Act 2006)
- Red List of British Butterflies; IUCN criteria: Near Threatened (NT)
- Butterfly Conservation priority: High

TRENDS AND DISTRIBUTION
UK rate of change in abundance
↓ 88% 1976–2018
South-west England rate of change in abundance
↓ 18% 2009–2018
↓ 60% 1976–2018
% of 1km squares occupied in Cornwall
22% 2009–2018 (84% 10km squares)
22% 1976–2008 (80% 10km squares)

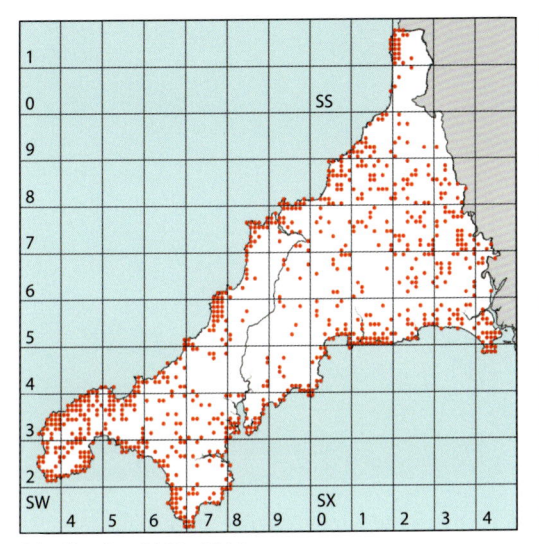

Distribution 2009–2018

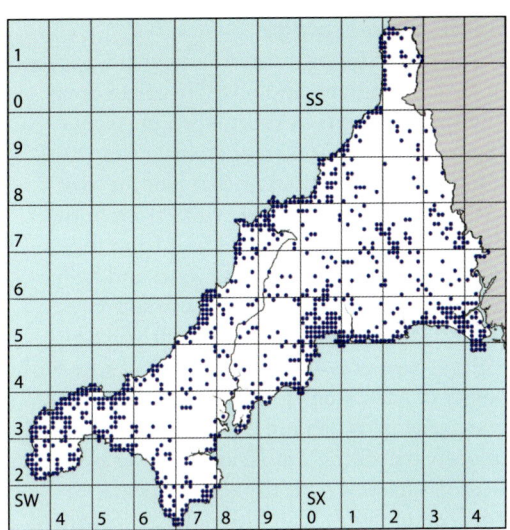

Distribution 1976–2008

Distribution in Cornwall

This species is widely distributed in Cornwall, usually in small numbers, and at one time or another has been recorded in every 10km square. However, in recent years it has become less common in the mid-Cornwall area, whilst it continues to thrive in West Penwith, on The Lizard and on the coast in general. Whether the maps represent the true distribution of the butterfly or reflect a disproportionate number of observers near the coast is a question that is raised for this and many other species.

Larval foodplants

Among the many grasses used by Wall larvae, those favoured in Cornwall are Cock's-foot *Dactylis glomerata*, Wavy Hair-grass *Avenella flexuosa* (which is mainly confined to Bodmin Moor), Common Bent *Agrostis capillaris*, Black Bent *A. gigantea* (which is scarce and is mostly found in arable fields in east Cornwall), Yorkshire-fog *Holcus lanatus*, Annual Meadow-grass *Poa annua* and False Brome *Brachypodium sylvaticum* (which is absent from the main granite areas).

Upperside, female

Habitat and ecology

The Wall can be seen almost anywhere, except in shady woodland and possibly urban areas. The species breeds in short, open grassland where the turf is broken or stony (Eeles, 2019). It is a sun-loving butterfly that likes to bask with wings open to gain warmth from both direct sunlight and radiated heat from the ground beneath it. It favours walls and hedges, particularly Cornish hedges, with south-facing, exposed stonework, as well as rough, unimproved grassland, old industrial sites with plenty of bare, stony ground, and bridleways and gardens. When not at rest, it can frequently be seen 'patrolling' alongside a wall or hedge, with its characteristically rapid, linear, purposeful flight. If an individual is disturbed along a footpath, it will often fly ahead and settle again, repeating this behaviour time after time, appearing to precede a walker for some distance.

The majority of adults emerge in May, although in Cornwall appearances in March and April are not uncommon. The male displays both perching and patrolling behaviour, investigating all passing insects in his search for a virgin female. In the process of courtship, the male showers the female with pheromones. Once mated, the female disappears into undergrowth in search of warm and sheltered places, where she lays her eggs, usually singly, on various parts of the foodplants or adjacent to them. The larvae that emerge are covered in long white hairs, although this reduces in later instars. Pupation takes place after the fourth instar, producing a second, often larger generation in July and August, sometimes extending into September. The offspring of the second generation usually go on to hibernate in the third instar phase, feeding in the winter during mild spells. In some years a partial third brood is seen from September to October or even as late as November.

Mating pair, showing undersides

Analysis of trends

Nationally, the Wall has suffered an 88% decline in abundance since 1976, the south of England being the most affected area. It does not appear to be in such dire straits in Cornwall. This species is subject to large annual fluctuations in abundance, but it generally seems to be doing well on most transects. Records show exceptionally high numbers in 2014 and again in the hot summer of 2018.

It has been suggested that an explanation for the butterfly's comparative success in Cornwall lies in the County's network of footpaths, bridleways, farm tracks and country lanes, which provide lines of communication as well as the bare ground that the butterfly favours. Perhaps the peninsula's extensive coastline and associated terrain also contribute to the butterfly's better fortune in Cornwall,

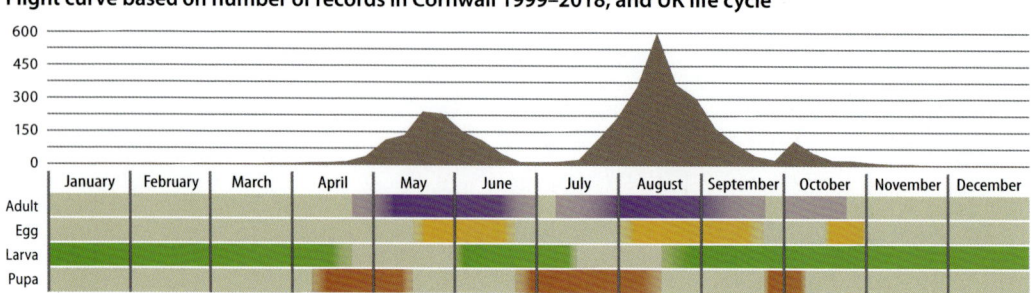

Flight curve based on number of records in Cornwall 1999–2018; and UK life cycle

especially as inland sites are increasingly developed or intensively farmed. However, there may be an additional explanation, based on research which suggests that the Wall, amongst other butterfly species, may be the victim of climate change, through a process known as the developmental trap. The second, and traditionally the last, summer brood produces well-fed caterpillars that normally overwinter. In hotter inland locations, researchers have found that all the second-brood Wall caterpillars studied became a third generation of adults. Larvae from this subsequent brood were then observed to be unable to reach the third instar required to survive the winter, and consequently perished. In cooler coastal locations, however, only 42.5% of second generation caterpillars developed into a third generation of adult butterflies in the autumn (Van Dyck *et al.*, 2014).

It may be, then, that the Cornish peninsula's summer temperatures, buffered from extremes by the surrounding sea, provide a more congenial and sustainable environment for the Wall than elsewhere in more land-locked southern England. Evidence that might support this explanation is in both transect and general records from 2014/2015. Because 2014 was an unusually warm year, there was a big increase in the third generation, confirmed by over 40 sightings of the butterfly in October, compared with just one or two individuals in other years. This was followed by a crash in numbers in 2015. It will be interesting to see whether a similar pattern emerges in the future, since 2018 was another very warm year with a large crop of third-generation Wall butterflies in Cornwall.

Conservation

The maintenance and extension of the Cornish system of footpaths and bridleways is important for this species, as is the preservation of Cornish hedges with exposed stone faces. Coastal grazing helps to keep areas open and scrub-free and to break up the sward for basking.

Best places to see
- Gear Sands (Penhale) SW772556
- Nare Head SW917371
- Pentire SW938800
- Levant Mine SW370346
- Struddicks (near Murrayton) SX293543

Coastal footpath, a habitat favoured by the Wall, Ardensawah Cliff to Gwennap Head

Upperside, male and female similar

Speckled Wood
Pararge aegeria

Sponsored by
Sonja Ede, Eileen Gamston
and Jackson Foundation

The Speckled Wood is an insect of shade, flying in dappled sunlight from early spring to late autumn. It is a common, medium-sized butterfly, easily identified by its dark velvety-brown colour with pale spots that are typically creamy-white or pale yellow. Background colour and size are also variable. In Cornwall, Speckled Wood usually have pale yellow spots, the summer generation commonly having lighter and smaller markings than the spring generation; these have been identified as being the subspecies *tircis*. The Isles of Scilly population, which has distinctively orange spots, belongs to the subspecies *insula*. There are occasional reports of butterflies on the Cornish mainland that resemble those on the Isles of Scilly.

Male and female are similar, although the female has brighter and more distinct markings and lacks scent scales on the forewings. Both sexes have eyespots on the apex of the forewings as well as a row of four eyespots towards the margin of each hindwing, the fourth spot being sometimes very small or even absent. The underside of the forewings is a paler version of the upperside, whilst the hindwings are a patterned greyish/brown bordered by a row of tiny white spots.

Resident in Cornwall
(resident on the Isles of Scilly)

CONCERVATION STATUS
■ Butterfly Conservation priority: Low

TRENDS AND DISTRIBUTION
UK rate of change in abundance
↑ 67% 1976–2018
South-west England rate of change in abundance
↓ 3% 2009–2018
↓ 3% 1976–2018
% of 1km squares occupied in Cornwall
56% 2009–2018 (84% 10km squares)
42% 1976–2008 (81% 10km squares)

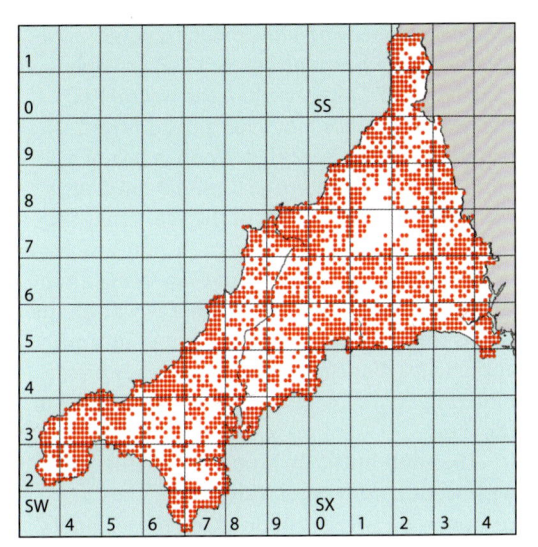

Distribution 2009–2018

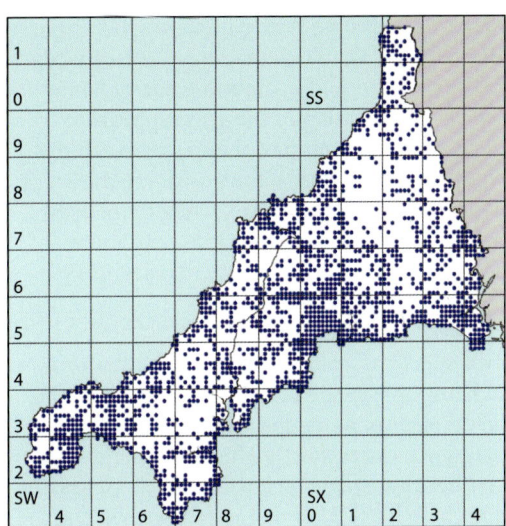

Distribution 1976–2008

Distribution in Cornwall

Speckled Wood are found throughout Cornwall, having been recorded in nearly every 10km square, the only clear gap being Bodmin Moor, where exposed Cornish hedges and remnant woodland seem to provide insufficient shelter (when looking at the distribution of the butterfly at 1km scale). This was also the case over 100 years ago when Clark described it in the Victoria County History for Cornwall in 1906 as "common and in many places abundant except on the moors" (Penhallurick, 1996: 124). The distribution maps indicate an increase in the butterfly's range in Cornwall over the last 10 years, although a growth in the number of recorders is a likely factor affecting this.

Larval foodplants

Speckled Wood larvae feed on a wide variety of grasses, with over 15 recorded. They tend to

Mating pair, showing undersides

95

prefer False Brome *Brachypodium sylvaticum*, Cock's-foot *Dactylis glomerata*, Yorkshire-fog *Holcus lanatus* and Common Couch *Elymus repens*. The location of the grass appears to be more important than the species, with the female thought to vary the position of her eggs depending on temperature, using more open positions during cooler periods, and more shaded places when it is hotter (Eeles, 2019).

Habitat and ecology

This species is unique among British butterflies in being able to spend the winter either in the larval stage or as a pupa. The female usually mates just once, shortly after she emerges. She lays her eggs singly on a wide variety of grasses, and these hatch after one to two weeks. The caterpillars start off pale yellow with a large black bulbous head, but become greener and longitudinally striped as they mature. Only third instar larvae are thought to be able to overwinter successfully; younger ones are usually killed by severe weather. However, some develop more rapidly, depending on temperature and exposure to light, and pupate suspended beneath a grass blade or other vegetation just above the ground. These overwintering pupae emerge the following spring, sometimes as early as February in Cornwall, to be followed slightly later by (and frequently overlapping with) those that overwintered as caterpillars.

Thereafter hardly a week passes without a Speckled Wood appearing somewhere in Cornwall, although it is possible to identify the emergence peaks illustrated in the flight curve. Despite its ubiquitous presence from spring to autumn, the adult butterfly lives for barely a week. In Cornwall, there are usually three overlapping generations; and fresh specimens seen in October and November, in good weather, have been interpreted as a possible fourth brood. In years when the weather is exceptionally poor, not even a third brood is produced. The Speckled Wood is one of a handful of Cornish butterflies that have been recorded in every month of the year. They are often the first and last to fly in the day and have been seen before sunrise and even flying after sunset, provided that the weather is warm. The adults feed on flowers and berries as autumn approaches, but earlier in the year they are more interested in aphid honeydew, high up in the trees.

The adult butterfly is normally associated with woodland fringes, rides and glades, but in Cornwall, while not forsaking these haunts, it has become a butterfly of hedges and shady lanes, railway embankments and gardens. Whilst the female is more likely to spend her time high up in trees and bushes, the male is often found perched or patrolling lower down in the sunshine, where, ever alert for a female, he fiercely defends his territory. Males are described as either 'perchers' or 'patrollers', depending on body shape and wing markings, which are adaptations to the two alternative strategies to attract a female (Thomas and Lewington, 2014).

Analysis of trends

The Speckled Wood has greatly expanded its range in Britain over the last few decades, following a marked decline in the early part of the twentieth century. Better weather and, latterly, climate change are thought to have contributed to the gradual expansion of its range from the 1920s onwards to reoccupy former areas and to move into new territory. South-west England has always been one

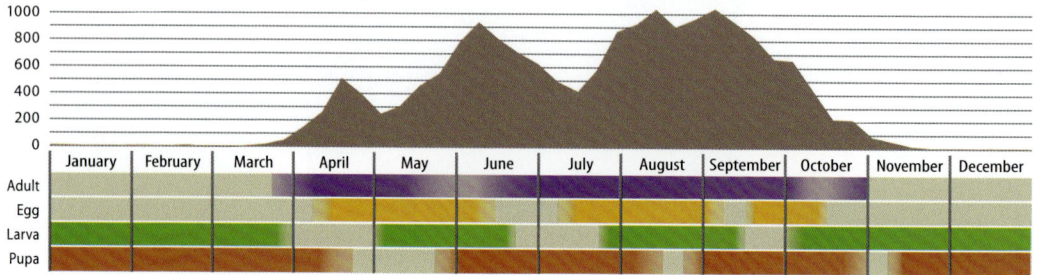

Flight curve based on number of records in Cornwall 1999–2018; and UK life cycle

A woodland track, typical habitat used by the Speckled Wood

of the butterfly's strongholds (Asher *et al.*, 2001), with its comparatively mild winters and relatively high night temperatures in summer (Penhallurick, 1996).

The decline in coppicing, which has adversely affected many other butterfly species, may have helped this butterfly to increase in abundance as well as range, because the decrease in woodland management has increased the habitat available for this shade-tolerant species (Asher *et al.*, 2001). Again, in contrast to some other species of butterfly, the Speckled Wood has adapted to thrive in areas of scrub if there is sufficient shade and when its preferred woodland habitat is unavailable.

Like all common butterflies, this species suffers periodic fluctuations, usually due to poor weather, but it seems to have no difficulty re-establishing itself once the weather becomes more benign. Although the trend in South-west England indicates a slight drop in abundance over the last 10 years, this is not enough to raise concern. A period of unseasonable weather can adversely affect individual broods, but after the very wet weather of 2012, which had a negative impact on most butterflies, data in Cornwall show that the Speckled Wood has been on an upward trend in abundance in the County.

Conservation
No specific conservation is required for this butterfly.

Best places to see
Wooded areas in most parts of Cornwall.

Underside, male and female similar

Small Heath
Coenonympha pamphilus

Sponsored by
*Maggie Goodere, Tony Phillips
and Gareth Ronson*

In the UK, the Small Heath is the smallest butterfly in the subfamily Satyrinae (the 'browns'), which contains thousands of species worldwide. It flies only in sunshine, its characteristic fluttering flight not far from the ground, showing flashes of its orange-coloured upperside, which is difficult to catch on camera. When it settles, its wings always closed and angled to catch the sun's rays, it resembles a diminutive Meadow Brown *Maniola jurtina*, with a prominent eyespot near the apex of its orange forewing acting as a decoy to predators. The underside of the hindwings has bands of brown, grey and cream. In poor weather, or when the butterfly is roosting, the distinctive forewings are usually tucked behind the hindwings in behaviour reminiscent of the Grayling. The male is slightly smaller and more intensely coloured than the female, but otherwise they are very similar.

Resident in Cornwall
(not on the Isles of Scilly)

CONSERVATION STATUS
- Section 41 Species of Principal Importance (NERC Act 2006)
- Red List of British Butterflies; IUCN criteria: Near Threatened (NT)
- Butterfly Conservation priority: High

TRENDS AND DISTRIBUTION
UK rate of change in abundance
↓ 54% 1976–2018
South-west England rate of change in abundance
↓ 40% 2009–2018
↓ 59% 1976–2018
% of 1km squares occupied in Cornwall
9% 2009–2018 (69% 10km squares)
11% 1976–2008 (71% 10km squares)

Distribution in Cornwall
The butterfly is fairly widespread, usually in small numbers, across the County, both along

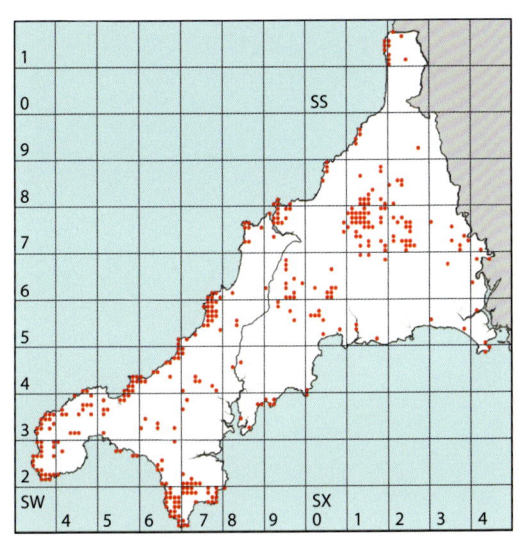

Distribution 2009–2018

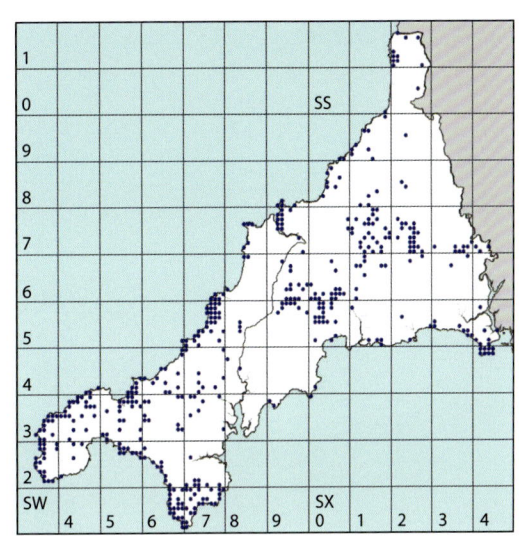

Distribution 1976–2008

the coast, particularly in west Cornwall, and inland on the Mid Cornwall Moors, Bodmin Moor and former mine sites. However, the maps and number of 1km squares where this butterfly has been seen in the two periods show a decline that is also recorded elsewhere in England, although not in Scotland and Wales. There is only a slight difference when comparing its presence in 10km squares, but the number of locations within some of these squares, together with the number of individuals, has decreased, even though more records have been submitted in the last 10 years.

Larval foodplants

Most fine-bladed grasses will serve as larval foodplants for the Small Heath, especially fescues *Festuca* spp., meadow-grasses *Poa* spp. and bents *Agrostis* spp., all of which are well represented in the County by one species or another. Sheep's-fescue *Festuca ovina* has been found to be the preferred larval foodplant in both English and Swedish studies (Eeles, 2019).

Habitat and ecology

The Small Heath occupies a variety of habitats. In Cornwall it can be found on grassland, dunes and grassy heaths, so long as the sites are dry and well-drained, with a short sward that is not too dense. It is noted as breeding at a higher altitude than most other butterflies (Asher *et al.*, 2001).

The breeding pattern is complex, with overlapping generations, and caterpillars developing differentially. It is reported that the caterpillar shows "a high level of 'plasticity' in terms of both the number of instars, and also the instar in which the larva can overwinter" (Eeles, 2019: 140). Hibernation always occurs in the larval phase. The first caterpillars, after hibernating, resume feeding in early spring and start to pupate in April, attached to blades of grass. The earliest that adults may be seen in Cornwall is April or

Underside

Newlyn Downs, one of the best places to see Small Heath

May, and they continue to emerge into July. Males live for around a week and focus their activities on mating, having gathered around suitable landmarks, where they compete for the best territory. The female lays each egg singly on a grass blade or on vegetation close to the foodplant, where it hatches in about two weeks. Larvae spend the daytime mostly tucked away at the base of their foodplant, feeding at night on the tender tips of the blades of their chosen grass. Some of these caterpillars will go through several moults before entering hibernation; others will enter an accelerated life cycle, developing into a second, late-summer brood, which may fly alongside the last stragglers of the first generation. In warm summers there may even be a third generation flying until October.

Analysis of trends

The decline in abundance of Small Heath has been so marked, both nationally and in South-west England, that the butterfly has been added to the Butterfly Red List (Fox *et al.*, 2010) and included as a 'species of principal importance' under the NERC Act (2006). Figures from the Cornwall database show some very large fluctuations. Unsurprisingly, the Small Heath does best in good summers, notably 2014 and 2018; it barely registered in the dire summer of 2012.

Transect data show a significant downward trend when comparing average numbers of Small Heath counted per transect over the last nine years. Conditions in a given year undoubtedly have an impact on numbers over succeeding years. A hot summer one year is

Flight curve based on number of records in Cornwall 1999–2018; and UK life cycle

followed in the next by a large emergence in August, succeeded in turn by a smaller emergence the following spring. Because of the Small Heath's staggered and overlapping emergences, it is hard to confirm any increase in a third generation in Cornwall.

It is difficult to attribute the decline in the number of sites where this butterfly has been seen over the last 10 years directly to changing land management, other than to note the inexorable increase in building development across Cornwall in recent decades, as elsewhere in the UK. The Small Heath has become much less abundant at the local scale, despite remaining widespread throughout the UK. This reduction may be due to "habitat fragmentation, overgrazing, or weather conditions, but much remains to be learned about the ecology and requirements… before the reasons for its current decline are understood and a prognosis can be given for its future" (Asher *et al.*, 2001: 283). The authors conclude that the decline of such a common species is cause for serious concern.

Conservation

Agricultural 'improvements' and changes to grazing regimes of grasslands, particularly inland, have no doubt led to the loss of colonies in the past. The reintroduction by landowners of regimes more sympathetic to the environment would ensure that suitable habitats continue to exist.

Best places to see

- Bodmin Moor including Garrow Tor SX145785 and Casehill/King Arthur's Downs SX128774
- Gear Sands SW769559 / Mount Field SW782568
- Porth Joke and Cubert Common SW777594
- Levant Mine SW369342 / Enys Zawn SW369345
- Porthgwarra SW368218
- Newlyn Downs SW834545
- Breney Common SX056610
- Kynance SW692130

Upperside, male and female similar

Ringlet
Aphantopus hyperantus

Sponsored by
*Tristram Besterman, Jenny Hill
and Sylvia Rose Oliver*

The Ringlet is a distinctive medium-sized butterfly that in flight can be confused with the male Meadow Brown, although it is generally darker and smaller, has no trace of orange, and has many more spots or rings on both the upper- and underside of its wings. At rest, particularly with undersides displayed, the butterfly is unmistakable. A freshly emerged individual has dark, chocolate-brown, velvety wings, the underside of which is prominently marked with the rings that give the butterfly its name. Each ring has three concentric colours: a bright outer rim of pale gold encloses a black disc with a small point of white at its centre. The rings vary in size, shape and number, but there are normally five on the hindwings and three on the forewings. On the upperside, the number of rings also varies, although towards the outer margins there are commonly two rings on the hindwings and three on the forewings. Both sets of wings are fringed with white, but this may not be so noticeable when the butterfly fades. The female is usually slightly larger and fractionally paler than the male.

Resident in Cornwall
(resident on the Isles of Scilly)

CONSERVATION STATUS
- Butterfly Conservation priority: Low

TRENDS AND DISTRIBUTION
UK rate of change in abundance
↑ 307% 1976–2018
South-west England rate of change in abundance
↓ 6% 2009–2018
↑ 94% 1976–2018
% of 1km squares occupied in Cornwall
40% 2009–2018 (82% 10km squares)
23% 1976–2008 (80% 10km squares)

Distribution in Cornwall
In Cornwall, the Ringlet can be seen flying in almost any rural location, typically along

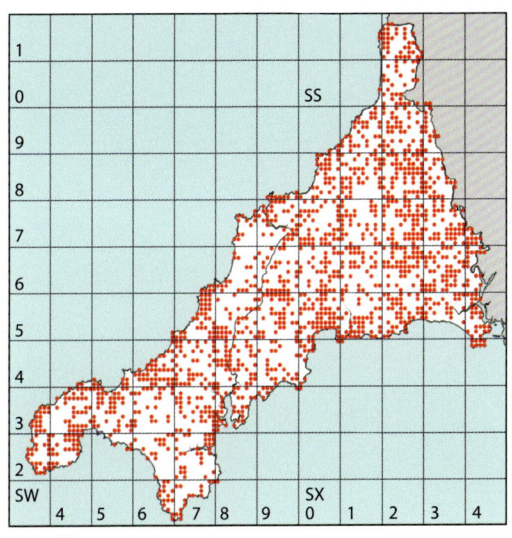

Distribution 2009–2018

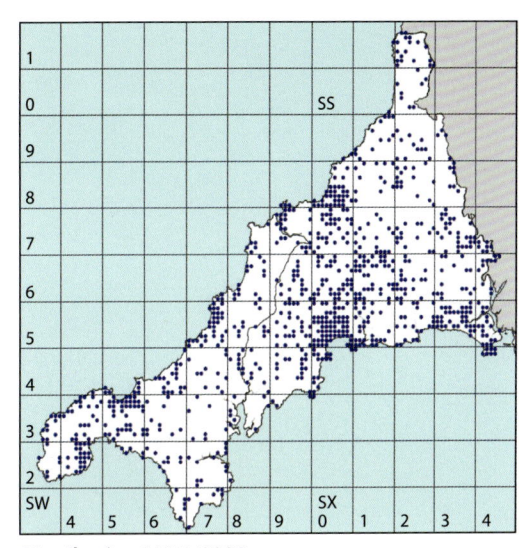

Distribution 1976–2008

hedges and rides and in meadows and woodland clearings. The distribution maps show a marked increase in the number of 1km squares occupied in the last 10 years; this is probably explained by a new generation of recorders more evenly distributed throughout the County, and confirms the suspicion expressed previously that the butterfly had been under-recorded (Wacher *et al.*, 2003). It can form sizeable colonies, and there are locations where hundreds have been recorded in a single session. There are also smaller colonies numbering no more than a few individuals extending thinly over a distance along bridleways and footpaths. Although this species is generally thought not to travel far, research from Switzerland suggests the butterfly "might reside in a metapopulation structure" and therefore a degree of connectivity and mixing between colonies may be expected (Eeles, 2019: 156).

Larval foodplants

The Ringlet favours coarse grasses such as Cock's-foot *Dactylis glomerata* and False Brome *Brachypodium sylvaticum*, although False Brome is missing from the main granite areas of the County, where its place is probably taken by Sheep's-fescue *Festuca ovina*. Tufted Hair-grass *Deschampsia cespitosa* and Creeping Bent *Agrostis stolonifera* have also been reported as foodplants. Regardless of the species of grass selected, the prime requirement is its presence in lush, uncropped tussocks.

Habitat and ecology

The adult butterfly can be expected almost anywhere, but it is more common in damp, half-shaded, sheltered places, be they hedgerows or woodland clearings. This species is particularly fond of areas with patchy scrub containing Bracken *Pteridium aquilinum* and

Underside, male and female similar

Mating pair

gorse *Ulex* spp. It nectars on a wide variety of flowers such as Bramble *Rubus fruticosus* agg., Hemp-agrimony *Eupatorium cannabinum*, Wild Privet *Ligustrum vulgare*, ragwort *Jacobaea* spp. and Common Fleabane *Pulicaria dysenterica*. It can often be found flying on cloudy days, and even sometimes in light rain, its dark colour enabling it to warm up more quickly than most other butterflies. The adult has a characteristic bobbing flight. The male emerges earlier than the female and patrols grassy areas on the look-out for a mate.

The Ringlet is single-brooded and is one of the most consistent of butterflies, seldom appearing before the last week of May and usually disappearing by mid-August, some battered individuals surviving until September.

The female ejects her eggs haphazardly, usually at rest high on a grass stem or sometimes on the wing above the tussocks. This behaviour may be triggered by quite specific scents from the grass, so it is less chaotic than it appears (Thomas and Lewington, 2014). The eggs are non-adhesive and fall directly into the warmer base of the foodplant, where they are concealed from both natural predators and the inquisitive human eye. After about two or three weeks the larvae emerge, and feed only at night on the most tender part of grass stems. The larval stage lasts for about 10 months, overwintering at the third instar stage and continuing to feed on mild days. Although this buff-coloured, hairy little caterpillar is very difficult to find in the autumn, it can be located

Flight curve based on number of records in Cornwall 1999–2018; and UK life cycle

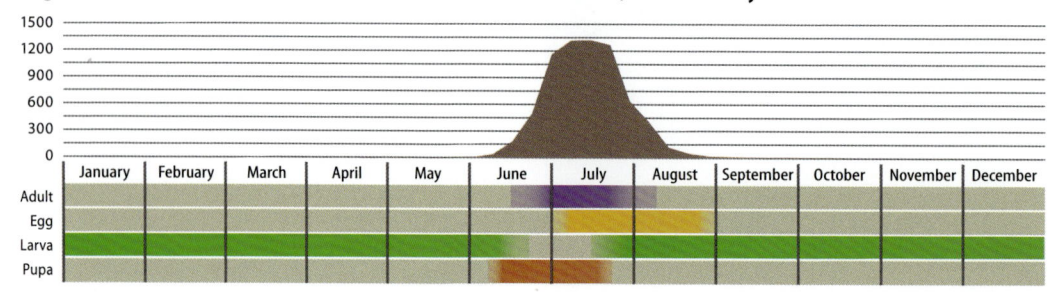

Unimproved grassland, habitat favoured by the Ringlet, Kelsey Head

with a torch on warm nights in May, when its back exhibits a characteristic longitudinal chocolate-brown stripe, which becomes a broken stripe in the final instar. The pupal stage lasts for around two weeks, the adult usually emerging in late June.

Analysis of trends

In terms of its distribution, the Ringlet is doing very well in Cornwall, and nationally distribution is also recovering, as well as abundance increasing dramatically. Opinions vary on the causes, including climate change, a decrease in industrial pollution, and an increase in the deposition of nitrogen, which favours the growth of coarse grasses. In Cornwall over the last 10 years there have been considerable fluctuations in the records and numbers submitted, with a generally upward trend. It is said that the Ringlet does worse in the year following an exceptionally hot and dry summer. The drought of 1976 is often cited as an example, but Cornwall rarely experiences the extreme conditions endured by other areas of the UK; moreover, the butterfly did quite well in Cornwall in the year following the hot summer of 2018.

Conservation

Although the Ringlet appears to have expanded its range in Cornwall, the same favourable conditions continue to be required as for many other butterflies. Recommendations for it to flourish include maintenance and extension of hedges bordering footpaths, bridleways and tracks throughout the County and ensuring that they are managed in a sympathetic manner so that saplings are allowed to grow and create shade. Road verges and field margins should increasingly be left uncut in summer.

Best places to see

Ringlet can be seen all over Cornwall.

Upperside, male

Meadow Brown
Maniola jurtina

Sponsored by
Lucy Hill, Jo Poland
and Kew Devon Cattle

The Meadow Brown is usually the most frequently recorded butterfly in Cornwall, flying throughout the summer months and into early autumn. It is medium-sized, the male slightly smaller than the female. Male and female are markedly dimorphic: the upperside of the male is mainly a dark chocolate-brown colour (fading with age), whilst the female is more orange, particularly on her forewings, although the orange patterning is quite variable. The presence of tiny black dots on the underside of the butterfly's hindwing is also extremely variable, some individuals being devoid of any, and others displaying up to five. (See **Habitat and ecology** below for further details.) Male and female undersides are quite similar, both having an eyespot in the tawny or light-orange background of the forewings, while the hindwings are a two-toned greyish brown, the tonal contrast between the bands of colour being more obvious in the female, whose eyespot is generally more prominent than the male's.

Resident in Cornwall
(resident on the Isles of Scilly)

CONSERVATION STATUS
- Butterfly Conservation priority: Low

TRENDS AND DISTRIBUTION
UK rate of change in abundance
↓ 3% 1976–2018
South-west England rate of change in abundance
↑ 39% 2009–2018
↓ 9% 1976–2018
% of 1km squares occupied in Cornwall
55% 2009–2018 (86% 10km squares)
39% 1976–2008 (80% 10km squares)

The black eyespot, which has a single white 'pupil', is visible on the upper- and underside of the forewings. This distinguishes it from the similar Gatekeeper *Pyronia tithonus*, which has an eyespot with twin white 'pupils'.

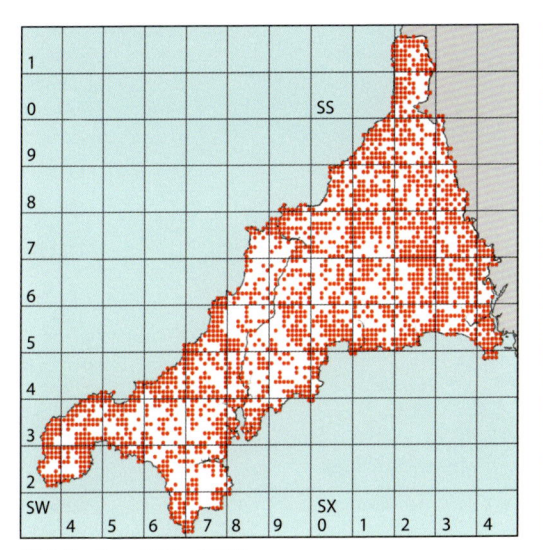

Meadow Brown

Distribution 2009–2018

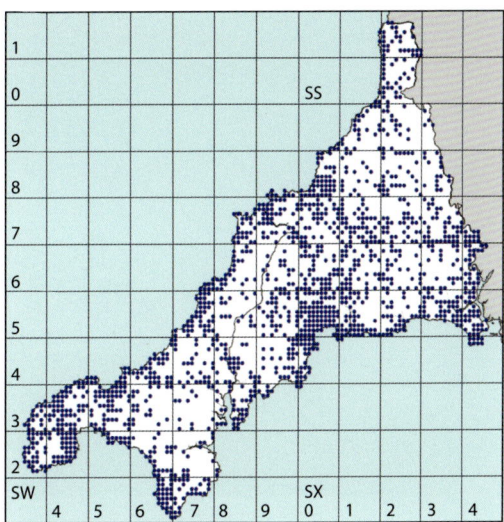

Distribution 1976–2008

However, it is not uncommon to come across a Meadow Brown with two white 'pupils', so care should be taken. There are other differences: the Gatekeeper is slightly smaller, and its upper hindwings are a brighter orange than the Meadow Brown's, again helping to tell the two species apart, as they both fly in July and August. In flight, the upperside of the newly emerged dark male can sometimes be confused with that of the Ringlet.

As with other 'browns', the adults are often active in dull weather and have even been seen in light rain.

Distribution in Cornwall

The maps show how widely distributed this butterfly is in the County. Its range appears to have increased slightly in the last 10 years, but this could well be a feature of having more recorders and ever-increasing data from the

Upperside, female

Underside, with the typical single white 'pupil' in the black eyespot

Underside, showing two white 'pupils' in the black eyespot, seen occasionally in Cornwall

Big Butterfly Count. Nevertheless, the picture looks quite optimistic for Meadow Brown in Cornwall and across the south-west of England, although there has been a slight decline in occurrence nationally. This species has always been regarded as widespread and abundant across Cornwall (Penhallurick, 1996).

Larval foodplants
A wide range of grasses is used, although those with finer leaves are preferred, particularly in the earlier stages of larval development, including the fescues *Festuca* spp. and meadow-grasses *Poa* spp. Larger larvae will consume Cock's-foot *Dactylis glomerata* and False Brome *Brachypodium sylvaticum* and possibly other coarser grass species.

Habitat and ecology
The ideal habitat for Meadow Brown is "warm, open grassland, 0.5m or so in height, containing an abundance of summer flowers and medium- or fine-leaved grasses" (Thomas and Lewington, 2014: 264). In such areas high numbers of Meadow Brown can emerge. However, the butterfly can be met almost anywhere, from urban parks and gardens to grassy road verges and hedgerows. Its more natural habitat is open grassland of almost any type that has not been too agriculturally 'improved', including grassy heaths, dunes and Cornish hedges. It also favours open woodland rides and glades.

The variation in the number of small spots on the underside of the hindwing, described above, along with other differences such as size, has led to the identification of four subspecies in the British Isles (Eeles, 2019). The subspecies *Maniola jurtina insularis* is thought to be the commonest subspecies in Cornwall, and indeed is widespread throughout the rest of England. The subspecies found on the Isles of Scilly is *cassiteridum*. However, there has been much speculation about which subspecies actually occurs in the County, with occasional recording in west Cornwall of individuals

Flight curve based on number of records in Cornwall 1999–2018; and UK life cycle

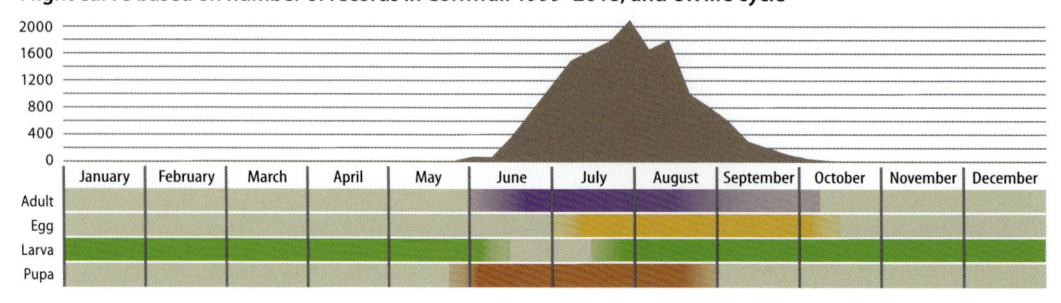

matching the appearance of the Isles of Scilly subspecies (as well as numerous transitional individuals between the two). Recent research suggests that the genes associated with the black dots also code for developmental differences that may be important drivers for survival under different circumstances (Thomas and Lewington, 2014).

The butterfly has a long flight period for a single-brooded species. In Cornwall this can run from mid-May until the end of October. This is caused by a lack of synchronisation in the development of the larvae, due partly to the long period of egg laying and possibly to different types of habitat.

The female Meadow Brown generally mates on the first day of her adult life, which lasts on average between five and twelve days. A single egg is laid on the blade of a wide variety of grasses or vegetation close by. After about three weeks, the initially brown larva hatches and devours its eggshell. The caterpillar is quite hairy and turns green once it has begun feeding on grasses, continuing to feed through high summer and autumn. Once the weather cools, the larva enters hibernation, although it still emerges to graze on mild winter days. In spring it is more easily found after dusk with a torch as it crawls up the blade from its daytime hiding place at the base of the grass. The larva moults five times before pupating. The chrysalis, which is green with variable markings, is attached to a grass blade, and the butterfly emerges after two to four weeks.

Analysis of trends

The Meadow Brown appears to be more than holding its own in Cornwall, as the 10 years to 2018 have seen its distribution expand throughout the County; indeed, trends across South-west England indicate an increase in abundance of 39% over this time period. Analysis of Cornish butterfly transect data is not straightforward, as in 2018 three times as many transects were walked where the Meadow Brown was seen as in 2009. When the number of Meadow Brown per transect was calculated as an average, this increased slightly year-on-year despite some weather-related fluctuations, with 2018 the best year in a decade. Transects are generally more likely to be created in areas with interesting semi-natural habitats, so they may not be entirely representative of the overall picture in the wider countryside. Some colonies have been lost as a result of habitat destruction by over-zealous mowing or spraying. Other agricultural 'improvements' to grassland, including the reseeding of some meadows with monocultural rye-grass *Lolium* spp., have also destroyed colonies.

Conservation

No specific conservation methods are needed, except to ensure that areas of semi-natural grassland are maintained.

Best places to see

Almost anywhere in Cornwall where a variety of grasses are present.

Upperside, male

Gatekeeper
Pyronia tithonus

Sponsored by
*Roy Hilder, Lynn Jeffries
and Kew Devon Cattle*

Despite its shorter flight period, the Gatekeeper is almost as widespread and abundant in Cornwall as its fellow grassland species, the Meadow Brown. It can sometimes be confused with the Meadow Brown when its wings are closed, revealing a brownish/grey and cream mottled hindwing with a few tiny white spots, and an orange forewing with an eyespot at the tip. The upperside of the wings is predominantly bright orange, apart from a broad brown border on both forewing and hindwing. The male tends to be smaller than the female and has a prominent sex brand of dark scales running through the orange on the upperside of the forewing. The Gatekeeper usually has two white 'pupils' within the black eyespots on both underside and upperside of the forewing. The Meadow Brown looks similar but it is larger and it can usually be distinguished by the presence of a single white 'pupil' in its eyespots and by the lack of white spots on the underside of its hindwing, where it may instead have black dots. Variations and aberrations in the Gatekeeper are quite common.

Resident in Cornwall
(vagrant on the Isles of Scilly)

CONSERVATION STATUS
■ Butterfly Conservation priority: Low

TRENDS AND DISTRIBUTION
UK rate of change in abundance
↓ 46% 1976–2018
South-west England rate of change in abundance
↑ 11% 2009–2018
↓ 55% 1976–2018
% of 1km squares occupied in Cornwall
47% 2009–2018 (84% 10km squares)
36% 1976–2008 (80% 10km squares)

Distribution in Cornwall
As with many of Cornwall's most common butterflies, distribution appears to have

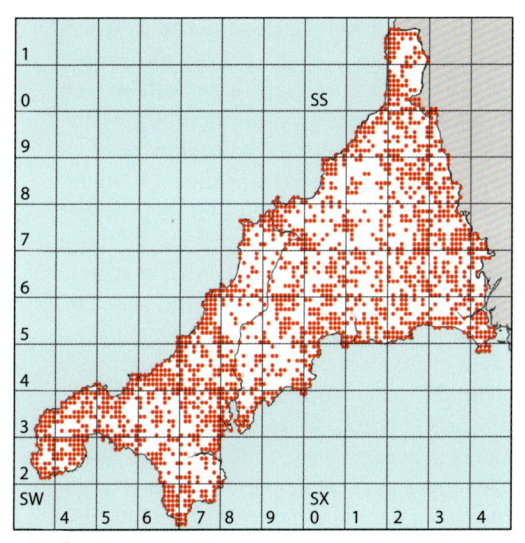

Distribution 2009–2018

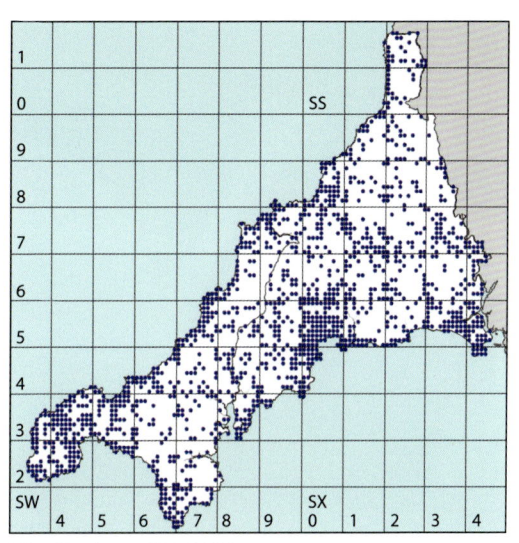

Distribution 1976–2008

increased over the last 10 years, coinciding with a growing number of recorders. The Gatekeeper is now widespread all over Cornwall, and its occurrence in 1km grid squares appears to have increased by over 10% in the most recent period, 2009 to 2018.

Larval foodplants

Various common grasses are the larval foodplant, with narrow-bladed species preferred, such as the fescues *Festuca* spp., of which Red Fescue *F. rubra* agg. and Sheep's-fescue *F. ovina* are widespread in Cornwall (French, 2020). Meadow-grasses *Poa* spp., bents *Agrostis* spp. and Common Couch *Elymus repens* are also used.

Habitat and ecology

As the distribution map implies, Gatekeeper can be seen in almost any habitat that

Upperside, female

Underside, male and female similar

contains nectar sources, of which Bramble *Rubus fruticosus* agg. is especially popular. The butterfly mostly favours hedgerows; hence its alternative name Hedge Brown. The ubiquitous Cornish hedge with lush growth at its base provides just the right habitat, so the Gatekeeper is common along roadsides, footpaths, field margins, bridleways and farm tracks. It is also found on scrubland and sand dunes and increasingly in urban gardens, but it tends not to penetrate too deeply into woodland, instead remaining on the fringes, along rides and in glades and clearings. The colonies are largely sedentary and vary greatly in size from a few dozen adults up to several hundred.

After emerging, Gatekeeper can be seen flying from late May, peaking in July, with a few still being seen at the beginning of September. The butterfly produces a single brood each year. Males will set up territory around a particular bush or shrub on the lookout for passing females. After mating, the female lays eggs singly on and near the larval foodplant, usually in shade, but she often ejects them into the air randomly over a suitable patch containing the foodplant. The larvae hatch two to three weeks later. The caterpillar eats its eggshell and feeds slowly on the tender parts of its preferred grass. It goes into hibernation in September or October at the base of a grass clump after moulting only once, remaining motionless until the following spring, when it starts to become active again. After a further three moults and continued feeding, it pupates, having spent about 240 days as a caterpillar. The butterfly emerges from the chrysalis after about three weeks (Eeles, 2019).

Analysis of trends

National trends indicate a significant long-term decline in abundance, which is even more pronounced in South-west England, although there seems to have been something of a recovery in the region over the last 10 years. The removal of hedgerows and the intensification of agriculture undoubtedly contributed to the loss of habitat in the second half of the twentieth century. Statistics for Cornwall, drawn from both the ERICA database and transect information, indicate considerable yearly fluctuations, but there seems to have been an overall upward trend

Flight curve based on number of records in Cornwall 1999–2018; and UK life cycle

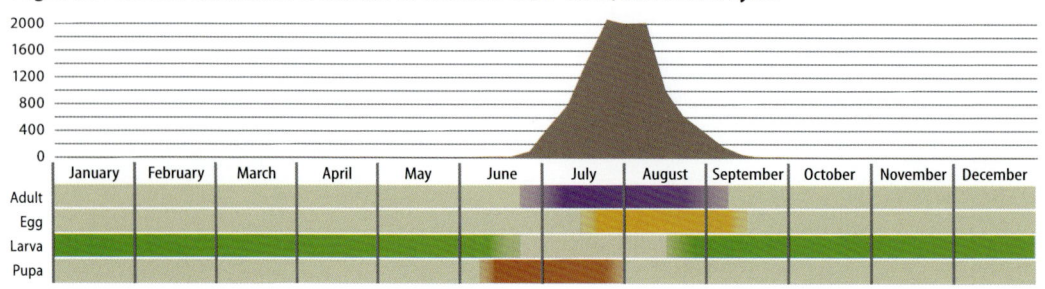

Coastal habitat for the Gatekeeper, near Fowey

over the last 10 years, echoing the picture in South-west England. By far the best year for the Gatekeeper was 2018, which was a good year for many butterflies.

Conservation

No specific conservation methods are considered necessary for this species, although there is the suggestion that "there is scope for populations to be increased by the planting of hedges and the growth of strips of native grasses alongside them" (Asher *et al.*, 2001: 271). The retention and appropriate management of hedgerows and Cornish hedges are also important.

Best places to see

The Gatekeeper is common throughout Cornwall and can be seen almost anywhere in the countryside on sunny days in high summer.

Upperside, showing distinctive marbled patterning

Marbled White
Melanargia galathea

Sponsored by
*Belinda Batt, Sarah Brennan
and Philip Hambly*

The wings of the Marbled White are marked with an unmistakable black and white chequered pattern, particularly vivid when freshly emerged. The marking distinguishes this medium-sized butterfly from all other UK butterflies and in particular from the other native 'browns', which are characterised by earth pigments, typically brown and ochre. Its striking appearance may be a warning to predators that it is unpalatable or may possibly act in a similar way to 'dazzle camouflage' once used by the military to confuse an attacker. Male and female are subtly different: the females are usually a little larger and have a brown hue to their underside and a very thin, pale brown rim along the leading edge of their forewings. Both sexes have a similar but fainter patterning on the underside, with a line of eyespots close to the borders of the hindwings. The butterfly often feeds and rests with wings closed, but under low light conditions, when

Resident in Cornwall
(not on the Isles of Scilly)

CONCERVATION STATUS
- Butterfly Conservation priority: Low

TRENDS AND DISTRIBUTION
UK rate of change in abundance
↑ 51% 1976–2018
South-west England rate of change in abundance
↑ 54% 2009–2018
↓ 24% 1976–2018
% of 1km squares occupied in Cornwall
4% 2009–2018 (51% 10km squares)
5% 1976–2008 (40% 10km squares)

it is overcast or at either end of the day, it rests with wings fully extended to optimise heat absorption, its contrasting tones clearly displayed.

Although it is not easily confused with other butterflies, a sighting of a perceived Marbled White is occasionally sent to

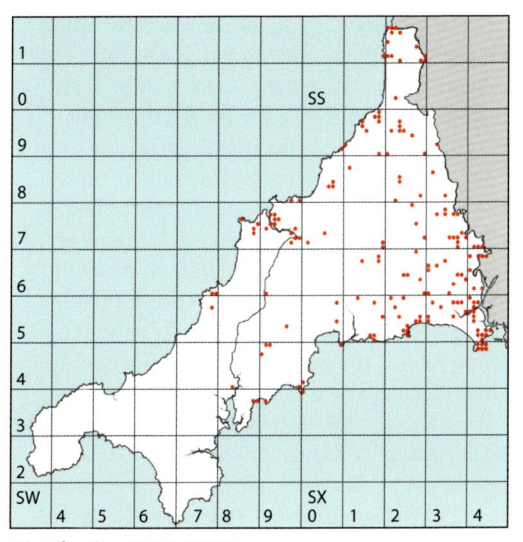

Distribution 2009–2018

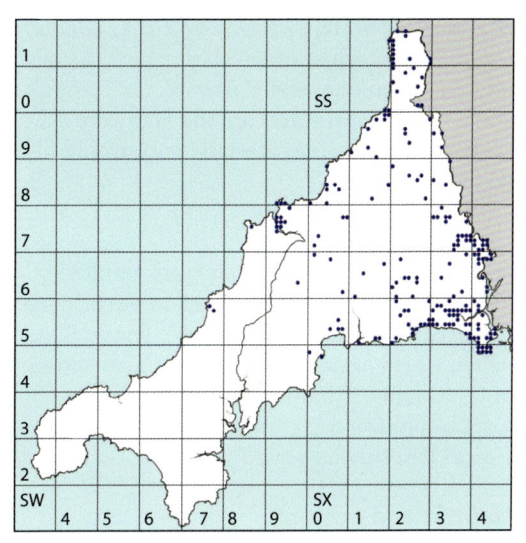

Distribution 1976–2008

the County Recorder accompanied by a photograph of a Magpie moth *Abraxas grossulariata*. It is therefore possible that the occasional sighting of this butterfly in an unexpected place might be a case of mistaken identity, since the moth and the butterfly fly at the same time. Although the markings on the Magpie moth can be variable, the open upperside of their wings always has some traces of bright yellowy-orange patterning, which is entirely absent from Marbled White and is the most obvious way to tell the difference for the uninitiated.

Distribution in Cornwall
In Britain, the strongholds of the Marbled White are the southern and south-western counties of England. However, the butterfly is common throughout central England and can be found as far north as Yorkshire, its range still expanding. It breeds primarily on unimproved grassland, but limestone and chalk grasslands are thought to support the biggest populations. In Cornwall, the explanation for its distribution is a bit of a mystery, although it is thought that the County straddles the western limit of the species' range (Penhallurick, 1996). Most colonies of Marbled White lie in the area east of the rivers Camel and Fowey up to the border with Devon. Small colonies exist along the north coast from the dunes at Rock to Marsland Nature Reserve at the north-eastern tip of Cornwall, with a reasonable sprinkling of colonies inland. A greater concentration lies in south-east Cornwall close to the upper reaches of the Tamar Valley and in the coastal region around Penlee Point and Rame Head, where large numbers have been recorded. This picture has not changed greatly since the 2003 Atlas (Wacher *et al.*, 2003) was published, although the number of 10km squares where this butterfly has been recorded has increased to a further 11% of Cornwall's total over the last 10 years. The butterfly's foodplant grasses are widespread in Cornwall, so there is no easy answer to why it has not spread any further south than Crantock on the north coast, although there was one reliable record from Zelah in 2019 and single sightings from the Truro and Roseland areas a few years earlier.

Magpie moth, sometimes confused with the Marbled White

However, the butterfly generally dislikes damp conditions, which might explain why it has not yet penetrated into west Cornwall: the "rainfall and winter temperatures are too high here for the overwintering caterpillars to survive" (Spalding, 1992: 9).

Larval foodplants
Several grasses serve as foodplants, but Red Fescue *Festuca rubra* agg. seems to be the most favoured. Alternatives are Sheep's-fescue *F. ovina*, which has a more restricted distribution in the County, Timothy *Phleum pratense*, Cock's-foot *Dactylis glomerata*, Yorkshire-fog *Holcus lanatus* and possibly others.

Habitat and ecology
The Marbled White inhabits areas of unimproved grassland with a tall sward, particularly in coastal areas. Colony size can vary greatly. This is usually associated with the area of available habitat, so in some areas just a few adults are seen, whilst in others hundreds have been observed. On south-facing, sheltered, grassy sites with a tall sward this species can become especially abundant. The adults are attracted to purple flowers such as thistles *Cirsium* spp. and knapweeds *Centaurea* spp. for nectar. The highest numbers occur on fields that are not grazed or are very lightly grazed, but they can also be seen in woodland clearings, where they breed along unmown grassy rides or in open glades. Although larger populations remain consistently local, the butterfly has excellent powers of dispersal and is able to colonise suitable uncut sites nearby, such as embankments and verges. Jeremy Thomas cites an example of a successful colony in the central reservation of a dual carriageway (Thomas and Lewington, 2014).

The butterfly is single-brooded and appears on the wing from mid- to late June, and, depending on the temperature, can be seen until mid- to late August. Its gliding flight is typically relaxed, the males patrolling constantly, often mating with females on their maiden flight; they are not particularly territorial. Most butterflies attach their eggs to specific plants, but the female Marbled White expels hers randomly onto the ground. The globular white eggs hatch after about three weeks, the emerging larvae devouring their eggshells and going straight into hibernation, eventually waking in late winter and feeding by day on one of the preferred fescues. As it matures, the caterpillar becomes more of a nocturnal feeder and varies its diet

Mating pair, female above

Flight curve based on number of records in Cornwall 1999–2018; and UK life cycle

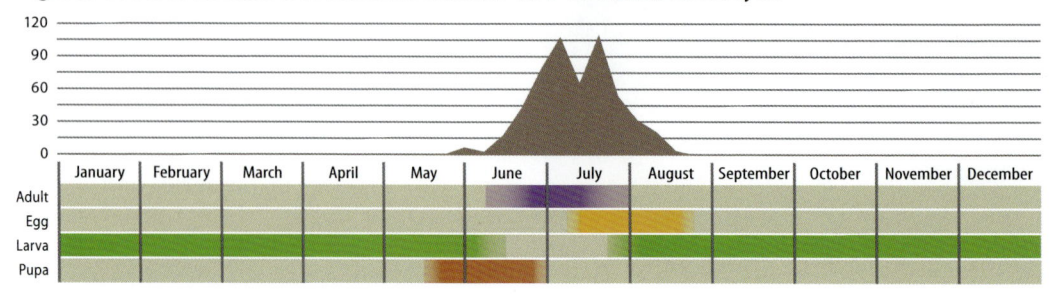

to include some of the coarser grasses, hiding away in the daytime at the base of a grass clump. The larva is initially 3mm long, pale brown with dark stripes running along its body, perfectly camouflaged when it hatches in withered winter fescue. From the second instar onwards it can be either light brown or light green, with correspondingly coloured longitudinal stripes. By the time it reaches its fourth instar it is 28mm long (Eeles, 2019). After three moults it will pupate on or just below the surface of the soil, the adult emerging three weeks later.

Analysis of trends

The national and south-west trends are confusing. Nationally there has been a healthy increase in abundance of the Marbled White since 1976, whilst South-west England showed a strong decline over the same period, albeit mitigated by a partial recovery in the last 10 years. Over the same period, the figures from Cornwall show a slight upward trend, mainly caused by two exceptional years in 2014 and 2015, which have not been repeated since. Indeed, each year since 2015 has shown a steady decrease in numbers, raising some concern and the need for increased monitoring. However, even in good years numbers submitted are not large and could be heavily influenced by differential reporting from the few main sites where hundreds are sometimes seen, thus influencing disproportionately the overall counts.

Grassland 'improvement' for agriculture, over-grazing, cutting too frequently and scrub encroachment all contribute to degrading the Marbled White's habitat and putting the species under pressure. Fragmentation of habitat can also be a problem, as for most species, but fortunately the Marbled White is a good traveller, so it is adept at establishing colonies on new sites or recolonising old ones.

Greena Moor, one of the best places to see Marbled White

A series of hot, dry summers may even lead to a temporary or permanent expansion of its range.

Conservation

It is important to ensure that fescue-rich grassland remains uncut and unimproved. Both over- and under-grazing should be avoided, as should cutting roadside verges too frequently in the vicinity of Marbled White colonies.

Best places to see

- Penlee Point SX438489 / Watch House Field SX438491
- Seaton Valley Countryside Park SX303550
- Greena Moor SX235961
- Rock SW930760
- Treburley SX342772

Underside, showing forewing, best seen when nectaring

Grayling
Hipparchia semele

Sponsored by
Jerry Dennis, Dave Thomas
and David F. Williams

The Grayling is the largest of the 'browns' in the UK, and perhaps the most challenging to see and identify. The open-wing markings have delicate light brown panels with pairs of eyespots on both upperside and underside, but this colourful aspect is usually glimpsed only during courtship or when the butterfly is in flight with its distinctive loops and glides, because the Grayling perches with wings closed. When in this pose, often with its wings tilted slightly to one side, it is hard to see, particularly in dry vegetation or on bare ground. The subtle off-white and brown marbling on the underside of the hindwings provides near-perfect camouflage.

The best chance of identifying this elusive butterfly is when it re-alights after being disturbed. The underside of the forewing can then be seen briefly, revealing one or both black-ringed eyespots. After a few seconds the forewing is lowered, showing only the

Resident in Cornwall
(not on the Isles of Scilly)

CONSERVATION STATUS
- Section 41 Species of Principal Importance (NERC Act 2006)
- Red List of British Butterflies; IUCN criteria: Vulnerable (VU)
- Butterfly Conservation priority: High

TRENDS AND DISTRIBUTION
UK rate of change in abundance
↓ 73% 1976–2018
South-west England rate of change in abundance
↓ 57% 2009–2018
↓ 70% 1976–2018
% of 1km squares occupied in Cornwall
3% 2009–2018 (46% 10km squares)
6% 1976–2008 (57% 10km squares)

underside of the hindwing. Distinguishing the sexes is possible when the butterfly is at rest or in copulation: the male has a more pronounced jagged white band across the

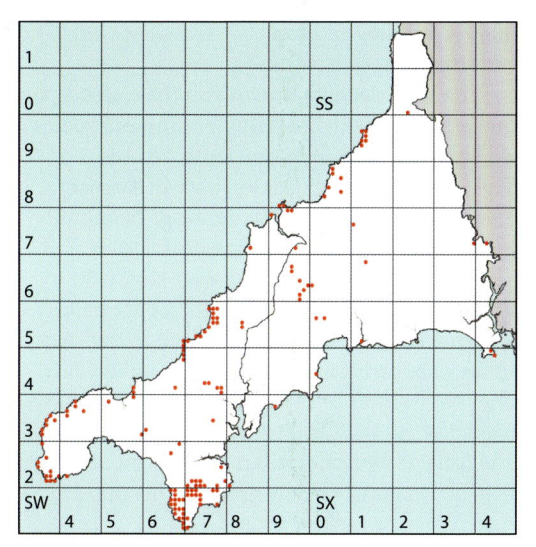

Distribution 2009–2018

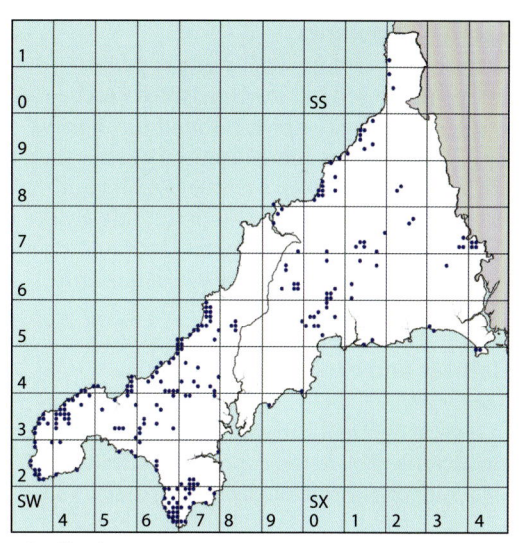

Distribution 1976–2008

lower underside of the wings and is smaller than the female.

Distribution in Cornwall

There are six named subspecies of Grayling in Britain and Ireland, but only the normal (nominate) subspecies *Hipparchia semele semele* is found in Cornwall. It is well-established in the County, particularly along the west and north coasts and on The Lizard peninsula, where it is one of the characteristic butterflies of the lowland heathland. It can be found less commonly inland, but often in former industrial locations such as disused mine sites, quarries and railway cuttings. It also inhabits unimproved moorland as well as heathland, notably west of Truro and in the environs of St Austell, Wadebridge and Bodmin. The butterfly is scarce along most of the south coast, especially in the east of the County, and it has been recorded at only a handful of locations along the border with Devon. The distribution maps appear to show a decline in sites, particularly inland. Grayling have been recorded in half as many 1km grid squares during the period 2009 to 2018 as from 1976 to 2008.

Underside, with forewing hidden

Larval foodplants

The commonest fine-bladed grass species in Cornwall used by Grayling is Red Fescue *Festuca rubra* agg., which is found throughout the County. Other, more localised grasses that are used include Sheep's-fescue *F. ovina* and Bristle Bent *Agrostis curtisii*. Early Hair-grass *Aira praecox*, Tufted Hair-grass *Deschampsia cespitosa* and Marram *Ammophila arenaria* may also be used.

Habitat and ecology

The Grayling lives in colonies and can be seen in several different habitats, all of which are characterised by nutrient-poor, well-drained soils supporting fine-bladed grass species and with areas of sun-exposed bare soil or rocky outcrop. Grazing is generally beneficial to Grayling habitat as it prevents the development of denser vegetation and the loss of bare ground. Heather species are often associated with its habitat in Cornwall, particularly Ling/Heather *Calluna vulgaris*, Bell Heather *Erica cinerea* and the more localised Cornish Heath *E. vagans*, which is found mainly on The Lizard peninsula. The lack of extensive heaths elsewhere on the southern coast of the County may explain why the butterfly has not colonised more widely there.

The species maintains optimal body temperature by tilting and orientating its body and closed wings in relation to the sun and to use the heat reflected off bare ground. It is therefore surprising that the butterfly thrives so well along the exposed cliffs of north and west Cornwall. Here the presence of heather may be useful in offering additional means of wind shelter and opportunities for basking. The Grayling spends less time than other butterfly species nectaring on flowers but is attracted to muddy puddles and sometimes feeds on sap exuding from tree trunks.

Like the Meadow Brown and the Gatekeeper, the Grayling produces a single summer brood, with the main flight period from July to September, peaking usually in early August. There have been rare early sightings from mid-June and late sightings up to mid-October.

The female lays her white eggs singly on fine-bladed grasses. The caterpillars are pale brown and striped and mainly feed in darkness, so they are rarely encountered during the day. They overwinter in the third instar stage and reach their final (fifth) instar during April/May. The fully grown caterpillars burrow into the soil in order to pupate below ground; this pupation stage lasts around one month.

Analysis of trends

In several inland counties of Britain the butterfly has disappeared completely, principally through loss of habitat. It is still fairly widespread on the coast and southern heaths, but significant declines in abundance have been recorded in both the UK and South-west England. Conservation work is being focused to save a dwindling number of sites in some counties still fortunate enough to have the butterfly.

The recording patterns in Cornwall mean that it is difficult to pinpoint exactly where the butterfly is struggling or indeed doing well. There are locations where abundance has fallen greatly or local extinction is thought to have taken place, such as Upton Towans and several sites on The Lizard peninsula. Conversely, there are locations where colonies have been found extant after many years without records, for example on the coastal section east of Land's End, where thriving colonies of Grayling were

Flight curve based on number of records in Cornwall 1999–2018; and UK life cycle

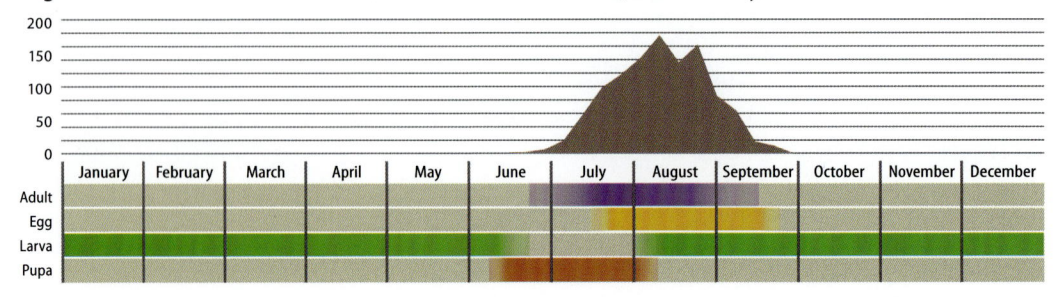

Goonhilly Downs, one of the best places to see Grayling

confirmed in 2019 after more than a decade when no records were submitted.

The butterfly's camouflage and behaviour make it difficult to see and identify in the field, yet there are more than 100 sightings each year, and in good years more than 400, and a few have been spotted in gardens seeking nectar. Good numbers of Grayling can still be seen and sightings of 20 or more occur in the butterfly's strongholds, such as on The Lizard heaths or at key coastal sites such as Porthgwarra. Penhallurick stated that the species is, "as Rollason described in 1910, 'local, but abundant where found'" (Penhallurick, 1996: 133).

There have been 13 Cornish UK Butterfly Monitoring Scheme (UKBMS) transects with Grayling records from 2009 onwards, but only Porthgwarra and Erisey Barton (on Goonhilly Downs), with their designated wildlife status as Sites of Special Scientific Interest (SSSIs), have significant colonies, which account for 93% of all transect sightings. The records at these two sites show that in 2018 and 2019 more Grayling were recorded than in any previous years. This may be in part a recording artefact, but there is no suggestion that the butterfly is struggling in its strongholds where habitat is not under threat.

In recent years, the number of butterfly recorders has increased markedly and the trends in casual records in the last few years reflect this change. Grayling are being seen in more places each year, but generally in small numbers. Most sightings are of one or a low number of individuals, and annually these sightings cover only about 10% of the known range of the butterfly, with just a few new places identified in most years. Many sightings are coastal, often close to the South West Coast Path and under the auspices of conservation organisations such as Natural England, the National Trust and Cornwall Wildlife Trust.

Conservation

The main threat to Grayling habitat in Cornwall is the development of denser vegetation and the resulting loss of fine-bladed grass species and bare ground. This may come about because of natural maturation of heathland, perhaps exacerbated by reduced grazing, where species such as Gorse *Ulex europaeus*, Western Gorse *U. gallii* and Bramble *Rubus fruticosus* agg. colonise. Agricultural intensification and the 'tidying up' or development of derelict mine sites have also resulted in the loss of essential habitat.

There is limited conservation work specifically for Grayling in the County. The management of sites such as the Porthgwarra SSSI for other species is favourable to Grayling and relies in part on targeted cattle grazing of coastal grassland and periodic localised refreshing of heathland.

Best places to see

- Porthgwarra: Gwennap Head SW366216 / Roskestal and Ardensawah Cliff SW363221
- Erisey Barton (Goonhilly Downs) SW711189
- Pentreath (Kynance Cove) SW688131
- Enys Zawn SW376354
- Rosenannon Downs SW954670
- Newlyn Downs SW835543

Upperside, male and female similar

Pearl-bordered Fritillary
Boloria euphrosyne

Sponsored by
James Board, Liz and Andrew Thomas
and in memory of Alison Norris

It is fortunate that the nationally rare Pearl-bordered Fritillary is still present in several areas of Cornwall. It is a medium-sized orange and black butterfly, slightly larger than its close relative the Small Pearl-bordered Fritillary *Boloria selene*. These two butterflies have some overlap in flight periods and habitats, so it takes a practised eye to distinguish between them correctly, as the patterning of the upperside of the wings is very subtly different in the two species. The main difference on the upperside is that the 'chevrons' on the borders of the Pearl-bordered Fritillary's wings are usually 'floating', whereas on the Small Pearl-bordered Fritillary they are joined to black patterning on the edges. On the underside of each hindwing of both species there is a line of seven silvery 'pearls' along the wing margin. However, other markings that are also visible when the wings are closed make it possible to tell the two apart. The Small Pearl-bordered

Resident in Cornwall
(not on the Isles of Scilly)

CONSERVATION STATUS
- Schedule 5 of the Wildlife and Countryside Act 1981 (as amended) (sale only)
- Section 41 Species of Principal Importance (NERC Act 2006)
- Red List of British Butterflies; IUCN criteria: Endangered (EN)
- Butterfly Conservation priority: High

TRENDS AND DISTRIBUTION
UK rate of change in abundance
 ↓ 78% 1976–2018
South-west England rate of change in abundance
 ↓ 59% 2009–2018
 ↓ 61% 1976–2018
% of 1km squares occupied in Cornwall
 <1% 2009–2018 (7% 10km squares)
 1% 1976–2008 (36% 10km squares)

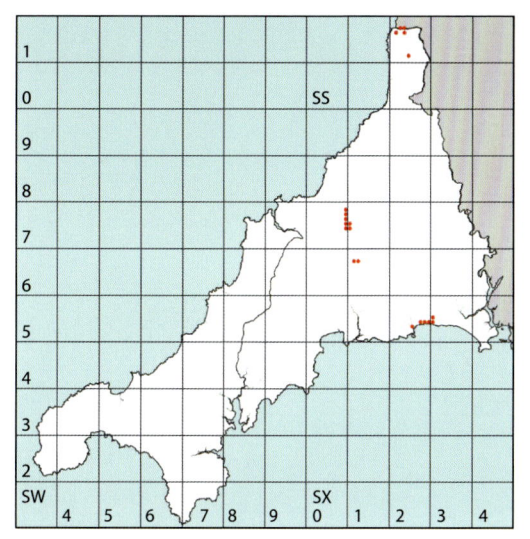

Distribution 2009–2018

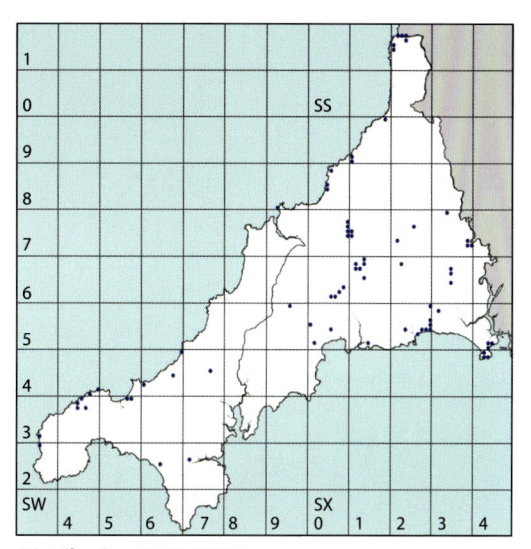

Distribution 1976–2008

Fritillary has a greater contrast in patterning on the underside of the wing, characterised by numerous silvery-white, cream and reddish-brown polygonal cells. The Pearl-bordered Fritillary has less contrasting colours on the underside, with an overall ginger appearance and only two silvery white cells, and it lacks the bold black markings of the Small Pearl-bordered Fritillary. Male and female Pearl-bordered Fritillary are similar, although the female tends to have less intense colour and a shorter, broader abdomen.

Although the Pearl-bordered Fritillary generally emerges from mid-April, there is often a significant overlap in the flight period with the first brood of the Small Pearl-bordered Fritillary, which emerges from mid-May. This affects the reliability of identification from a few sites.

Distribution in Cornwall

Historically, the Pearl-bordered Fritillary was more widespread in the County, although Penhallurick states that the "paucity of historical records makes it impossible to know the extent to which the Pearl-bordered Fritillary has declined in Cornwall" (Penhallurick, 1996: 106). Nationally, in terms of occurrence, this species has declined by 95% between 1976 and 2014, a rate only just exceeded by the High Brown Fritillary *Fabriciana adippe* (Fox *et al.*, 2015). The 1976–2008 Cornish distribution map may not be wholly reliable, as it includes past records from habitats that are unlikely ever to have provided suitable conditions for this butterfly. The Towans area near Hayle is an example of this, so it is probable that the Small Pearl-bordered Fritillary, which is still relatively common in this area, has been wrongly identified as a Pearl-bordered Fritillary in the past. Although records of Pearl-bordered Fritillary were submitted from the Zennor and St Just areas in 1989 and 1999 by a well-respected recorder, some doubt has been expressed about the authenticity of past records from the whole West Penwith area (Penhallurick, 1996). In the last 20 years, there have been no further records submitted from west Cornwall.

The areas of the County where the butterfly is still found are in the far north at Marsland, along the south coast between Millendreath (near Looe) and Seaton Valley, and in central Cornwall, with the western edge of Bodmin Moor supporting several populations. The All the Moor Butterflies project run by Butterfly Conservation from 2017 to 2020 confirmed two metapopulations on Bodmin Moor: one to the north-west of St Breward focused around Fellover Brake, and the other further south of the village centred around De Lank Quarry and Pendrift Downs.

The largest known colony is at the Marsland Nature Reserve, near the Devon border, which

Underside, male and female similar

is a site specifically and very successfully managed for this butterfly. Colonies have undoubtedly been lost in the last 40 years from wooded areas in east Cornwall. The isolated colonies along the coastal path east of Looe, between Millendreath and Seaton Valley, rarely produce high numbers; however, records span several 1km grid squares. Unfortunately, this site has become increasingly inaccessible because of coastal erosion, but there is access to the eastern end of the colony on National Trust land at Struddicks.

Sadly, the colony discovered at Bunny's Hill, near Cardinham, Bodmin, at the end of the twentieth century now appears to be extinct, with the species last recorded there in 2014.

Larval foodplants
Common Dog-violet *Viola riviniana* is the most commonly used foodplant, although other violets may occasionally be used, including Marsh Violet *V. palustris*, which is mainly used by the Pearl-bordered Fritillary in Scotland. In Cornwall, however, Marsh Violet *V. palustris* subsp. *juressi* is an alternative favoured by the Small Pearl-bordered Fritillary, which can cope with damp, humid conditions and seems to be more adaptable to a variety of habitats in Cornwall than the more environmentally sensitive Pearl-bordered Fritillary.

Habitat and ecology
Pearl-bordered Fritillary used to thrive principally in woodland areas throughout Britain, where the butterfly was known as 'the woodman's friend' for its opportunistic habit of colonising freshly cleared areas after coppicing or felling had taken place. The decline of coppicing and woodland management has rendered much of this old habitat unsuitable. The primary habitat of Pearl-bordered Fritillary in Cornwall is now bracken-covered, south-facing, warm, sheltered hillsides that are lightly grazed and adjoin old remnant woodland. Where this species occurs on the south-east coast near Looe, the habitat is open steep slopes created by landslips with loose shale and sheltered from the strong sea breezes by Sycamore *Acer pseudoplatanus* scrub.

The Pearl-bordered Fritillary is a sun-loving butterfly that prefers the protection of dells, glades and combes with low ground cover and an abundance of its larval foodplant; more exposed sites are seldom chosen unless they are facing south or south-west. A mosaic of Bracken *Pteridium aquilinum*, light scrub, bare ground and short grass is important in these situations. Dry, dead Bracken enables the larvae to raise their body temperature, aiding their development early in the year, when this warm microclimate is essential.

The butterfly can fly reasonable distances. Although it usually does not venture far from its own territory, Eeles, quoting from Matthew Oates' 2004 *British Wildlife* article, states that "one marked female travelled 4.5km, across arable land and a deep wooded valley," evidence that the butterfly has the potential to colonise suitable habitat several kilometres away from known populations (Eeles, 2019: 186). One appeared in 2018 in a garden in West Looe whose owner, who had a special interest in this species, realised that it must have travelled nearly 3km from the nearest known colony along the coast.

In most years, the adult butterfly starts to emerge in mid-April in Cornwall and is therefore the first of the fritillaries on the wing, its flight period lasting until the end of May and occasionally going into June. There is said to be a partial second brood in Cornwall in August, although there has been little evidence of this in the last 10 years. It is important that a second brood not be discounted and that searches continue, especially in view of climate change.

A good time to observe the butterfly is in the early morning, when it can be seen nectaring avidly, particularly favouring Bugle *Ajuga reptans* amongst a variety of other spring flowers. Males are more active as they patrol, flying low above their breeding ground and indulging in brief skirmishes with rivals. Mated females will wait a few days until their eggs mature and then take their time searching for suitable patches of foodplant. They rarely lay in grassy areas, preferring either bare ground or heat-absorbing leaf litter, and will usually lay single eggs either directly on the foodplant or close by on dead vegetation. Eggs take about two weeks to hatch and the larvae then feed during the day on the freshest leaves or seedlings of violets.

The banded caterpillars become darker and spinier as they mature, and overwinter after their third moult, usually inside a twisted, dead leaf. They shrink considerably during this period and emerge early in March, small and hungry, feeding for three weeks on the leaves of violets until they have doubled in size before their next moult. The resulting larvae are largely black, with a pale band running longitudinally on either side of their bodies, black-tipped yellow spines, and a light-brown underside. They can move quickly and at this stage are relatively easy to find, still feeding in the daytime as well as basking in the sunshine. A fully grown larva is 25mm in length. It spins a silk pad after two to four weeks, suspending itself head-downwards on withered vegetation, before moulting one last time to reveal the pupa, from which the adult emerges after about a fortnight, depending on the temperature (Eeles, 2019).

Egg on underside of Tormentil leaf

Fifth instar larva with its black-tipped yellow spines

Struddicks, one of the best places to see Pearl-bordered Fritillary

Analysis of trends

Decline in both distribution and abundance is extremely worrying nationwide, with this once widespread species now surviving in scattered, isolated colonies (Eeles, 2019). In South-west England, abundance decreased by 59% during the period 2009–2018. Since the species prefers drier terrain with sparse vegetation, a succession of wet springs and summers can be detrimental, whereas a hot spring promotes a temporary increase in the number of adults. The assessment of abundance on a yearly basis between 2009 and 2018 in Cornwall is not reliable, as variable recording can change the statistics significantly for such a rare species, which is on the wing for only a limited period of time. Figures are also heavily influenced by how well the species is faring at the Marsland Nature Reserve, as this area contributes by far the greatest number of Pearl-bordered Fritillary to the County's yearly total. In the last 10 years, there have been no transects monitoring this butterfly species in Cornwall.

Decline in most parts of Britain (excluding Scotland) has undoubtedly been brought about by the gradual loss of habitat through reduced traditional woodland management, and particularly through the replacement of broadleaved woodland with conifers after 1945.

Flight curve based on number of records in Cornwall 1999–2018; and UK life cycle

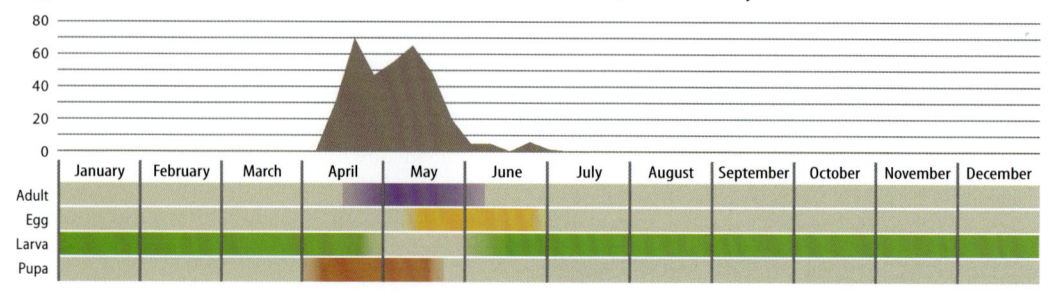

In more open habitats, the cessation of grazing leads to the uncontrolled encroachment of scrub and occasionally too much shading by Bracken. Equally, overgrazing and trampling on grassland can destroy the larval foodplants. Rapid destruction of habitat can be caused by large-scale and repeated burning, or by Bracken eradication.

Conservation

Conservation for this species now depends on management, which varies according to the habitat type. In woodland, the butterfly does best in newly cleared areas, and it declines as more shading occurs, so periodic coppicing is necessary, coupled perhaps with the control of any excess herbage. Felling of trees in adjacent areas, with the consequent growth of ground flora, allows this not particularly mobile butterfly to move from an older clearing shaded out by tree growth to a newer area where the ground flora has just been renewed.

In Cornwall, the butterfly mainly occupies a bracken-rich habitat, where scrub, especially gorse *Ulex* spp., also occurs. Conservation measures are urgently needed here to provide sheltered glades and rides where the larval foodplant can grow in abundance, in conjunction with sufficient dead Bracken and leaf litter, on which most of the eggs are laid, to provide warmth for the larvae. Habitat is best managed by light grazing with ponies or stock, or by regular manual clearance, to ensure that Bracken does not become dominant and to control invasion by scrub (Spalding with Bourn, 2000). Conservation in the Bodmin area has been inconsistent. Earlier this century, annual habitat conservation work was carried out by Cornwall Butterfly Conservation (CBC) in conjunction with other members of the Cornwall Fritillary Action Group, but this ceased for a few years after 2010. Work restarted prior to the commencement of the All the Moor Butterflies project (2017–2020) in the St Breward and De Lank Quarry areas and has significantly improved the habitat for this species. It is essential to the conservation of the Pearl-bordered Fritillary for this work to continue on an annual basis. As part of the legacy of the All the Moor Butterflies project, CBC has committed to carrying out conservation work at the De Lank Quarry site from 2020 until 2030.

Best places to see
- Marsland Nature Reserve SS235171 (Permits from the site warden are required)
- Struddicks SX290543
- Fellover Brake SX091774
- Pendrift Downs SX099746

Upperside, male and female similar

Small Pearl-bordered Fritillary
Boloria selene

Sponsored by
Rowan Board, Dr Simon Wearne
and Natural England

Cornwall is fortunate to have the Small Pearl-bordered Fritillary resident in quite a wide variety of habitats and locations, because it has disappeared from much of the rest of England. This small- to medium-sized, vividly orange and black butterfly is named for the string of white 'pearls' that run along the outer margin of the underside of its wings. Male and female are similar, although the female is usually a little larger and less brightly coloured.

Despite its 'small' name, there is an overlap in size with the very similar Pearl-bordered Fritillary, and when they are flying together the two species are difficult to distinguish, particularly when only the upperside of the wings is seen. To the practised eye, small differences of detail can become apparent, but the undersides of the two species have clearer features that make it easier to tell them apart. Although they both have a row of seven white 'pearls' along the edge of each hindwing, the

Resident in Cornwall
(not on the Isles of Scilly)

CONSERVATION STATUS
- Section 41 Species of Principal Importance (NERC Act 2006)
- Red List of British Butterflies; IUCN criteria: Near Threatened (NT)
- Butterfly Conservation priority: High

TRENDS AND DISTRIBUTION
UK rate of change in abundance
↓ 70% 1976–2018
South-west England rate of change in abundance
↓ 77% 2009–2018
↓ 84% 1981–2018
% of 1km squares occupied in Cornwall
5% 2009–2018 (48% 10km squares)
7% 1976–2008 (67% 10km squares)

Small Pearl-bordered Fritillary has a more strongly contrasted mosaic of white, orange, brown and black patches, in which white predominates. The underside of the Pearl-

Small Pearl-bordered Fritillary

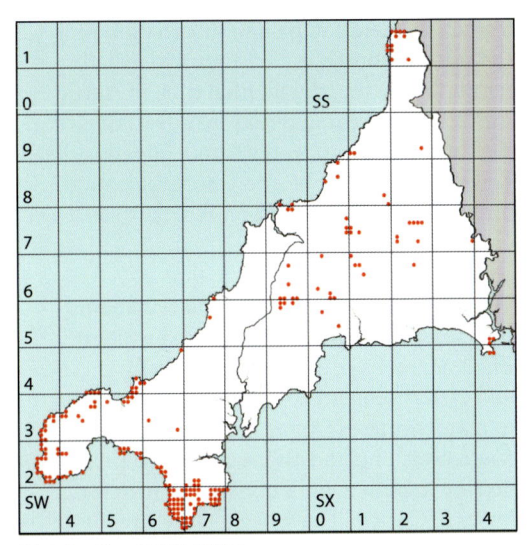

Distribution 2009–2018

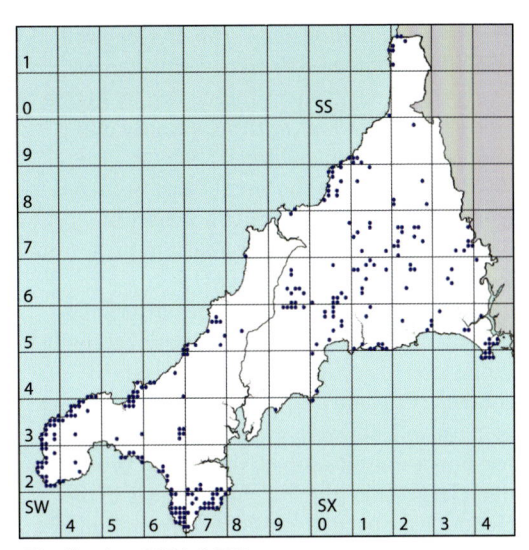

Distribution 1976–2008

bordered Fritillary's wings is a more subdued pattern of pale orange and yellow. There are only a few sites in Cornwall where these two species can be seen together, as the Pearl-bordered Fritillary has a much more limited range.

Distribution in Cornwall

The Small Pearl-bordered Fritillary is now the most abundant fritillary in Cornwall. The number of locations where this species has been reported has fallen in recent years; nonetheless, the butterfly is still locally common and widespread in parts of the County, notably The Lizard peninsula and West Penwith. There are several colonies along the north coast on dunes and clifftops and in the valleys (although records show fewer than formerly), and a few in the Rame Head area. Inland, the sites are more scattered, with the principal concentration stretching eastward from Goss Moor by way of Breney Common to beyond Bodmin. Additionally, the butterfly is found in large numbers at Marsland. In west Cornwall, the distribution is predominantly coastal, whilst inland sites are favoured in the east of the County.

Larval foodplants

The most commonly used foodplant is Common Dog-violet *Viola riviniana*, which occurs widely in Cornwall. Marsh Violet *V. palustris* subsp. *juressi* is used in damper areas and is the preferred foodplant on the Mid Cornwall Moors and Bodmin Moor.

Habitat and ecology

Small Pearl-bordered Fritillary populate a variety of habitats where there is a good supply of the larval foodplant violets *Viola* spp. and nectar sources, as they are avid feeders. In areas of southern England, the butterfly is thought to favour deciduous woodland with clearings. However, this is not generally the case in Cornwall. Although some woodland colonies such as those in Cabilla Wood remain, many others have died out. On the coast in west Cornwall and on The Lizard, it inhabits damp grassland, open bracken-scrub (which replicates woodland and heathland conditions) and sand dunes. Inland, the butterfly favours grassland, especially damp grassy heath with Bracken

Underside, male and female similar

Pteridium aquilinum and scrub. In Cornwall, Small Pearl-bordered Fritillary are often seen flying in the boggy areas frequented by Marsh Fritillary *Euphydryas aurinia*, because damp conditions encourage Marsh Violet to grow vigorously. Perhaps the Small Pearl-bordered Fritillary's ability to adapt to a variety of habitats in Cornwall will enable it to persist into the future within the County.

In other parts of the UK, the butterfly is usually single-brooded, with a partial second brood said to occur sometimes in parts of southern England and South Wales. In the west of Cornwall, however, on The Lizard and in a few north coast locations, the second brood is often larger than the first. Other areas of the County have partial second broods; but inland, and especially on higher ground, they are often absent (Dennis, 2021). The first brood appears in May or June (with very occasional reports from late April), with a second brood, sometimes overlapping the first, being seen from July until early September. Colony numbers are hard to estimate. Most records report between one and five individuals, but some of these may appear at a distance from their colony, indicating a wandering habit. Numbers in the low hundreds have been recorded in a few areas.

Eggs are laid singly, sometimes on the underside of the larval foodplant, but the preferred oviposition location appears to be on nearby vegetation. Medium-sized plants growing in damp, open, sunny situations are usually chosen, in contrast to those favoured by the Pearl-bordered Fritillary, which prefers drier conditions (Thomas and Lewington, 2014). Larvae are initially pale yellow but soon change in colour and appearance as they moult. Together with the offspring of the second brood of butterflies, those that do not produce a second generation go into hibernation in the third or fourth instar. They will usually moult for the fourth and final time in April the following year and feed mainly in the daytime for several weeks before pupating, head downwards, in leaf litter or short vegetation. This stage usually lasts two to three weeks.

Analysis of trends

The Small Pearl-bordered Fritillary remains widespread in Scotland, Wales and western England but has declined dramatically in central and eastern England, where it is now extinct in many counties (Asher *et al.*, 2001). Despite South-west England also registering a dramatic decline in both abundance and occurrence, losses in Cornwall have been less severe (Butterfly Conservation, 2018). The distribution maps for Cornwall show that there has been a decline in range of nearly 30% when comparing the last 10 years to the previous 32 years, although after reviewing the occurrence data for the decade 1999–2008 it seems that there has been a slight recovery in the most recent period, 2009–2018 (possibly due to increased recorder effort).

Destruction of habitat is the primary cause of the loss of colonies, whether by agricultural 'improvement' and fragmentation, overgrazing, growth of scrub or changes to woodland management. It is thought that the reason the Small Pearl-bordered Fritillary is faring better in Cornwall than in the rest of South-west England is that the species has its strongholds in the rough coastal grassland and dunes, which are largely untouched by intrusive development. Elsewhere, loss of habitat has occurred in woodlands where management has ceased, resulting in a decrease in the woodland glades and clearings favoured by the butterfly (Asher *et al.*, 2001).

Flight curve based on number of records in Cornwall 1999–2018; and UK life cycle

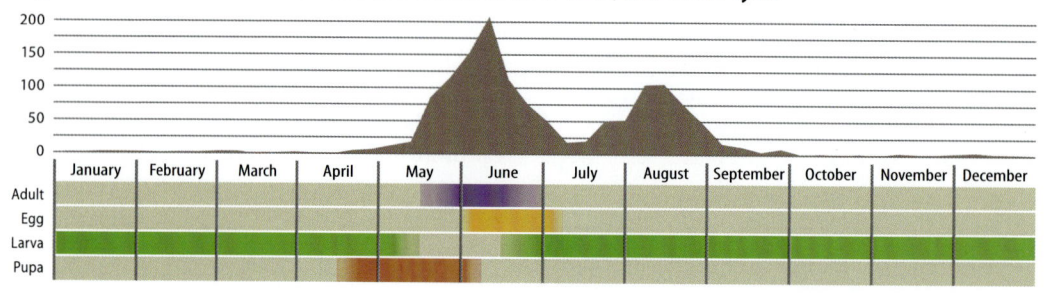

Cudden Point, one of the best places to see Small Pearl-bordered Fritillary

Weather undoubtedly affects this species, and over the 10 years prior to 2018 there was considerable fluctuation in records and numbers. The warm summer weather of 2018 raised hopes that the Small Pearl-bordered Fritillary might have recovered slightly in Cornwall, despite the overall downward 10-year and longer-term trends in South-west England as a whole. However, trends in relation to less common species are more susceptible to artificial distortion because of the smaller populations involved. This can be caused by an area being heavily recorded in some years and not recorded at all in others; this appears to be the case with Small Pearl-bordered Fritillary numbers. It is also a reflection of the numbers of recorders who are contributing records and walking transects, both of which have increased recently.

The transect data reveals the difficulty of assessing the true picture. There were far fewer transects monitored at the beginning of the last 10-year period, and only two of those walked in 2010 recorded Small Pearl-bordered Fritillary, compared with 18 transects in 2018. Of the two transects that were walked in both those years, Roskestal Cliff in West Penwith recorded three times as many Small Pearl-bordered Fritillary in 2010, and Upton Towans twice as many, as in 2018, a year that was acknowledged as exceptionally good for butterflies.

The fact that counts appear to have decreased in some key areas, combined with the noted decline in range, gives cause for concern and requires careful monitoring in the future. However the species is potentially under-recorded in several areas in the County, including the Mid Cornwall Moors and Bodmin Moor.

Conservation

Fortunately for this butterfly, many of the areas it now inhabits are owned or leased by conservation bodies, such as Natural England, Cornwall Wildlife Trust and the National Trust (which owns many of the coastal sites where it is found). A certain amount of grazing to check the uncontrolled growth of Bracken and scrub is essential. Recent habitat management by the National Trust specifically for Small Pearl-bordered Fritillary in the Hell's Mouth/Hudder Down area (near Godrevy) shows how successful strategically targeted conservation (including scrub clearance and reinstatement of conservation grazing) can be: the establishment of a transect at this location, following the conservation work, has already recorded encouraging numbers of the butterfly.

Best places to see

- Marsland Nature Reserve SS234171/ Marsland Mouth SS215173
- Bunny's Hill SX117675
- Breney Common SX055610
- Pendrift Downs SX098744
- Kenidjack SW356323 to Roskestal Cliff SW364223
- Cudden Point SW551277 / Rinsey Cove SW592270
- Poldhu SW664200 / Predannack Head SW661163
- Windmill Farm SW691151
- Chynhalls Cliff SW780169

Upperside, male, showing radiating sex brands on the forewing

Silver-washed Fritillary
Argynnis paphia

Sponsored by
*Pete Bousfield, Jim Cooper
and Roger Hooper*

The Silver-washed Fritillary is the largest and most spectacular fritillary in Cornwall. It can be confused with the Dark Green Fritillary *Speyeria aglaja*, which is a similar-sized, strong flier, although they are generally found in different habitats: the former in woodland glades, and the latter on open grassland. Occasionally both species can be seen flying together, but Silver-washed Fritillary can be distinguished by the faint silver streaks on the underside of their wings, whereas Dark Green Fritillary have bright silver spots. Both sexes of the Silver-washed Fritillary are orange with black spots and have slightly more pointed forewings than the Dark Green Fritillary. The male is uniquely identifiable by having sex brands that radiate as black lines from the thorax outwards across the forewings. These contain pheromones, which he expels during courtship as he flies in loops around a female (Eeles, 2019). The female is less intensely coloured and lacks the sex brands. The most common aberration is the sex-linked *valezina* form found particularly in Cabilla Wood, where the bronze-green female has been reported several times during recent summers.

Resident in Cornwall
(vagrant on the Isles of Scilly)

CONSERVATION STATUS
■ Butterfly Conservation priority: Low

TRENDS AND DISTRIBUTION
UK rate of change in abundance
↑ 127% 1976–2018
South-west England rate of change in abundance
↓ 6% 2009–2018
↑ 99% 1976–2018
% of 1km squares occupied in Cornwall
11% 2009–2018 (65% 10km squares)
8% 1976–2008 (61% 10km squares)

Silver-washed Fritillary

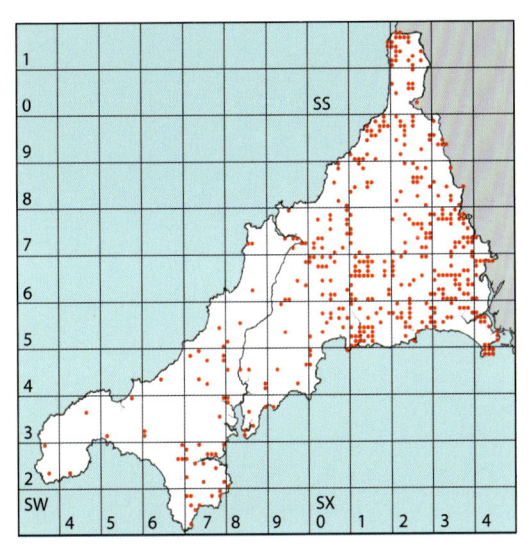

Distribution 2009–2018

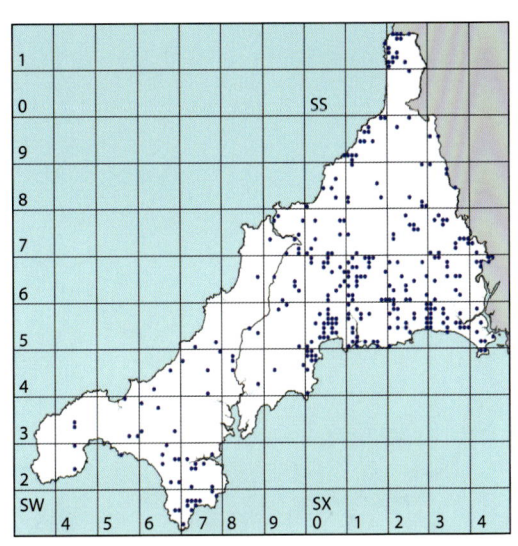

Distribution 1976–2008

Distribution in Cornwall

The number of 1km squares where Silver-washed Fritillary have been seen has increased slightly in the last 10 years. The butterfly continues to be found in suitable woods and nearby lanes and hedgerows, although sightings tend to diminish towards the west of the County, particularly in West Penwith and the western half of the north coast, mainly due to the lack of suitable habitat, although the odd stray butterfly has been spotted there beyond its usual territory. The main concentration of colonies is most often found in wooded river valleys, including the Fowey. There are also colonies in the lower reaches of the Seaton Valley and West Looe River in east Cornwall, and within the Penlee Battery reserve. The butterfly is also found in good numbers in

Upperside, female

Underside

Aberration *valezina*

the Valency and Tidna valleys close to the north coast. In the last 10 years, the number of records received has increased along the entire length of the Devon border in the Tamar Valley region, and there have been sightings in the Fal valley and on the Roseland peninsula, following a period when none were recorded in that area. They still maintain a presence on The Lizard, and within the woodlands fringing the Helford River.

Larval foodplants

Common Dog-violet *Viola riviniana* is the preferred larval foodplant; whether the butterfly is able to use other violet species is currently unknown in the UK (Emmet and Heath, 1990).

Habitat and ecology

The Silver-washed Fritillary is essentially a butterfly of woodland, preferably oak *Quercus* spp., and it can tolerate more shaded conditions than other fritillaries, provided that there are sufficient open spaces for sunlight and warmth to penetrate to allow the growth of the larval foodplant. It is less strictly confined to woods in Cornwall than elsewhere, being common also along lanes, hedges and banks close to woodland. The butterfly is most frequently found nectaring on Bramble *Rubus fruticosus* agg. and thistles *Cirsium* spp. in clearings and along rides. When it rests in cooler weather, a keen eye may spot it roosting in trees. Increasingly Silver-washed Fritillary are reported in small numbers in gardens. The male in particular has a strong, gliding flight when patrolling his territory and while awaiting a passing female, a habit that makes identification easier. The courtship flight is described as being unlike that of any other UK butterfly. Upon encountering a virgin female's pheromones, the male repeatedly loops around the straight-

Flight curve based on number of records in Cornwall 1999–2018; and UK life cycle

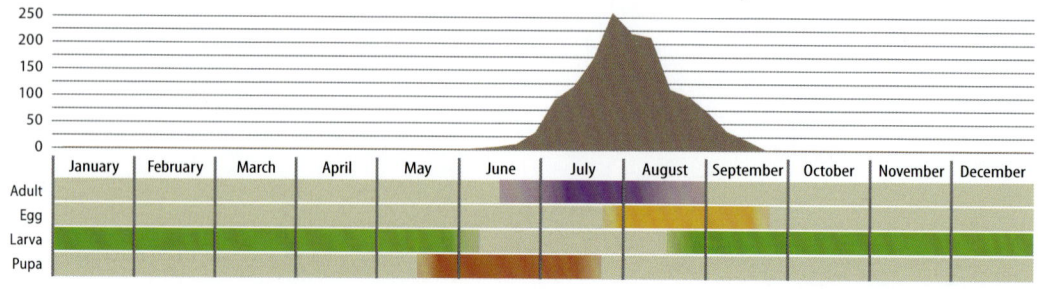

Cabilla Wood, one of the best places to see Silver-washed Fritillary

flying female, showering her with his own pheromones. Later, while mating, he injects her with a repellent to deter other males (Eeles, 2019).

The Silver-washed Fritillary is single-brooded, emerging in early July, although it has been seen in mid-June in Cornwall, and in recent years this seems to be a more common occurrence. The flight period normally lasts until late August, but adults have occasionally been seen in September. The female lays her eggs singly in the crevices of rough bark or on moss on the cooler, shaded side of her chosen tree, a few metres above the ground. Around two weeks later the caterpillar emerges, spins a silk pad as an attachment in a crevice of the tree, and almost immediately goes into hibernation until the following spring, when it re-emerges and crawls down the tree in search of its foodplant. The larva changes appearance quite significantly at each moult. In the final stage it is brown and spiny, and although very distinctive it is well camouflaged amongst leaf litter. After around 10 months in the larval stage, pupation begins. This involves spinning another silk pad and attaching itself to a twig, where it looks very much like a shrivelled leaf. The pupal stage lasts around two to three weeks (Eeles, 2019).

Analysis of trends

The national trends are now very positive for this species, following substantial declines during the twentieth century. South-west England is one of the butterfly's strongholds in the UK (Asher *et al.*, 2001). The principal cause of any decline is change in woodland management, often leading to increased shade and fewer larval foodplants. In central and eastern Cornwall, the species was quoted at the beginning of this century as holding its own (Spalding with Bourn, 2000). This would still seem to be the case. However, numbers of Silver-washed Fritillary continue to fluctuate year-on-year. One explanation for this could be the variation in how many people send in reports each year from the few principal areas where the butterfly is seen in large numbers. It is also likely to be affected by weather, with population size growing in hot summers.

Conservation

It is essential that woodland clearings and rides be kept free of overshadowing canopy, and scrub should be periodically cleared to encourage the growth of larval foodplants.

Best places to see
- Marsland Nature Reserve SS228172
- Cabilla Wood SX133653
- Greenscoombe Wood SX391726
- Seaton Valley SX302555
- Valency Valley SX103912
- Goss Moor SW949601

Upperside, male

Dark Green Fritillary
Speyeria aglaja

Sponsored by
Alex Ashburne, Dick Goodere
and Dave Thomas

The Dark Green Fritillary is a large butterfly with striking orange and black colouring and a powerful flight. It is locally common in a few parts of Cornwall but can be a rare treat to see elsewhere. The male is slightly smaller than the female and, compared with the intense orange colour of the upperside wings of the male, the female looks slightly washed-out despite having strong black markings. As the male fades towards the end of the flight period, the sexes are not so easy to distinguish.

The uppersides of Dark Green Fritillary and High Brown Fritillary can be difficult to tell apart; however, since the latter is now thought to be locally extinct, it is highly unlikely to be seen in Cornwall. Another species the Dark Green Fritillary may be confused with is the Silver-washed Fritillary, which is also a large orange and black butterfly, whose flight period overlaps with that of the Dark Green Fritillary. The Silver-washed Fritillary is larger,

Resident in Cornwall
(vagrant on the Isles of Scilly)

CONSERVATION STATUS
■ Butterfly Conservation priority: Medium

TRENDS AND DISTRIBUTION
UK rate of change in abundance
↑ 163% 1976–2018
South-west England rate of change in abundance
↑ 37% 2009–2018
↑ 88% 1976–2018
% of 1km squares occupied in Cornwall
3% 2009–2018 (50% 10km squares)
4% 1976–2008 (59% 10km squares)

and the male has prominent radiating black ridges along the veins on the upperside of the forewings, a unique identifying characteristic. The underside of the wings also differs markedly between the two species. Dark Green Fritillary have large, clear-cut silver spots on the hindwings, set in a predominantly creamy

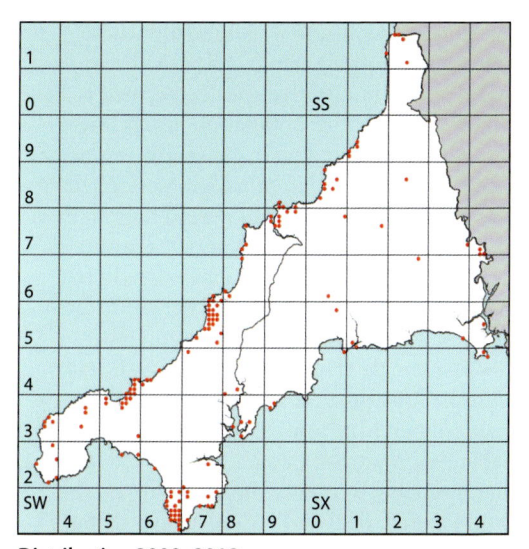

Distribution 2009–2018

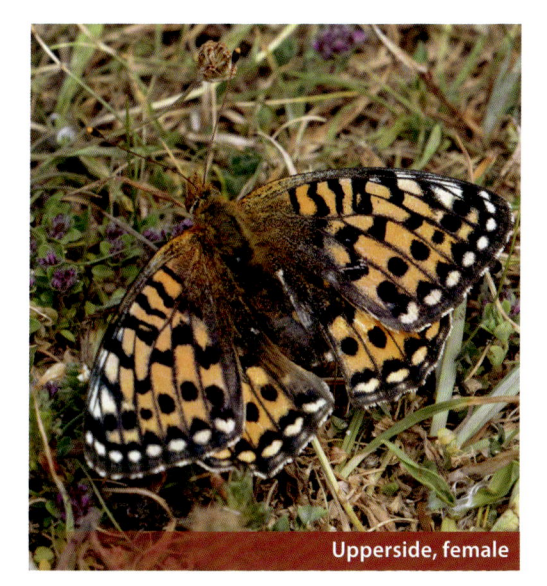

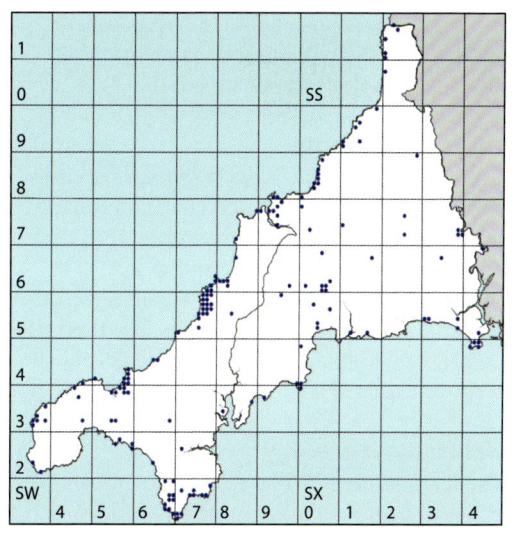

Distribution 1976–2008

Upperside, female

Underside, male and female similar

orange background with a wash of green. Silver-washed Fritillary have less-defined patterning on the underside of pale green hindwings, and in place of silver spots there are faint diagonal silver streaks, and green spots near the wing margins.

Distribution in Cornwall

In the last Cornwall Atlas it was stated that the Dark Green Fritillary was the commonest fritillary in Cornwall (Wacher *et al.*, 2003). This is no longer true in terms of either abundance or range, as the Small Pearl-bordered Fritillary is currently more numerous and the Silver-washed Fritillary more widespread. There has been a slight increase in reported locations in the last 10 years, compared with those identified in the first part of this century. The Dark Green Fritillary's stronghold remains the coastal regions, with inland colonies becoming rarer and more isolated. The distribution maps and the number of both 1km and 10km squares where the butterfly is recorded confirm that there has been a decline in range when the last 10 years are compared to the previous 32 years. The butterfly is still most commonly seen on the dune systems of the north coast, notably on the Towans between Hayle and Godrevy, and at Penhale and Gear Sands, where many dozens are seen during the flight season. It is also recorded regularly and in reasonable numbers in various locations on The Lizard peninsula and in West Penwith, as well as on many sites along the whole of the north coast.

Larval foodplants

Common Dog-violet *Viola riviniana* seems to be the most commonly used larval foodplant; other violet species may prove attractive, such as Marsh Violet *V. palustris* subsp. *juressi* or the rare Hairy Violet *V. hirta*, which is found on some of the north coast dunes, but their use as a larval foodplant by Dark Green Fritillary is so far unconfirmed in Cornwall.

Habitat and ecology

The Dark Green Fritillary seems to enjoy more exposed habitats than any other fritillary found in the County. In Cornwall it is typically found on sand dunes and coastal grassland. The species is less frequently encountered on moorland, heathland and inland flower-rich grassland, and in managed woodland. Favoured breeding areas are usually situated adjacent to scrub and are rich in larval foodplants, with a sward height of around 8–20cm (Eeles, 2019).

Dark Green Fritillary are said to occupy discrete colonies, although the adults are strong fliers and tend to occur at low densities over large areas. The butterflies are attracted predominantly to purple flowers such as Common Knapweed *Centaurea nigra*, thistles *Cirsium* spp., teasels *Dipsacus* spp. and particularly Buddleja *Buddleja davidii*. Another preferred nectar source is Red Valerian *Centranthus ruber*, as illustrated by an observation at Upton Towans in July 1999 (and similarly observed in subsequent years) when over 50 butterflies were seen feasting on a patch of Red Valerian no more than 2m by 20m in size. None were seen anywhere else on the Towans that day; Red Valerian, a naturalised, non-native species, seemed to have attracted the entire population, providing a valuable source of nectar, especially important in dry summers when other sources of nectar are hard to find. The flight period nationally is mid-June to mid-August, but in Cornwall, in warm years, the butterfly can emerge in late May, and it can still be on the wing into late August and occasionally early September.

The female Dark Green Fritillary is less conspicuous than the male because of her habit of perching low in tussocks of grass, where she waits for patrolling males to locate her, probably by scent. The males are on the move a great deal and are consequently hard to photograph unless nectaring, usually in the early morning or late afternoon. According to Jeremy Thomas, "more than any violet-feeding fritillary, except perhaps the Small Pearl-bordered", the female is extremely choosy in her selection of location and plants on which to lay her eggs, preferring "lush, cool or humid spots" (Thomas and Lewington, 2014: 213). The eggs are often laid on plants, dead leaves or dead Bracken *Pteridium aquilinum*, close to the larval foodplant rather than on the foodplant itself (Asher *et al.*, 2001). The larvae hatch in a few weeks, and after eating their eggshells they immediately hibernate in leaf litter and do not start feeding again until the following spring. Pupation commences from mid-May, and after three or four weeks the adult butterfly emerges.

Analysis of trends

The Dark Green Fritillary is the "most widespread of the fritillaries… in Britain and Ireland, although it is rarely found in great numbers" (Asher *et al.*, 2001: 226). National

Flight curve based on number of records in Cornwall 1999–2018; and UK life cycle

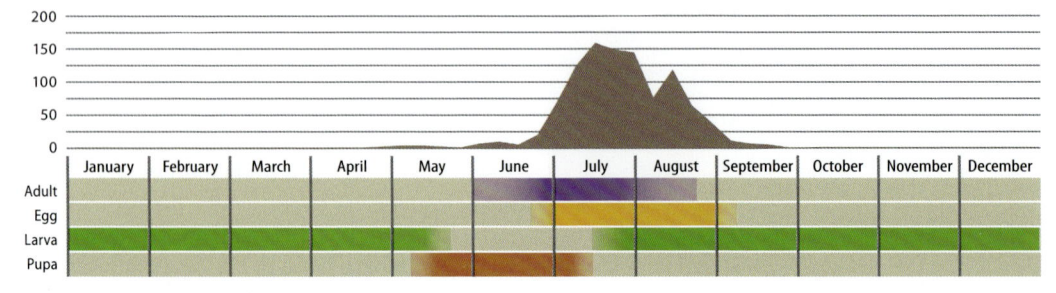

Porth or Polly Joke, one of the best places to see Dark Green Fritillary

and regional trends indicate an increase in abundance but a decline in distribution of 33% nationally over the period 1976 to 2014, although there are signs of a recent recovery (Fox et al., 2015). In Cornwall more records than ever were submitted in the last 10 years because of increased recording effort. However, abundance fell over that period, and the species was seen in fewer locations. Unfortunately, transects cannot provide us with answers regarding abundance in the County, as there are only a few where reasonable numbers of Dark Green Fritillary are present; Upton Towans is the only transect that has been consistently recording this butterfly in good numbers for the last 10 years. Records from here have fluctuated, with 2018 as the best year. Average transect numbers and *ad hoc* records show a similar pattern.

It is hard to know why Cornwall should be losing some of its colonies; targeted recording is required to check historical sites to determine whether the butterfly has truly disappeared from these locations. It is thought that the decline of Dark Green Fritillary may be associated with overgrazing, mainly by rabbits, which eat the larval foodplants. Other causes may include destruction of habitat through the agricultural 'improvement' of grassland and through encroaching development.

Conservation
Preservation of coastal dunes, heathlands and flower-rich grassland with rotational scrub management and moderate rabbit or cattle grazing would seem to be the main conservation management requirements to maintain healthy populations of this butterfly.

Best places to see
- Mexico Towans SW559388 and Upton Towans SW576397
- Gear Sands SW769561 and Mount (Penhale) SW782570
- The three bays south-west of Newquay, including Holywell Bay SW767597 and Polly Joke SW770605 and Cubert Common SW784597
- The Lizard, including Higher Predannack Cliff SW663171 and Kynance SW691129

Upperside, male and female similar

Red Admiral
Vanessa atalanta

Sponsored by
*Conor Allen, Rachel Bickerton
and Della*

The Red Admiral is one of the most widely recognised of UK butterflies; it can be seen almost anywhere in Britain and is a familiar garden visitor. This large, active butterfly has distinctive scarlet (or orange) and black colouring. The forewings have prominent white markings on their black tips, and the velvety-black hindwings have red borders containing four small black spots. The undersides are a mottled brown on the hindwings, and the forewings echo the patterning on the upperside but with a faint streak of blue. These cryptic undersides make it difficult to spot the butterfly when it is roosting on tree trunks. Male and female are alike, but the female is slightly larger.

Distribution in Cornwall
The Red Admiral is one of the most widespread species in Cornwall. The distribution map for the last decade shows that this butterfly has been recorded in more 1km and 10km squares than ever before. This is probably as a result of an increase in recorders and, in particular, a significant contribution from the Big Butterfly Count, which began in 2010.

Resident and migrant in Cornwall
(migrant on the Isles of Scilly)

CONSERVATION STATUS
■ Butterfly Conservation priority: Low

TRENDS AND DISTRIBUTION
UK rate of change in abundance
↑ 212% 1976–2018
South-west England rate of change in abundance
↑ 68% 2009–2018
↑ 127% 1976–2018
% of 1km squares occupied in Cornwall
57% 2009–2018 (86% 10km squares)
43% 1976–2008 (82% 10km squares)

Red Admiral

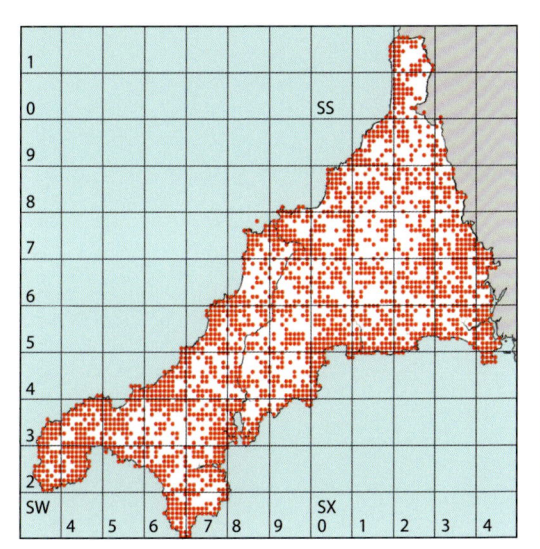

Distribution 2009–2018

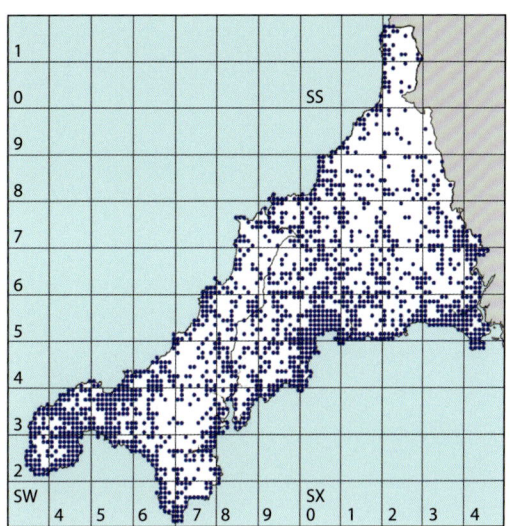

Distribution 1976–2008

Larval foodplants

The Common (Stinging) Nettle *Urtica dioica* is most frequently used, but Small Nettle *U. urens* will suffice, although this plant is infrequent in Cornwall and largely restricted to the west of the County. The caterpillars also feed on another member of the nettle family, Pellitory-of-the-wall *Parietaria judaica*, and on Hop *Humulus lupulus*; the former is confined mainly to coastal regions in the County, while the latter is scarce and is found mainly on Cornish hedges.

Habitat and ecology

Because the Red Admiral is such a common butterfly and not prone to the kind of major periodic invasions of the closely related Painted Lady *Vanessa cardui*, it can come as a surprise to learn that it is not a true UK native species, despite the fact that the butterfly is

Underside, male and female similar

Egg on nettle

Larva

often seen in winter. Those recorded in this country are almost entirely yearly migrants, together with their progeny, which are only temporary residents. Three main waves of immigration have been described. The first, relatively small batch of immigrants arrives from southern Europe or North Africa early in the year. A second, larger wave appears in May or June from Spain and Portugal, and a third, which arrives in August from central Europe, appears to contribute to the largest population peak in Cornwall. This is probably an over-simplification, since immigrant Red Admiral continue to arrive in successive waves from various locations throughout the year until November, their offspring contributing to the overall numbers. These butterflies embark in their thousands on the return journey to their homelands from August onwards. Emigrant butterflies flying south frequently cross with new immigrants heading north-east (Eeles, 2019).

Despite many coastal sightings, when butterflies are seen 'refuelling' on their way through, this species can be found anywhere but is commonly seen in woodland, orchards, hedgerows and gardens. It particularly favours the flowers of ivy *Hedera* spp., sap of oak *Quercus* spp., rotting fruit, and a variety of garden nectar sources such as buddleja *Buddleja* spp., sedum *Sedum* spp., and Argentine Vervain *Verbena bonariensis* in late summer and September.

Newly arrived males, as well as British-bred individuals, soon establish territories, in which they look for females. Strangely enough, the mating process is very rarely observed. The female lays her green-ribbed eggs singly, usually on the upper surface of young nettle leaves. They can hatch in a week in warm weather, although development slows down in colder conditions. The caterpillar lives singly in a series of tents, which it constructs by tethering the edges of leaves together with silken threads.

Flight curve based on number of records in Cornwall 1999–2018; and UK life cycle

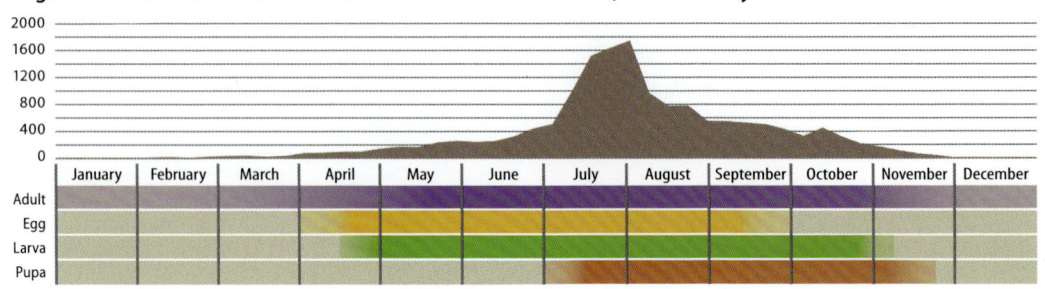

The spiny caterpillars are quite variable in colour: they are primarily black, but there are many different forms, particularly in the final instar. In warm weather the larval stage lasts for only two to three weeks, succeeded by a similar period as a pupa.

Because of the complications caused by the various waves of migration and the fact that the Red Admiral is a continuously brooded butterfly whose rate of development is affected by temperature, this species' life cycle and flight pattern remain the subject of interpretation. Nonetheless, the consensus has emerged that individuals can complete their life cycle over the winter in some parts of the UK, despite the majority migrating. The slightest warming by the sun can bring out adults that have settled in crevices, and these have occasionally been seen laying eggs during the winter months, resulting in adults in late spring or even earlier. In recent years there have been few mild, sunny, calm days when a Red Admiral cannot be seen somewhere in Cornwall. The late Roger Lane, who was for many years Migration Officer for Cornwall Butterfly Conservation, made observations during the mild winters of 2006/7 and 2007/8, when he distinguished between what he took to be overwintering adults and at least 10 pristine, newly emerged butterflies in mid-January 2007, an account of which he published in the Cornwall Butterfly Observer (Lane, 2009).

The proposition that this butterfly increasingly overwinters in southern counties of the UK is challenged by Jeremy Thomas, who states, "Although a few Red Admirals may seem to hibernate in the British Isles during mild winters, these generally settle in exposed places, such as tree trunks, or under branches, and usually perish. It is likely that these are late-emerging adults that become trapped after the sudden onset of conditions too cold for flight. The few adults that are seen on sunny days from December to February are believed to result from late caterpillars that were able to develop in warm spots" (Thomas and Lewington, 2014: 179). Despite this very well-respected, dissenting voice, evidence now points to the species having the capacity to overwinter in Cornwall.

Analysis of trends

The annual British populations of this largely migrant butterfly are controlled mainly by the varying weather in their countries of origin. As is the case with most butterflies, long, warm summers in Britain should generate larger populations, swelled by a greater survival of migrant offspring. However, in recent years this does not appear to have been the case. In 2017, which was not a particularly good summer, the highest number of Red Admiral in the last decade was recorded in Cornwall, whilst in the warm summer of 2018, when the majority of butterfly species prospered, the Red Admiral suffered a large drop in numbers (reflected also in local transect results). This slump was also recorded nationally.

Against the upward long-term and 10-year abundance trends nationally and in South-west England, Cornwall seems to have received more records but lower numbers per record in the last 10 years. Average transect numbers are stable, showing neither an upward nor a downward trend.

Conservation

This butterfly is largely dependent on conditions in other countries.

No specific methods of conservation are needed in Cornwall for this species, although it would be aided if gardeners left a patch of nettles uncut in a sunny corner of their plots. Since the butterfly prefers fresh, young leaves, cutting a section of unoccupied nettles to the ground once in midsummer will ensure a good supply.

Best places to see

Anywhere in Cornwall, especially gardens.

Pupa

Upperside, male and female similar

Painted Lady
Vanessa cardui

Sponsored by
Anne Banks, Christine Salter
and Ecological Surveys Ltd

The Painted Lady is a highly mobile, successful migrant that arrives in Cornwall in variable numbers each year. It is easily identified by its characteristic colour, which is somewhere between pale orange and salmon pink, fading as the butterfly ages. The upperside of the slightly pointed forewings has a black tip with prominent white spots, whilst the hindwing has a row of black dots near the outer margin. The underside of the forewing is a paler version of the upperside, and the underside of the hindwing is an elaborately reticulated brown and white mosaic with dark eyespots. Male and female are similar. This butterfly could at first glance be mistaken for a Small Tortoiseshell *Aglais urticae*, but the Painted Lady is larger and less intensely coloured and lacks the scalloped bright blue border of the Small Tortoiseshell.

Distribution in Cornwall
Although a considerably less frequent

Migrant in Cornwall
(migrant on the Isles of Scilly)

CONSERVATION STATUS
■ Butterfly Conservation priority: Low

TRENDS AND DISTRIBUTION
UK rate of change in abundance
↑ 46% 1976–2018
South-west England rate of change in abundance
↓ 69% 2009–2018
↓ 13% 1976–2018
% of 1km squares occupied in Cornwall
34% 2009–2018 (81% 10km squares)
28% 1976–2008 (81% 10km squares)

immigrant than the Red Admiral, and with much more variable numbers, in most years the Painted Lady is not uncommon in the County. In lean years, its numbers may be swelled by the recent fashion of releasing this species at weddings, as well as it being the most popular butterfly to rear in primary

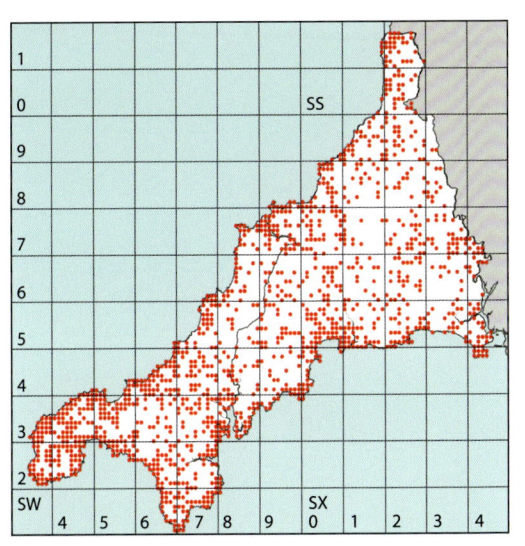

Distribution 2009–2018

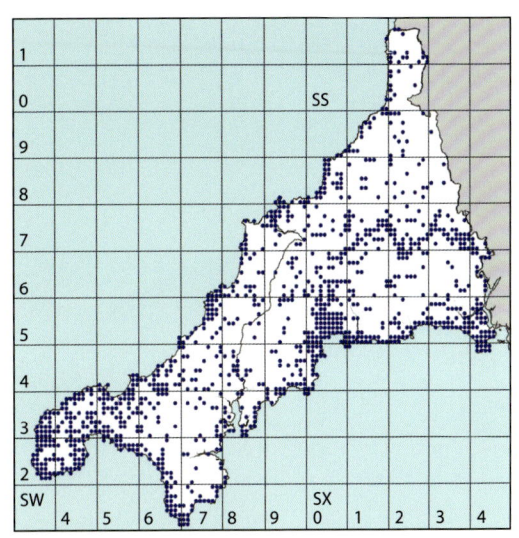

Distribution 1976–2008

schools! This may lead to a slight distortion of numbers.

The butterfly is widely distributed in varying abundance across the County in most years. However, every 10 years or so, the UK experiences a spectacular mass immigration. In 1996, 2009 and 2019, millions of Painted Lady butterflies flew north across the English Channel into Britain on a broad front, overflying or touching down in just about every part of Cornwall. In 2009 there were estimates of more than 10 million butterflies crossing the southern coastline (Thomas and Lewington, 2014). National statistics are not available for 2019 at the time of writing, although it was the butterfly that topped the charts in the UK for the Big Butterfly Count. In Cornwall, approximately twice as many records for the Painted Lady were submitted in 2019 as in 2009, although the total number of butterflies seen was lower.

Like the Red Admiral, to which the Painted Lady is closely related, this immigrant from the south also produces a brood in Cornwall, but unlike the Red Admiral it is unable to overwinter in the UK. Consequently, despite its annual appearance in the County, the Painted Lady is not a native species. As with the Red Admiral, Cornish broods of this butterfly tend to join the outward flights of emigrants in the autumn, when they can intermix with late immigrants. In 1996, hundreds were seen flying south and gathering in coastal sites before setting off on their journey across the English Channel (*contra* Asher *et al.*, 2001).

Another reasonably good period was 2000–2001, when several Painted Lady were sighted during the winter, with indications of immigration as late as December (Lane, 2001).

Since the butterfly tends to follow coastlines on both inward and outward migration, the

Underside, male and female similar

Feeding on Hemp-agrimony

distribution maps reflect a greater abundance on the coast than inland. In summary, however, over a 10-year period, every area of Cornwall is likely to be visited.

Larval foodplants

The larval foodplants most favoured by Painted Lady are Spear Thistle *Cirsium vulgare*, Creeping Thistle *C. arvense* and Marsh Thistle *C. palustre*, all of which are common in Cornwall, but almost any species of thistle will suffice. The butterfly is also known to lay eggs on a variety of other species, including burdock *Arctium* spp., mallow *Malva* spp. and nettle *Urtica* spp. Viper's-bugloss *Echium vulgare* is also used, but other than a stronghold on the Hayle dunes it is sparsely distributed throughout Cornwall.

Habitat and ecology

There has been a great deal of research in recent years into the migratory habits of Painted Lady. The butterfly begins its journey south of the Sahara. It is believed that with favourable winds a few can arrive in Cornwall directly from places such as Morocco, but the majority are the result of several generations of reproduction as they travel through mainland Europe in countries including Spain, Portugal and France on their way to Britain and further north to the Arctic Circle. Isotope analysis of the butterflies' wings shows that the annual distance travelled

Flight curve based on number of records in Cornwall 1999–2018; and UK life cycle

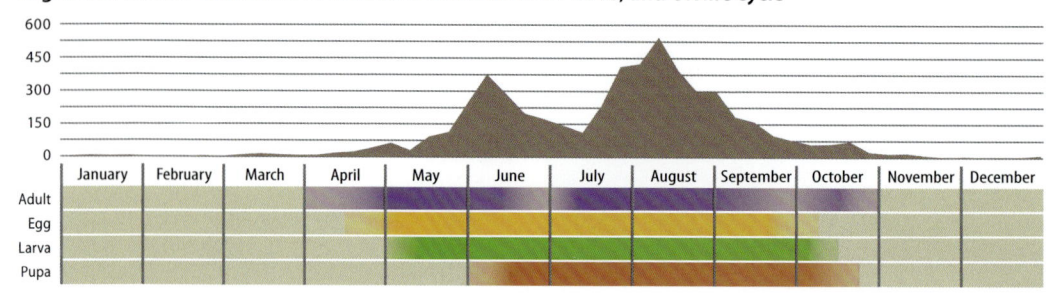

by the successive generations, crossing desert and seas twice, may reach about 12,000km (Talavera et al., 2018).

Painted Lady can be found almost anywhere in their relentless search for nectar, larval foodplants and mates. Any warm, flower-rich area can produce a sighting, be it clifftop, urban garden or moorland tor, especially after the first British brood has emerged in years of abundance.

After mating, the female frequents thistles, first as larval foodplants, where she deposits single eggs, usually on the upper surface of the leaves, and, later in the season, as a nectar source. The butterfly is tolerant of wind and can be drawn to thistles on exposed hillsides. Other attractive nectar sources for both sexes are Bramble *Rubus fruticosus* agg., Buddleja *Buddleja davidii*, Common Fleabane *Pulicaria dysenterica*, Red Valerian *Centranthus ruber*, Hemp-agrimony *Eupatorium cannabinum*, knapweeds *Centaurea* spp., Wild Privet *Ligustrum vulgare* and Atlantic Ivy *Hedera hibernica*, indigenous to Cornwall and the County's most widespread ivy.

The flight pattern of the butterfly, as with the Red Admiral, is complicated, owing to its continuous breeding habit; the whole life cycle can be completed in only six weeks (Eeles, 2019). No hibernating period is recognised and the species has been recorded in every month of the year in Cornwall. Normally, however, the main wave of immigrants arrive in May or June and produce a British brood, which emerges in July and August, when the butterfly is most abundant in the County. The egg, which has pale longitudinal ribs, is light green initially, turning grey as the larva inside develops. The larva is greyish brown with rows of hairy warts and pale, broken stripes running along each side of its body, although its appearance can be variable. It makes a silken tent on the underside of the leaf, in which it lives as it eats. The larval stage lasts just under a month, and then the caterpillar pupates on the foodplant or on suitable vegetation nearby. The pupal stage lasts around 10 days and can be a variety of colour forms (Eeles, 2019).

In very favourable summers, a second – and even possibly a third – brood also reaches maturity in Cornwall; the adult progeny tend to join any survivors of the first brood that attempt to emigrate. It is traditionally asserted that they cannot survive the British winter, and certainly there is no evidence for an indigenous spring brood. Nevertheless, in late October 1997, a freshly emerged individual was captured, marked and released by Professor John Wacher, close to his home near Hayle. He reported its re-emergence in early April the following year and its intermittent sighting until the middle of May (Wacher, 1998). Although such reports are extremely rare, it would not be surprising, for a species that is a prolific migrant constantly pushing at the edges of its distribution, if overwintering were possible somewhere across its range. Climate change is likely to propel shifts in such behaviour. There is no evidence for eggs, larvae or pupae overwintering in the UK.

Analysis of trends
The variation in numbers of Painted Lady from year to year is related to climatic conditions in North Africa and the Mediterranean region. A population explosion there can result from high rainfall causing luxuriant growth of larval foodplants. On the other hand, many thousands of butterflies can be killed in flight by an unseasonable thunderstorm.

Conservation
Leaving an uncut growth of thistles in field corners, on headlands and in gardens is about as much as can be done in Britain to support the conservation of this migratory species.

Best places to see
Anywhere in Cornwall, especially along the coast.

Upperside, male and female similar

Peacock
Aglais io

Sponsored by
Arthan-Davey, Tamara Hicks
and Deborah Soady

Because the Peacock is so common and well known, it is easy to take it for granted. It needs little description, as almost everyone can recognise these large, uniquely marked, rich red-coloured butterflies, a familiar sight in both town and countryside, possessed of spectacular 'eyes' on the upperside of both fore- and hindwings. Reminiscent of the 'eyes' on the tail plumes of a peacock, these markings are highly effective in deterring predators. The underside, by contrast, looks almost entirely black, providing excellent camouflage when the butterfly is hibernating or at rest, especially against a tree trunk. The butterfly is able to make hissing sounds audible even to human ears by rubbing its wings together, further deterring predators. Male and female are very difficult to tell apart.

Distribution in Cornwall
The Peacock is widespread throughout Cornwall. As with many other common

Resident in Cornwall
(resident on the Isles of Scilly)

CONSERVATION STATUS
■ Butterfly Conservation priority: Low

TRENDS AND DISTRIBUTION
UK rate of change in abundance
  ↓ 5% 1976–2018
South-west England rate of change in abundance
  ↓ 21% 2009–2018
  ↓ 46% 1976–2018
% of 1km squares occupied in Cornwall
  40% 2009–2018 (81% 10km squares)
  29% 1976–2008 (81% 10km squares)

species, the distribution maps show that within the last 10 years it has come to inhabit more 1km squares, likely as a result of more recorders reporting from a wider variety of places, a factor no doubt helped by the Big Butterfly Count. However, it is interesting to note that despite the increased number of

Peacock

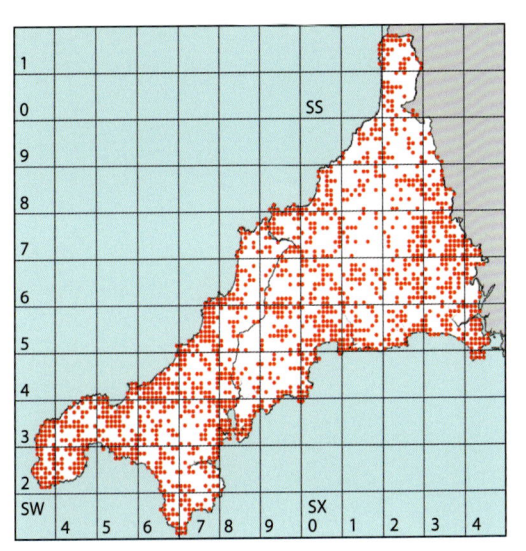

Distribution 2009–2018

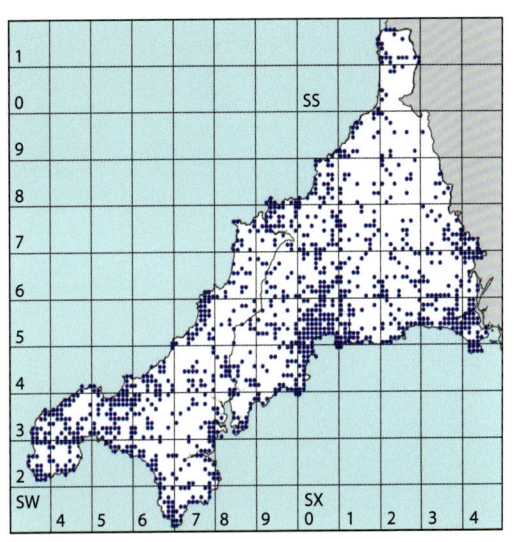

Distribution 1976–2008

records in the two time periods shown, the number of butterflies per record has declined by over a third (possibly due to more precise recording, with recent records collected at 100m, 10m or even 1m grid-square level rather than at the 1km scale).

Penhallurick states, "All writers have described the Peacock as 'common' in Cornwall" (1996: 97), and Clark in the *Victoria County History, Cornwall* has summarised its status as "common throughout the county and in some years… locally very abundant" (Clark, 1906).

Larval foodplants

Common (Stinging) Nettle *Urtica dioica* is known to be the principal foodplant. Occasional use of Small Nettle *U. urens* and

Underside, male and female similar

Larvae on their larval web

Pupa

Hop *Humulus lupulus* have been recorded, but both have a limited distribution in Cornwall.

Habitat and ecology
Peacock frequent woodland glades and rides, scrubland, waste ground, hedgerows, Cornish hedges and gardens – in short, wherever there are nectar sources and larval foodplants. Favoured sources of nectar are Bramble *Rubus fruticosus* agg., thistles *Cirsium* spp. and several cultivated flower species, including sedum *Sedum* spp., buddleja *Buddleja* spp., Red Valerian *Centranthus ruber* and michaelmas-daisies *Symphyotrichum* spp. A particular favourite is Hemp-agrimony *Eupatorium cannabinum*, a single clump of which can host a dozen or more butterflies.

Peacock can be encountered in every month of the year. The butterfly is normally single-brooded, although it can produce a partial second brood in very hot summers. It can live up to 11 months as an adult, overwintering in this stage. It tends to enter hibernation as early as August, and spends the winter in any dark place, such as a hole in a tree or in a garden shed. It is less commonly seen in houses than

Flight curve based on number of records in Cornwall 1999–2018; and UK life cycle

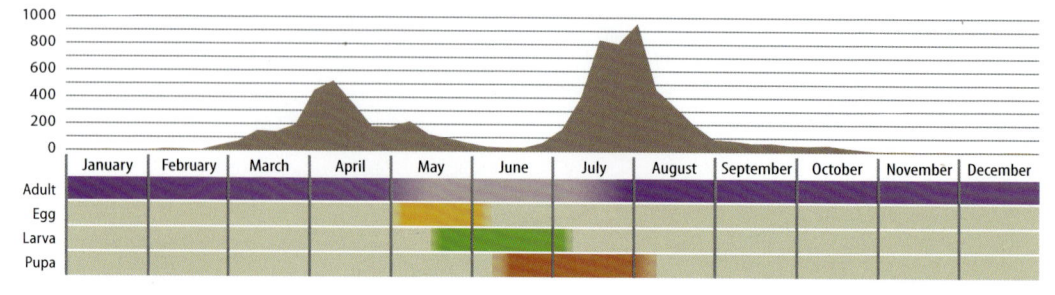

the Small Tortoiseshell. Any warm, sunny winter day in December, January or February can reactivate the butterflies, especially in Cornwall, where such days are not infrequent, but the peak in emergence from hibernation occurs in late March and April, when mating and egg laying take place. The male is fiercely territorial at this time and aggressively challenges any rival male that enters his territory. The female lays one or more clusters of up to 400 eggs in a dense mass on the underside of a nettle leaf. Large nettle clumps are chosen, in sheltered positions in full sun, situated on woodland edges or in south-facing hedgerows or Cornish hedges.

The eggs hatch in about 14 days and the caterpillars build a communal web, from which they emerge to bask and feed. At this stage they are very noticeable. They can be confused with early instar larvae of Small Tortoiseshell, which look similar and create webs on nettles in much the same way. However, in their later instars the jet-black larvae of Peacock are easier to distinguish from Small Tortoiseshell larvae, which have distinctive longitudinal yellow stripes. When disturbed, members of the caterpillar colony wave about in unison, a ruse to deter predators by giving the appearance of a single large, undulating organism. The larvae disperse as they become fully grown and wander off to find sites in which to pupate. They occasionally use their foodplant, but more often they choose branches, twigs or leaf stalks on trees or bushes, which can be some distance away from the nettles.

In very hot summers, development of a partial second brood is thought to occur, with adults from this brood emerging in late autumn. This has been backed up by laboratory research finding that longer periods of daylight hours could enhance the development speed of the first-brood larvae, thus enabling time for a second brood (Asher *et al.*, 2001). Under normal conditions, though, the adult progeny of the hibernators start flying from late July and eventually seek their own winter quarters. They may wander quite far afield looking for a suitable place, and once they find somewhere to their liking they tend to stay around that area. Since this is a mobile species that can fly considerable distances, it is in theory capable of migration. Jeremy Thomas, however, cites research that establishes the Peacock as "a nomad rather than a true migrant, for while the adults have a tendency to fly north in spring, and south in late summer, the furthest any individual had been shown to travel is about 95km" (Thomas and Lewington, 2014: 192).

Analysis of trends

The butterfly has extended its range over the UK as it has moved northwards, and there was a 16% increase in distribution between 1976 and 2014 (Fox *et al.*, 2015), although UKBMS data show that there has been a slight decline in abundance of 5% between 1976 and 2018.

Despite considerable fluctuations in the abundance of Peacock in Cornwall over the last 10 years, there has been an upward trend, with the highest numbers in 2014 and 2015. The butterfly is seen on almost every transect, and the results from these show a similar pattern. The situation in Cornwall seems to buck the declining trend in South-west England over both the longer period and the last 10 years. Also unexplained is the fact that regional decline is more severe than the national picture, but this may be connected to climate change. Early emergence from hibernation in mild winters in the wider south-west may also be a factor if those butterflies are unable to survive later cold spells, but because of its maritime climate Cornwall is less likely to experience such contrasts in temperature.

Conservation

Little needs to be done, as the supply of nettles in the County seems more than plentiful. Intensive agriculture is a potential threat, particularly where sprays are used to eliminate the foodplant, but this is probably more than offset by increased levels of nitrogen in the environment from vehicle emissions and agricultural fertilisers, creating ideal conditions for nettles.

Best places to see

Any area where nectar flowers are abundant.

Upperside, male and female similar

Small Tortoiseshell
Aglais urticae

Sponsored by
Murchadh Keohane, Shaun Poland
and Russell Williams

The Small Tortoiseshell is easily recognised by its attractive orange, yellow and black patterning, white tips on the forewings, and bright blue edging to the scalloped wings. The undersides of the wings, by contrast, are a dull mixture of dark and light brown patterning. Although becoming increasingly scarce, it is still often seen in gardens. Male and female are similar.

Distribution in Cornwall
The Small Tortoiseshell is widespread throughout most of Cornwall, although given to periodic fluctuations in numbers. Despite declines in abundance both nationally and locally, it has expanded its range in recent years, although this may only reflect the increasing number of recorders who now contribute from most parts of the County.

The indigenous butterflies of this species can be reinforced by migrants from across the

Resident and migrant in Cornwall
(resident on the Isles of Scilly)

CONSERVATION STATUS
■ Butterfly Conservation priority: Low

TRENDS AND DISTRIBUTION
UK rate of change in abundance
↓ 78% 1976–2018
South-west England rate of change in abundance
↓ 59% 2009–2018
↓ 84% 1976–2018
% of 1km squares occupied in Cornwall
44% 2009–2018 (85% 10km squares)
30% 1976–2008 (81% 10km squares)

English Channel, but, unlike the influx of Red Admiral and Painted Lady, high numbers are not thought to arrive. Most Small Tortoiseshell seen in the UK will be home-grown (Eeles, 2019). The butterfly is a great traveller and has been recorded up to 150km from the point of release (Roer, 1968).

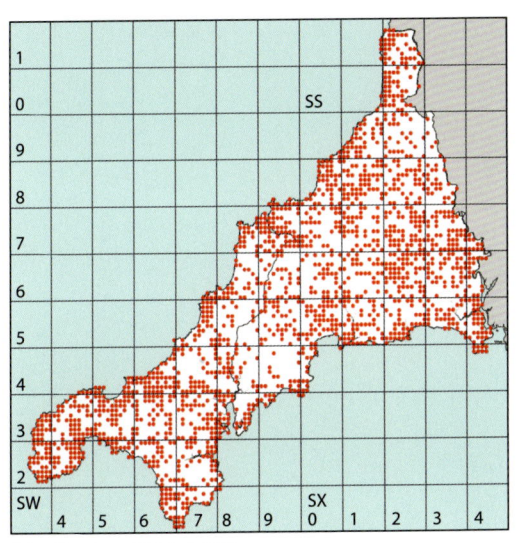

Distribution 2009–2018

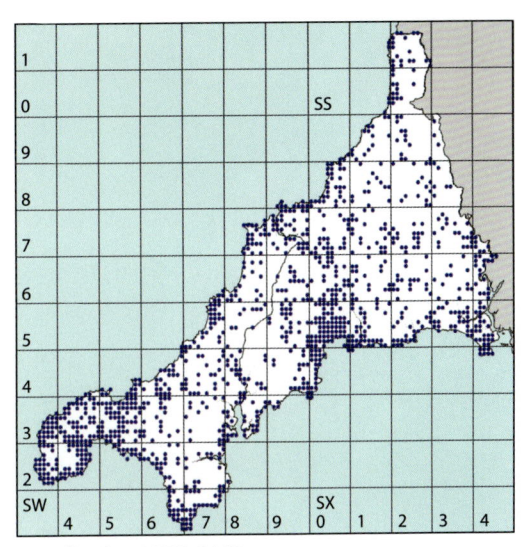

Distribution 1976–2008

Larval foodplants

Common (Stinging) Nettle *Urtica dioica* and Small Nettle *U. urens* seem to be the only larval foodplants used. Young plants on the sunniest and warmest side of large clumps are most favoured. Small Nettle is mainly restricted to west Cornwall, where it is infrequent.

Larva on Common Nettle

Habitat and ecology

The adult butterfly can be met with almost anywhere, including urban gardens, lanes, fields, waste land and dunes, wherever there are larval foodplants and an abundance of nectar sources. Its wandering habits can take it into town centres and to the top of the highest moors in Cornwall.

The Small Tortoiseshell's complicated flight periods are affected by both climatic conditions and geographical situation. It overwinters as an adult and can often be found sheltering in houses and sheds during winter months, although an unseasonably warm day or domestic heating can reactivate the butterfly, when it again takes to the wing. It is doubtful whether these butterflies, awakened so prematurely, can survive for long. Normally, hibernating adults emerge in Cornwall from March onwards, depending on the weather. They then mate and the female lays large clusters of lime-green eggs under young nettle leaves in a sunny, sheltered position. The black and yellow caterpillars with stumpy spines form a silken tent and live gregariously until almost fully grown, when they disperse to nearby nettles to pupate. The colour of the chrysalis can vary from golden to brown.

Temperature and day length affect the number of broods in a year. In Cornwall, the first-brood adults emerge in June and July, their more numerous offspring appearing from early

Underside, male and female similar

August, and can be seen on the wing as late as October. These second-brood adults usually go into hibernation by the end of September, reappearing the following year on warm spring days to breed, often surviving until mid-May. However, if spring is late and temperatures remain cooler, many first-brood adults will also enter hibernation, saving their eggs for the following year rather than laying them and producing a second brood during the same year. In very favourable years it is thought a third brood may even be possible.

The butterfly's flight pattern is further complicated by incoming migrants from continental Europe in early spring and summer, which presumably begin to breed on arrival, creating a life cycle that may or may not synchronise with butterflies developing in Cornwall. While there is evidence that autumn shows a southward drift of the species, there appear to be no observations that show whether immigrants can successfully overwinter in Britain, or whether they can return to the continent.

Analysis of trends

National statistics show that this once very common butterfly has declined by 78% in abundance since 1976, and in recent years it has from time to time given grave cause for concern, as the numbers have fallen steeply. The last 10 years (2009 to 2018) show a

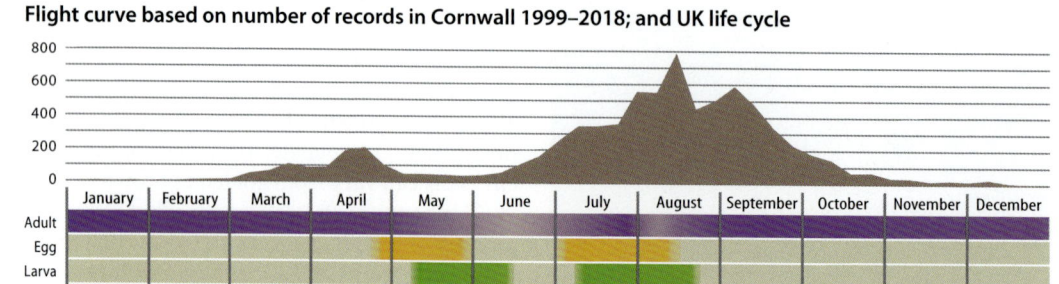

Flight curve based on number of records in Cornwall 1999–2018; and UK life cycle

considerable fluctuation in both numbers of individuals and records after an all-time low in 2008 when only 218 adult Small Tortoiseshell were recorded in Cornwall. The butterfly made a good recovery, in 2013, however, and even more so in 2014, the best result in this decade, with over 4,000 recorded in the database. The Cornish trends for this species have been derived from the records submitted to the County Recorder, mirrored by transect information. Unfortunately, since the revival of fortunes for this butterfly, numbers have declined again, but with a slight increase in the good summer of 2018.

It is interesting to note that in Cornwall, for each Small Tortoiseshell record submitted before 2009, an average of just over four individuals were seen, whereas in the last decade this has gone down to only two butterflies per record, possibly confirming the drop in abundance reported nationally. This is puzzling because nettle, the caterpillar's foodplant, is more widespread now than at any point in the past, nourished on nitrate-rich land caused by vehicle emissions and fertilisers.

The Small Tortoiseshell has always been vulnerable to various parasitoids that are likely to contribute towards annual fluctuations. An example of this is the tachinid fly *Sturmia bella*, which arrived in Britain in the late 1990s and has spread rapidly, eventually reaching Cornwall. Since it was first recorded in England, the Small Tortoiseshell has declined considerably, especially in areas where *S. bella* was known to be present. However, the jury is still out regarding how much this tachinid has contributed to the decline of the butterfly. Weather is also thought to be a possible factor influencing abundance of the Small Tortoiseshell, with numbers observed to drop substantially after a dry, warm summer (Thomas and Lewington, 2014). Research continues!

Conservation
As with other closely related species, the maintenance of good patches of the larval foodplant is a necessary aid to conservation.

Best places to see
Anywhere in Cornwall, especially gardens.

Pupa on Common Nettle

Upperside, male and female similar

Comma
Polygonia c-album

Sponsored by
Anne Green, Gilli Keohane
and Xanny Stride

This eye-catching orange butterfly with dark brown spots and blotches on the upperside of its wings is named after the small white comma or 'C' mark on the underside of its hindwings, a feature expressed in both its common and scientific names. Another unique distinguishing characteristic of the Comma is the scalloped outline of its wings. It is a master of disguise in all its stages. In profile, with the marbled brown, tan and grey colours of the underside of its wings exposed, it looks like a dead leaf, which is perfect cover for a winter-hibernating butterfly. The pupa also resembles a shrivelled leaf, and the larvae can be mistaken for bird droppings. Male and female are similar, although there are small differences in the intensity of colour, the male being slightly darker than the female. It is sometimes confused with a ragged Small Tortoiseshell or one of the species of fritillary found in the County.

Resident in Cornwall
(resident on the Isles of Scilly)

CONSERVATION STATUS
■ Butterfly Conservation priority: Low

TRENDS AND DISTRIBUTION
UK rate of change in abundance
↑ 130% 1976–2018
South-west England rate of change in abundance
↓ 7% 2009–2018
↑ 63% 1976–2018
% of 1km squares occupied in Cornwall
31% 2009–2018 (80% 10km squares)
16% 1976–2008 (76% 10km squares)

There are two different forms of the adult Comma. The normal (nominate) form is darker and hibernates over winter, and the *hutchinsoni* form (named after the Victorian entomologist Emma Hutchinson, who first identified it), does not usually hibernate and is paler on both the upper- and the underside of the wings.

Comma

Butterflies of Cornwall

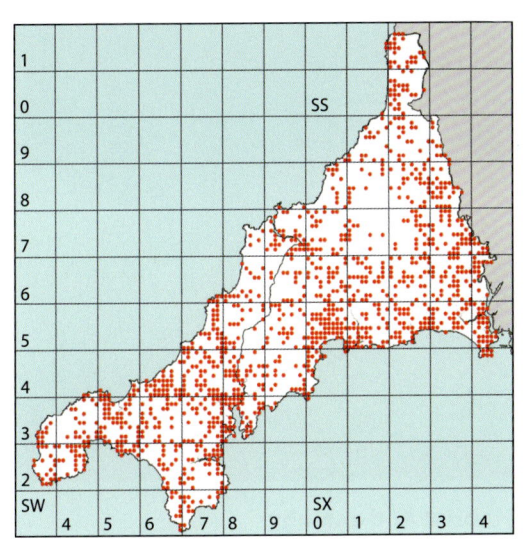

Distribution 2009–2018

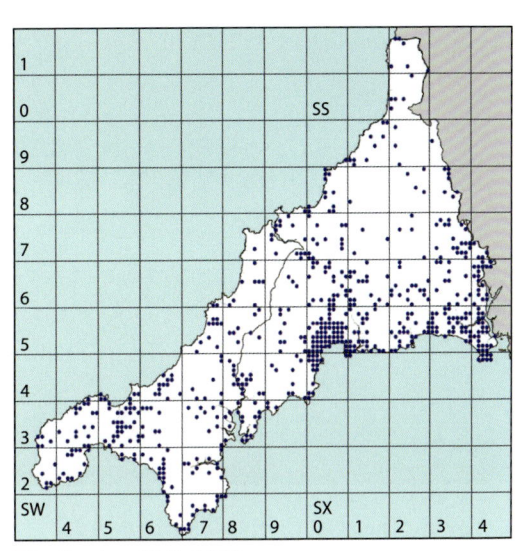

Distribution 1976–2008

Distribution in Cornwall

Starting in the early nineteenth century, the Comma suffered a serious decline, eventually being restricted to a few counties along the southern Welsh Marches. During this period the butterfly was unknown in Cornwall; indeed, it is thought that it was never a Cornish resident, although rare occurrences may have been overlooked (Penhallurick, 1996). Recovery began in about 1914 and the Comma gradually spread over the whole of southern England, moving northwards and eventually colonising Scotland in 2000 (Eeles, 2019). After a prolonged march through Devon in the late 1920s, the butterfly was first recorded in Cornwall at Polyphant in 1933 (Bracken, 1936). It continued its westward journey, reaching Redruth by 1935, followed by Hayle and Marazion the following year. Since then it has gradually become widespread throughout

Underside of the normal darker form, showing the small white mark after which the species is named

The paler *hutchinsoni* form

157

Larva on Common Nettle flowers

Cornwall. The 2003 *Cornwall Butterfly Atlas* refers to the Comma still being scarce in the west of the County (Wacher *et al.*, 2003), but the most recent distribution map shows that this is no longer the case. Indeed, the number of 1km squares where it has been seen has almost doubled in the last 10 years. It is also interesting to note that since the last Atlas was published the Comma has also become locally common on the Isles of Scilly.

Larval foodplants

During the nineteenth century the principal larval foodplant was Hop *Humulus lupulus*. However, with the decline of hop-growing in England in the later twentieth century, the butterfly has adapted opportunistically, and its caterpillars now feed mainly on Common (Stinging) Nettle *Urtica dioica*. Comma larvae have also been found on elms *Ulmus* spp., currants *Ribes* spp., willows *Salix* spp. and Hazel *Corylus avellana*. Prominent stands of Common Nettle in the shelter of woodland edges and hedgerows are favoured.

Habitat and ecology

The Comma is essentially a species of sunny rides and glades in wooded areas, but due to its strong flight and wandering habits it can also be seen in tree-lined lanes, on trackways and in scrubland. It is a frequent visitor to gardens and orchards, where it is attracted by Buddleja *Buddleja davidii*, michaelmas-daisies *Symphyotrichum* spp. and rotting fruit. Pre-hibernation individuals also enjoy Bramble *Rubus fruticosus* agg., ivy *Hedera* spp. and Hemp-agrimony *Eupatorium cannabinum*.

The flight pattern is complicated. Comma butterflies overwinter as adults and can reappear as early as January/February in Cornwall. They are, however, mainly recorded between late June and early August. There is an earlier, smaller peak in late March and April, when those that emerge from hibernation mate, usually high up in a tree or a bush, the male having established a specific territory and intercepted a passing female. The female mates with multiple males to fertilise her eggs. She can select males that have fed on high-quality

Flight curve based on number of records in Cornwall 1999–2018; and UK life cycle

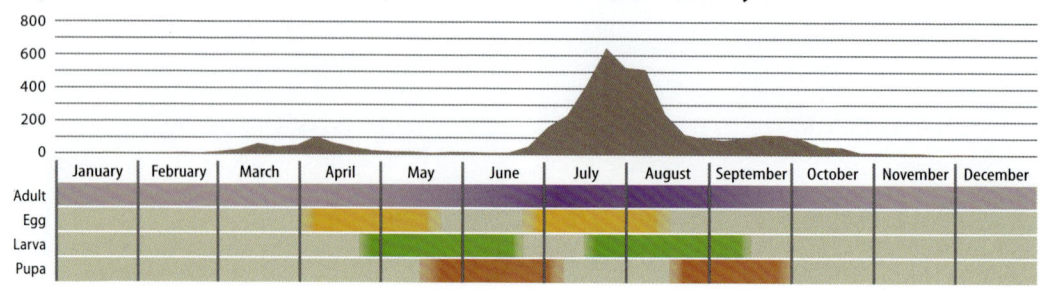

food that enables them to provide superior 'nuptial gifts' in the form of nutrients along with their sperm. These nutrients can increase the female's longevity and reproductive success (Wedell, 1996). The green, ribbed eggs are laid singly, usually on the upperside of a nettle leaf, and change colour to yellow and then grey. The larvae moult four times, changing in appearance at the various stages and developing differently coloured spines. The fifth (and final) larval instar has a very unusual appearance with a clear distinction between the orange-brown colouring of its first few segments and the startling white of the rest of its body. Before pupation, the larva spins a dense silk pad on its foodplant or nearby vegetation, from which it hangs in a 'J' shape while it pupates. The butterfly emerges from the chrysalis after about two weeks (Eeles, 2019).

The first generation, which results from the spring mating of overwintering adults, usually emerges in late June. Some of this brood are the darker normal form; these adults survive for about 10 months and overwinter to mate the following spring. Others are the pale *hutchinsoni* form, which are short-lived and mate soon after emergence to produce "normal-looking adults" in August and September (Thomas and Lewington, 2014: 195). Several factors have been thought to control this twin-track life cycle, including temperature, when the egg is laid and quality of foodplant (Asher *et al.*, 2001). Laboratory research, however, has found that the key determinant of which form emerges is the length of daylight hours, whether these are increasing before midsummer or decreasing after it has passed, and the effect this has on the developing caterpillar. Lengthening, longer daylight hours result in the *hutchinsoni* form, whereas decreasing, shorter hours correlate with the dark form (Nylin, 1989).

Analysis of trends
Cornwall has seen a very definite increase in geographical distribution in recent decades, although there is some evidence from both Cornish data and South-west England trends that abundance has slightly decreased in the last 10 years. The UK increase in abundance over the whole period 1976–2018 is an unrivalled 130%. This spectacular nationwide recovery over the last century has been attributed to various factors, particularly a warming climate and the switch to alternative foodplants from this butterfly's initial favourite, Hop. However, at present the reasons for the recovery of the Comma are still not fully understood.

Conservation
No specific methods of conservation are considered necessary for this species, although continuing the management of woodlands to encourage glades and rides would be beneficial.

Best places to see
Wooded areas and gardens throughout Cornwall.

Butterflies of Cornwall   Nymphalidae

Upperside, male and female similar

Marsh Fritillary
Euphydryas aurinia

Sponsored by
*Alan Stapleton, Simon Tagney
and in memory of Ian Pryor*

This medium-sized butterfly is the most colourful of the British fritillaries, and one of the rarest. Its protected status today is in stark contrast to the vast swarms reported in the nineteenth century, when the butterfly and its caterpillars occasionally reached huge proportions in Ireland (Thomas and Lewington, 2014).

The variable orange, brown and cream chequered pattern of the upperside of the wings makes Marsh Fritillary relatively easy to identify. The colours fade after a few days, so that the butterfly appears rather shiny, earning it the older, unflattering names of 'Dishclout' and 'Greasey Fritillaria' (Eeles, 2019). The undersides are patterned with cream spots on a pale orange background, with a row of tiny black spots near the border of the hindwings. Male and female are similar, but the female is larger than the male and slightly paler.

This species is unusual because it is not only the adult that is recorded and monitored, but

Resident in Cornwall
(not on the Isles of Scilly)

CONSERVATION STATUS
- Annex II Species (EU Habitats Directive 1992)
- Fully Protected under Schedule 5 of the Wildlife and Countryside Act 1981 (as amended)
- Section 41 Species of Principal Importance (NERC Act 2006)
- Red List of British Butterflies; IUCN criteria: Vulnerable (VU)
- Butterfly Conservation priority: High

TRENDS AND DISTRIBUTION
UK rate of change in abundance
 ↓ 13% 1981–2018
South-west England rate of change in abundance
 ↓ 28% 2009–2018
 ↓ 65% 1981–2018
% of 1km squares occupied in Cornwall
 2% 2009–2018 (25% 10km squares)
 3% 1976–2008 (46% 10km squares)

Marsh Fritillary

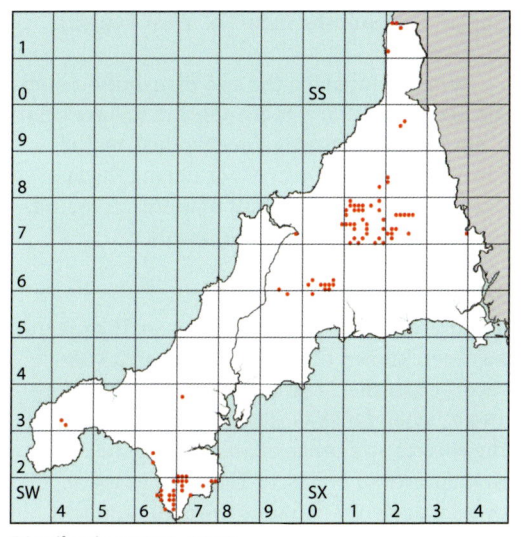

Distribution 2009–2018

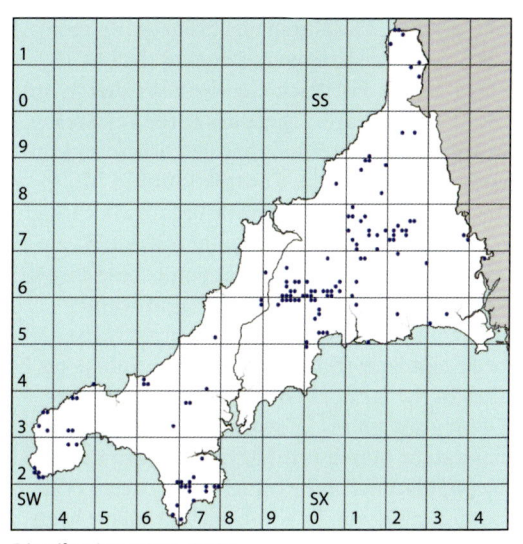

Distribution 1976–2008

also the webs that are spun by the gregarious larvae. Taken together, larval webs and adults provide good evidence of the strength of a colony.

Distribution in Cornwall

The Marsh Fritillary is one of the most rapidly declining butterflies in Europe. A similar pattern of decline has taken place in Britain, where the butterfly is mainly restricted to South-west England and parts of Wales and Scotland. Today the South-west accounts for one fifth of the UK population, of which a small proportion survives in Cornwall. There are several localities where the species was recorded prior to 1976 and not seen again. These sightings may represent gradual losses of populations to agricultural intensification after the second world war.

The 1976–2008 map shows a wider scattering of sites, covering 3% of 1km squares

Underside, male and female similar

in Cornwall. Despite the fact that many more sites have been discovered during the last 10 years, Marsh Fritillary are now recorded in only 2% of 1km squares. There are three key areas in which the butterfly is consistently found in good numbers in the County: Bodmin Moor, The Lizard and Mid Cornwall Moors (although distribution here is principally restricted now to Breney Common). Beyond these three areas the species has declined significantly in recent years, albeit that the butterfly has been regularly recorded in small numbers on sites in the far north of Cornwall, where rare Culm grassland is known to be an important habitat for Marsh Fritillary and is also used by populations of the butterfly just across the border in Devon. In addition, there have been a number of unusual and unexplained lone sightings from locations including Stithians, Wadebridge, Greenscoombe Wood and the north of The Lizard, some backed up by photographs. There is always the possibility that some of these individuals have been captive bred and released.

Extensive surveys in West Penwith indicate that the Marsh Fritillary has been lost from this area (having last been seen in 2010), despite the existence of a number of sites with suitable habitat, searched by Cornwall Butterfly Conservation (CBC) volunteers and confirmed by Patrick Saunders on behalf of Natural England, within the moors of West Penwith in 2014.

When comparing the two distribution maps for Marsh Fritillary, there appears to have been a substantial increase in occurrence on The Lizard and on Bodmin Moor during the last decade, probably the result of more surveying, monitoring and conservation work.

The Lizard

The presence of Marsh Fritillary on The Lizard has been known of since 1935, when it was recorded in the 'Lizard area', and then in the 1940s, when Dr W.G. Tremewan recorded the species at Kennack Sands, along the coast to the north of Poltesco. This species is still present at Kennack Sands in small numbers. However, the strongest colonies on The Lizard are towards the west coast, focused on the cliffs of Predannack and Mullion, and around the eastern fringes of Predannack Airfield (including Hayle Kimbro, Windmill Farm and Lower Predannack Downs). Numbers at some of these sites (including their associated sub-sites) range from 20 to 100 adults, with a similar number of larval webs. The central heathlands of Goonhilly Downs support smaller colonies of the butterfly around the southern and western fringes, where Devil's-bit Scabious *Succisa pratensis* grows on the side of, and around, Purple Moor-grass *Molinia*

Predannack Head, The Lizard, coastal grassland, an alternative habitat for the Marsh Fritillary

caerulea tussocks, the primary habitat used on The Lizard. It is thought that in years with hot summers individuals of the adult butterfly spread out across the Downs, with both adults and larval webs recorded around scabious plants scattered across the heathland. Smaller colonies also persist towards the east coast of The Lizard, focused mainly to the north and west of Coverack, with populations at Crousa Downs and the coastal stretch between North Corner and Lowland Point. Main Dale lies within the East Lizard Heathlands Site of Special Scientific Interest (SSSI); the citation, last revised in 1995, notes "a strong breeding colony of the marsh fritillary butterfly". The last Marsh Fritillary record for this site, however, was in 2011. The site was searched in 2018 and 2019, but no adults or larval webs were found.

Natural England produced an action plan for surveying and monitoring the legally protected and notable butterfly species found on The Lizard covering the period 2018 to 2022. This plan includes the annual monitoring of the known sites for Marsh Fritillary on the peninsula, as well as pinpointing potential new sites through the identification of suitable grasslands/heaths with Devil's-bit Scabious. Coordinated by Natural England in collaboration with the National Trust, Cornwall Wildlife Trust and CBC, the aim of the work is to improve our understanding of the Marsh Fritillary on The Lizard and provide data that will inform better habitat management. The University of Exeter has been conducting research on Marsh Fritillary on The Lizard since 2014.

Bodmin Moor

On Bodmin Moor there has been the most dramatic increase in identified Marsh Fritillary sites since 2009. This has been mainly due to the ambitious All the Moor Butterflies project, which ran from 2017 to 2020 across Dartmoor, Exmoor and Bodmin Moor. The project focused on surveying and conservation, with the aim of safeguarding a select number of rare and threatened butterfly and moth species. In autumn 2015, in the developmental stage of the project application, 35 sites were surveyed on Bodmin Moor for Marsh Fritillary larvae. Larval webs were found on a total of 17 sites, including six entirely new locations.

This initial survey generated a list of sites that merited further investigation, as well as locations that would benefit from practical conservation work. The All the Moor Butterflies project enabled this programme of work to be delivered, and resulted in the discovery of 29 Marsh Fritillary sites, of which 23 were previously unknown and six were rediscovered sites with historical records. The resulting number of sites recorded in the

Bodmin Moor, typical damp grassland habitat for the Marsh Fritillary

project for Marsh Fritillary on Bodmin Moor is 49 grouped into 12 metapopulation networks (Phelps, 2019). A metapopulation is defined as "groups of local populations being connected by occasional dispersal" (Ellis *et al.*, 2012: 11). These metapopulations stretched from St Breward in the north to Penkestle Moor in the south, and Harpur's Downs in the west to Hawkstor Marsh in the east. Each site was very variable in the number of records it generated, with six sites producing only single sightings each of either a web or an adult. Carkeet in the Fowey valley recorded the highest numbers in both 2018 and 2019: 145 webs were recorded in the final year of the project.

Mid Cornwall Moors
By contrast, the outlook for Marsh Fritillary on the Mid Cornwall Moors is worryingly bleak, with only a handful of sites on the east side of the moors where the butterfly is still known to be extant. In 2017 six SSSIs were amalgamated and expanded to create the Mid Cornwall Moors SSSI, which lies within the Hensbarrow and Cornwall Killas National Character Areas. (See chapter 2.) This incorporated known and potential Marsh Fritillary habitat in the region to help ensure that it would be best protected.

As a result of careful management and monitoring by Cornwall Wildlife Trust, Breney Common has a relatively stable population, with consistent records of varying numbers every year. The surrounding areas of Red Moor and Lowertown Moor both have records for the species in 2018 and 2019, giving cause for hope that the metapopulations on the east side of the Mid Cornwall Moors may occupy a larger area than previously thought. This may aid expansion into nearby habitat that is suitably managed.

On Goss Moor National Nature Reserve (NNR) the Marsh Fritillary metapopulation has collapsed. The last natural record of the species there was a single larval web in 2011. Numbers peaked between 2003 and 2005, when the species was widespread across the NNR, followed by a steady decline until the butterfly's disappearance by 2012. The reason for this decline is unknown, but substantial regrowth of scrub across the site is probably a major factor. This would have fragmented the populations on the moor, isolating smaller colonies from each other. Reduced grazing by livestock in some areas has enabled scrub to grow and Purple Moor-grass to dominate, out-competing the larval foodplant, Devil's-bit Scabious. Bad weather in the summer of 2012 could have been the final straw.

In 2016, Natural England started a project to reintroduce the species to Goss Moor NNR, and multiple egg batches were translocated from the Hayle Kimbro population on The Lizard the following year. This reintroduction resulted in four larval webs being recorded in the autumn of 2018, followed by a maximum count of two adults in spring 2019; no larval webs were found in searches during autumn 2019. This unsuccessful reintroduction was halted after concerns were raised about habitat sustainability and impact on the donor population.

There are vast numbers of other sites that historically held Marsh Fritillary populations on the Mid Cornwall Moors, notably Carbis, Criggan Moors, Molinnis Downs and Retire Common. These areas have all been searched in recent years, with no evidence of Marsh Fritillary found. Sites were observed to be suffering from the same scrub encroachment and grazing issues as Goss Moor, resulting in habitat fragmentation and a loss of connectivity. Further monitoring and assessment of habitat condition, followed up by appropriate management, will be vital to restoring Marsh Fritillary metapopulations on a landscape scale across the Mid Cornwall Moors. Natural England, which manages Goss Moor NNR, attracted European Regional Development funding for Growing Goss. This project, ending in 2022, aims to improve habitat on the NNR through work such as scrub clearance and controlled grazing. Natural England is working closely with CBC in the long term to restore the area for Marsh Fritillary in the hope that the butterfly will return to Goss Moor if conditions allow.

Larval foodplants
The larvae feed almost exclusively on Devil's-bit Scabious, which is widespread throughout most of Cornwall, although it is more sporadic along

the south coast, perhaps explaining the rarity of the butterfly there. The species has been known to use Field Scabious *Knautia arvensis* and Honeysuckle *Lonicera periclymenum*, although these are unlikely to be used in Cornwall unless the preferred foodplant is scarce.

Habitat and ecology

In Cornwall, Marsh Fritillary occupy a variety of habitats. The primary habitat type comprises damp grassland dominated by tussock-forming Purple Moor-grass interspersed with ample larval foodplant; this is typical of the majority of sites in the County, from the wet heaths of The Lizard and the mires of Bodmin Moor to the Culm grassland in the far north of Cornwall. For the butterfly to flourish, it requires a varied sward structure 5–25cm in height. Notably, a favourable habitat is coastal grassland, as at Higher Predannack Cliff/Mullion Cliff on The Lizard. The colonisation of this habitat appears to mirror the fairly recent trend for the species to use areas of shorter turf containing the larval foodplant, similar to that of the chalk and limestone downs of southern England.

The butterfly is single-brooded, the adults usually emerging towards the middle of May and reaching a peak in early June. The male emerges first, followed a few days later by the female. The male sets up small territories, darting around to investigate any passing butterfly. After mating, he seals the genital opening of the female with a substance that he secretes, to prevent further insemination (Eeles, 2019). The female then searches out suitable plants for egg laying, being very particular about the location, but not too concerned if others have laid eggs nearby or even on the same large leaf. She lays her eggs in a single batch on the underside of a leaf of the foodplant, usually on the day of mating. Egg laying takes several hours, as there can be up to 400 yellow eggs. Further eggs develop in the female as she feeds, and these are later laid in smaller batches. Eggs hatch in about four or five weeks, and the larvae immediately start creating a communal web, in which they live, feed and moult. In addition to being protective, the web provides a humid atmosphere that stops the caterpillars from drying out and helps to retain heat when they huddle together. The caterpillars go through several moults, changing slightly in appearance at each stage, from a golden-brown initially, to jet black with tufts of spines and a peppering of white in their final instar. After the third moult, the larvae construct a new web low down in the vegetation, where they hibernate over winter. There is speculation that the sight of large

Eggs, freshly laid on the underside of a Devil's-bit Scabious leaf

Maturing eggs

Larval web with caterpillars

Pupa attached to the leaf of Devil's-bit Scabious

numbers of prickly caterpillar bodies adjacent to each other may deter predators.

Larvae emerge from the web in early February to bask communally on warm and sunny days (at which time they are easier to find), and they resume feeding after about a week. They go through a fourth moult in late March or early April, now living in much smaller groups. After the fifth moult in April, each larva for the first time lives as a solitary individual, searching for food alone and sometimes, where this is in short supply, varying its diet. It pupates low down in the vegetation, where it suspends itself head down from a silken pad. The pupa is very striking, being largely white with black, brown and orange markings (Eeles, 2019).

Marsh Fritillary metapopulations comprise a network of self-contained colonies from which, in warm summers, females sometimes disperse to recolonise vacant areas or to establish new colonies, if suitable habitat can be found close by. The abundance of the butterfly fluctuates substantially from year to year. A number of interacting factors cause metapopulations to crash and recover, such as adverse weather, availability of the larval foodplant, and infestation by parasites. *Cotesia bignelli* is a parasitic wasp that can cause considerable fluctuations to its sole host, the Marsh Fritillary. The wasp is thought to be present in the County, having been observed in recent years at Carkeet, one of the core sites for Marsh Fritillary on Bodmin Moor. However, to be certain of identification, a wasp needs to be examined by an expert. The standard distance adult Marsh Fritillary fly is thought to be less than 100m (Eeles, 2019), but movements of over 10km are possible, presenting opportunities to colonise unoccupied habitat if the right circumstances occur. This exemplifies the importance of the metapopulation, with its network of high-quality core sites located within flight distance of lower-quality sites. For a metapopulation to be sustainable, the colonisation rate in the region must be at least equal to the rate of extinction. The larger the area occupied by a metapopulation, the greater its potential chance of survival.

Analysis of trends

In England there has been a 66% decline in abundance of Marsh Fritillary between 1982 and 2018 (Brereton *et al.*, 2019). In the UK as a whole, however, the decline in abundance was 13% between 1981 and 2018. The trend in South-west England is more closely aligned to that of England, although the decline has begun to flatten out in the last 10 years. It is difficult to know whether this butterfly is genuinely doing slightly better in Cornwall or whether the All the Moor Butterflies project on Bodmin Moor has merely brought to light long-established sites. However, in 2000, 89 sites were investigated for the species as part

Flight curve based on number of records in Cornwall 1999–2018; and UK life cycle

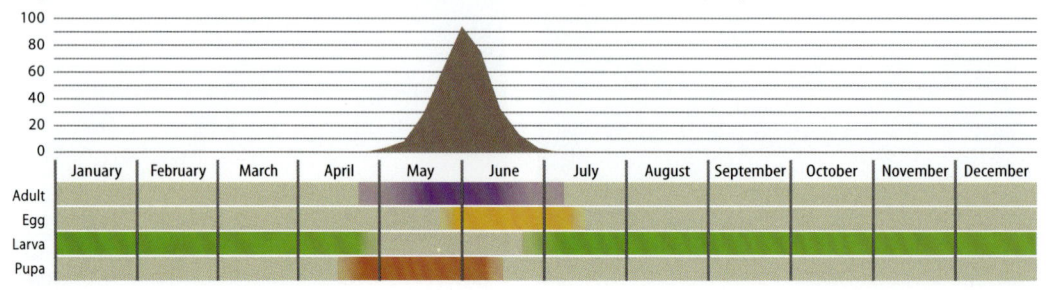

of the creation of a 'site dossier' for Marsh Fritillary in Cornwall. These searches only found larval webs on five of the 27 sites that were surveyed in detail. The butterfly was thought to exist at low density on Bodmin Moor; however, many sites were noted to have potential and it was suggested that more populations may be found (Hobson and Budd, 2001). Out of 49 sites identified in the All the Moor Butterflies project, 37 were covered by agri-environment schemes in the development phase of the project. Sites that were previously unsuitable were noted to improve in habitat quality after entering into these schemes (Phelps, 2019; Foster, 2015). The expansion of Marsh Fritillary on Bodmin Moor, therefore, may well be the result of incentivised habitat enhancement schemes, allowing the species to expand into suitably managed areas.

Measures of abundance rely on consistent transect data. Regrettably, only one transect in Cornwall has recorded Marsh Fritillary (in very small numbers), and that has not been walked regularly, so we have no reliable data on which to assess abundance trends. Numbers in the ERICA database fluctuate annually, in large part a result of the amount of surveying focused on the species in any one year. What does seem to have changed significantly from when the 2003 Atlas (Wacher *et al.*, 2003) was published, is the resurgence of populations on The Lizard. Together with the metapopulations confirmed on Bodmin Moor, this might compensate numerically for the loss of sites in West Penwith and on the Mid Cornwall Moors.

Draining wetlands, ploughing fields, fertilising meadows and over-grazing have all contributed to the decline of this species. In Britain, as in much of Europe, it has not been profitable for farmers to support non-intensive grazing regimes, so either grasslands are over-grazed, often with the addition of fertilisers, to the extent that the larval foodplant is lost or is present in such low quantities that it can no longer support colonies, or they are simply abandoned and left to become overgrown with scrub.

Although the main metapopulations of Marsh Fritillary in Cornwall are looking reasonably healthy at the moment, many of the remaining pockets of suitable habitat are now so small and fragmented that the chances of the butterfly's long-term survival away from the main colonies is slight.

Conservation

Conservation must be aimed at the maintenance of suitable habitats, on a scale that can sustain metapopulations. This can best be achieved by financially incentivised management agreements with landowners to prevent over-grazing, under-grazing or draining of land, tailoring the land-management regimes to deliver the specific habitat requirements of the species. Recent surveys focused on Bodmin Moor and The Lizard have yielded crucially important, site-specific data. This information means that resources can be accurately targeted on habitat management and the recruitment of landowners as partners to help conserve the Marsh Fritillary in its Cornish strongholds. Many sites that contain populations of the butterfly or suitable habitat within colonisation distance of known colonies now have some sort of statutory designation protecting them, such as being a SSSI. However, there are still several areas that do not, such as the vast majority of the south of Bodmin Moor. Ensuring that these vital habitats are awarded protection will be key to further safeguarding the Marsh Fritillary's future in Cornwall.

Best places to see
- Higher Predannack Cliff SW661164 and Mullion Cliff SW664172
- Breney Common SX054608
- Carkeet SX216729
- Kerrow SX109744

Upperside, male and female similar

Heath Fritillary
Melitaea athalia

Sponsored by
Tamara Allardes, Mark Wash
and in memory of Ian Pryor

The Heath Fritillary is today found at only one site in Cornwall, where it was reintroduced and persists because of careful habitat management. It now occupies only four areas in the UK – the Cornwall/Devon border, Exmoor, Kent and Essex – and is thus one of the UK's rarest butterflies. The species has been present in Cornwall since at least the 1850s, although for a few years, just after the publication of the 2003 Atlas (Wacher *et al.*, 2003), its future was by no means assured. Male and female are similar, but the female tends to be a little larger and paler. The upperside of the wings has orange and dark brown chequered patterning, with a fringed border of white hairs, but patterns can be variable, as can the amount of dark colouring, occasional aberrations showing very little orange at all. The underside of the forewings is predominantly light orange with variable small black markings. The underside of the hindwings, by contrast, which is normally

Resident in Cornwall
(not on the Isles of Scilly)

CONSERVATION STATUS
- Fully Protected under Schedule 5 of the Wildlife and Countryside Act 1981 (as amended)
- Section 41 Species of Principal Importance (NERC Act 2006)
- Red List of British Butterflies; IUCN criteria: Endangered (EN)
- Butterfly Conservation priority: High

TRENDS AND DISTRIBUTION
UK rate of change in abundance
 ↓ 91% 1981–2018
South-west England rate of change in abundance
 ↓ 31% 2009–2018
 ↓ 94% 1981–2018
% of 1km squares occupied in Cornwall
 <1% 2009–2018 (1% 10km squares)
 <1% 1976–2008 (3% 10km squares)

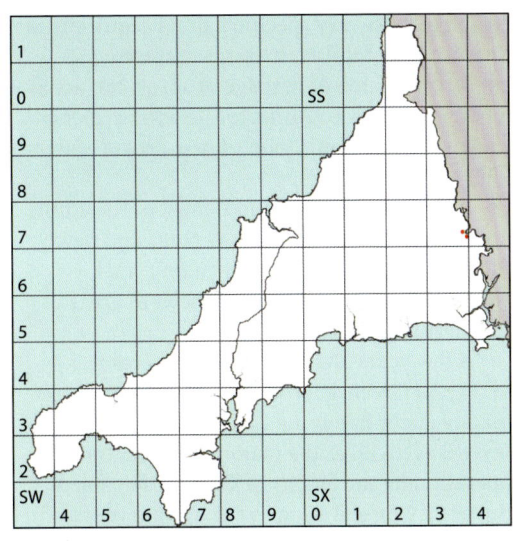

Heath Fritillary

Distribution 2009–2018

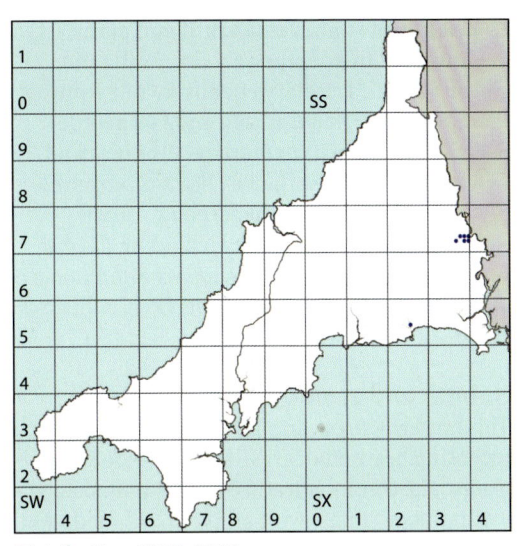

Distribution 1976–2008

all that is seen when the wings are closed, has a bold pattern of orange and white cells.

Distribution in Cornwall

The Heath Fritillary is now restricted to a single location, Greenscoombe Wood (sometimes known as Luckett Wood), a small area of woodland and meadow on two steep-sided valleys near the village of Luckett, in the Tamar Valley, close to the Devon border. The area is designated a Site of Special Scientific Interest (SSSI) for Heath Fritillary and is owned and managed by the Duchy of Cornwall, which receives woodland management advice from Butterfly Conservation to conserve the species.

Underside, male and female similar

The management programme has been financed by Natural England in partnership with the Duchy of Cornwall. In 2002, the Greenscoombe Wood population became extinct, probably because of shading by conifers. This setback was temporary, and the present population is the result of a reintroduction four years later, the continuing success of which depends on careful habitat management and monitoring. During the four years when no Heath Fritillary were recorded at Greenscoombe, the only remaining colony in Cornwall was a short distance away, to the west of Luckett, at Deer Park Wood; this colony became extinct in 2008.

There are few historical records in other areas of Cornwall to supplement those from the Tamar Valley, but the Heath Fritillary is thought to have become extinct in these other areas by the beginning of the twenty-first century. Smith (1997) makes reference to a few small colonies in river valleys near Looe that had disappeared by the 1960s apart from a single record in the 1990s. Penhallurick (1996) also mentions the existence of colonies present near Herodsfoot and Callington that survived up to the 1960s, as well as a record of the species in the Falmouth area in 1850.

Larval foodplants

Common Cow-wheat *Melampyrum pratense* is often said to be the favoured foodplant, and it does grow around Luckett, although it was almost absent from Deer Park Wood at the beginning of this century. In Cornwall (and Devon), Heath Fritillary equally favour Ribwort Plantain *Plantago lanceolata* and Germander Speedwell *Veronica chamaedrys* (Thomas and Lewington, 2014). Foxglove *Digitalis purpurea*, Ivy-leaved Speedwell *V. hederifolia*, Yarrow *Achillea millefolium*, Lesser Celandine *Ficaria verna* and Wood Sage *Teucrium scorodonia* have also been recorded as foodplants (Penhallurick, 1996).

Habitat and ecology

This butterfly uses various habitat types, including heathland, grassland and woodland glades, the common feature being an unshaded warm microclimate with ample larval foodplant. In recent years, the requirements of the Heath Fritillary have become better understood. This species is more widespread in warmer parts of the continent; in Britain the Heath Fritillary's need for heat means that it is extremely intolerant of any shading and therefore has far fewer areas of available habitat. A traditional name for the butterfly was 'woodman's follower' for its opportunistic habit of colonising freshly cleared areas after coppicing or tree felling. Well-drained, warm, sheltered, sunny woodland clearings with sparse vegetation but an abundance of foodplants provide an ideal habitat for the butterfly while the area is in the early stages of succession, but the understorey becomes too dense to support a colony as succession progresses, and the butterfly is forced to move on.

Heath Fritillary have been extensively studied, particularly by Butterfly Conservation, because of the threat of extinction in the UK. Although this species appears to be an accommodating butterfly in terms of its choice of foodplants, research by Martin Warren established its very specific habitat requirements and limited mobility. It seems that even an overgrown patch of herbage or scrub can act as a barrier, which accounts for its poor dispersal and prevents colonisation of new areas (Warren, 1987).

In Greenscoombe Wood, most of the adults are recorded in woodland clearings and newly managed coppice habitat and along track edges, with relatively few now seen in areas of unimproved grassland.

A few years after the butterfly became extinct at Greenscoombe Wood, extensive management involving consultation with several agencies and a major tree-clearance operation by the Duchy of Cornwall enabled Butterfly Conservation to carry out a successful licensed reintroduction. When the habitat was assessed to be suitable, stock was taken from Lydford in Devon to breed large numbers of the adults, which were released in several locations in the Greenscoombe area in 2006. The Lydford colony had originally come from Greenscoombe stock, so they had a common gene pool. It was reported that when the adults were released, the areas most used by the butterflies were the sunny parts of the rides and the places where deciduous trees had recently been felled or trimmed. Eggs were laid in these latter areas as well as the south-east-facing walls and banks, in clearings and at the sides of tracks. The females chose sunny areas with patches of bare ground and abundant foodplant but limited vegetation.

Prior to this introduction, Cornwall Butterfly Conservation volunteers carried out regular clearance work at the Deer Park Wood site, where the butterfly had been recorded over many years. In 2006, it was noted in the Cornwall Butterfly Observer that the habitat was 'looking good', with 25 adults seen on the annual

Flight curve based on number of records in Cornwall 1999–2018; and UK life cycle

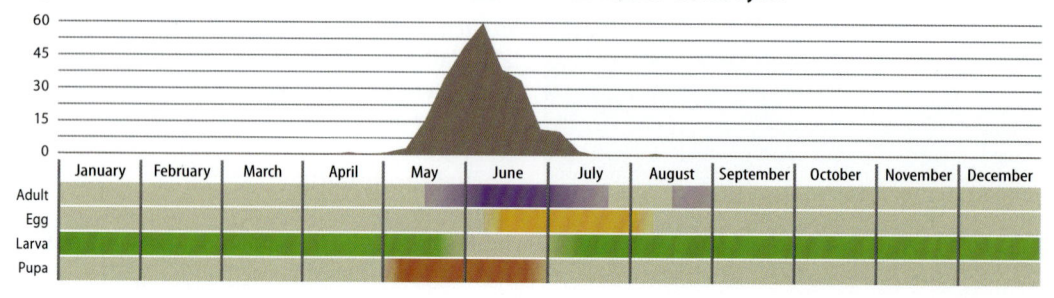

field trip. The butterfly was last recorded at this locality in 2008. Unfortunately, after a series of wet years it was thought to have become extinct due to extensive growth of Rosebay Willowherb *Chamaenerion angustifolium* and Bracken *Pteridium aquilinum*, which took over the cleared area. Fortunately, by then the Greenscoombe colonies had become well-established.

Important lessons have been learnt from both Deer Park Wood and Greenscoombe Wood about the importance of sustained and scientifically informed habitat management to maintain a viable population of Heath Fritillary. Despite the setbacks, this nationally rare butterfly has had an unbroken presence in Cornwall for at least the last 200 years, a tradition that Butterfly Conservation is determined to continue.

The Heath Fritillary is single-brooded, and in Cornwall the adult butterfly appears in the last two weeks of May or early June. The male is more conspicuous, patrolling close to the ground looking for females. Once fertilised, the female waits a few days before laying large batches of eggs in sunny spots on leaves or dead vegetation close to suitable larval foodplants. After hatching several weeks later, the larvae eat their eggshells and then congregate on a nearby foodplant, spinning a web within which they bask and feed. After the third moult the larvae become less gregarious and can be identified by their distinctive spiky two-toned appearance (said to resemble the flowers of a plantain, particularly when mature). In August, when the larva is in its fourth instar, it will start searching for a suitable place to overwinter, usually within a dead, rolled-up leaf held together by the web it spins. The fourth moult takes place within this safe environment, just before the larva emerges in March or April to start feeding again intermittently during warm spells; it is thought that the heat aids digestion. After two further moults, pupation occurs amongst leaf litter, the chrysalis being a very distinctive pearly white with black and white markings (Eeles, 2019). It is thought to be particularly susceptible to predation by small mammals at this stage (Warren, 1987).

Analysis of trends

The long-term data for the Heath Fritillary show a catastrophic decline in both abundance and occurrence in the UK. This decline has slowed down in South-west England over the last 10 years, but at the time of writing Cornwall presents a more optimistic picture. There is fluctuation in transect data, but it seems to show an upward trend. Even more encouraging, since the reintroduction at Greenscoombe Wood, where timed counts for the two main colonies have been carried out on a yearly basis by Butterfly Conservation, is that the estimated peak in abundance has increased by 418% over the period 2009–18 (Plackett, 2018). The year 2018, with its warm, sunny weather, was a recent high point for abundance, although counts in the following year were considerably down from the numbers recorded in preceding years, so there is no room for complacency.

Conservation

The Heath Fritillary requires constant vigilance in terms of its habitat requirements, as its decline is strongly linked to the cessation of traditional coppicing in woodland during the twentieth century (Asher *et al.*, 2001). The situation can change very quickly, as was illustrated at the Deer Park Wood site, which rapidly became overgrown. The most important requirement for the species is warmth, which means that cutting and coppicing regimes must be carefully planned and closely monitored. Extreme weather events, which are becoming more frequent with climate change, can result in waterlogging and rapid overgrowth. Butterfly Conservation keeps the habitat management regime under constant review in partnership with the Duchy of Cornwall and other bodies. In addition to rotational coppice management, necessary measures include the removal of Bracken and scrub barriers, keeping tracks open, management of unimproved grassland, and the creation of nectar-rich field margins. Ultimately, the aim is to increase connectivity between a number of new or historical sites in the Tamar Valley. The creation of a more extensive metapopulation would provide a more secure future for this butterfly in Cornwall.

Best places to see

- Greenscoombe Wood SX391724 (west colony) and SX392725 (east colony)

Upperside, male and female similar

Small Copper
Lycaena phlaeas

Sponsored by
*David Fryer, Kathy Hicks
and Alec Mackonochie*

At rest, it is hard to confuse the Small Copper with any other butterfly, because of its diminutive size and the fiery orange of its forewings. Male and female are similar, with the female slightly larger. Aberrations and variations are not uncommon, particularly in relation to the extent of the black markings. The upperside of the hindwings is a deep chocolate brown, with a strip of orange along the rear margin. The attractive aberrant form *caeruleopunctata*, which has a row of blue spots just inwards from the orange marginal band on the hindwing, is quite frequently reported in Cornwall. On the underside, which is only seen when the butterfly is at rest, the forewings are creamy orange with a sprinkling of black dots, whilst the hindwings are a light brown with fainter dots and a barely discernible orange border at the rear.

Distribution in Cornwall
This small, very territorial butterfly is one of the most widespread in Cornwall, occurring on all except the highest ground. The species prefers open areas, either flat or undulating, hill slopes, coastal dunes and former mining sites, where there are both flowers for nectar and open patches of ground suitable for basking. It usually occurs in low numbers,

Resident in Cornwall
(resident on the Isles of Scilly)

CONSERVATION STATUS
■ Butterfly Conservation priority: Low

TRENDS AND DISTRIBUTION
UK rate of change in abundance
↓ 42% 1976–2018
South-west England rate of change in abundance
↓ 43% 2009–2018
↓ 73% 1976–2018
% of 1km squares occupied in Cornwall
30% 2009–2018 (81% 10km squares)
21% 1976–2008 (82% 10km squares)

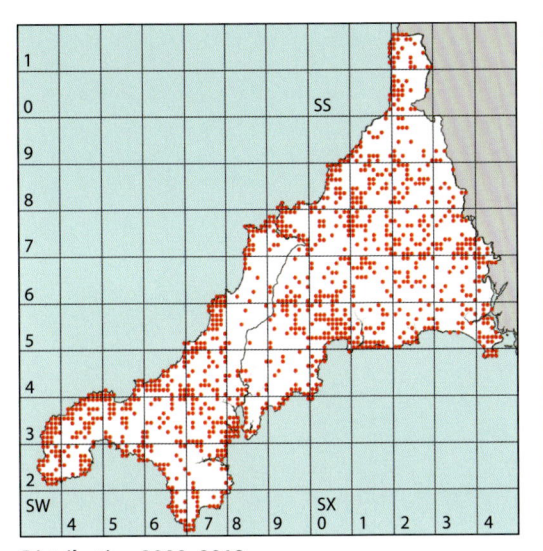

Small Copper

Distribution 2009–2018

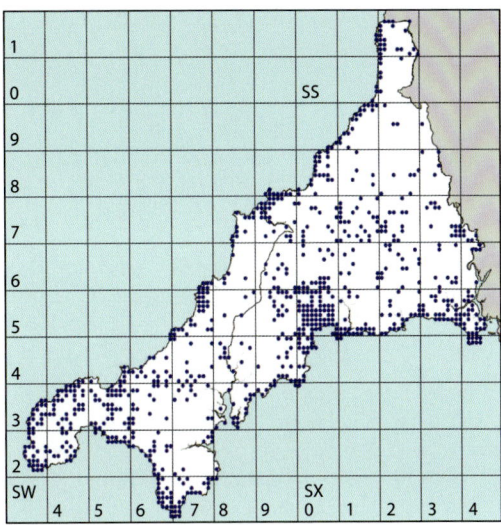

Distribution 1976–2008

although occasionally larger numbers have been recorded. The distribution maps show a widening of the butterfly's range in recent years, although this could be explained by the increasing number of recorders covering further areas of the County.

Larval foodplants
The favoured larval foodplants are Common Sorrel *Rumex acetosa* and Sheep's Sorrel *R. acetosella*, both abundant in Cornwall, although Broad-leaved Dock *R. obtusifolius* is occasionally used

Habitat and ecology
The adults seek out sheltered, wind-free, dry and warm corners in woodland clearings, on coastal sand dunes, on former mining sites, on hill slopes, on roadside verges and railway embankments, and even in gardens. In these locations the butterfly will bask on a stone, bare ground or low-growing flowers and vegetation, the male fending off any species of intrusive butterfly it encounters. Where there are ample larval food-plants, Small Copper tend to form small colonies, although the mobility of the butterfly means that they can be seen almost anywhere. The female lays single eggs, usually on the upperside of leaves, resembling tiny white golf balls when viewed through a hand lens. Once hatched, the caterpillars feed on the leaves, creating a distinctive 'windowpane' effect. The colour of the caterpillars is variable, some developing deep-pink markings, whilst others are consistently green in the later instars. October is normally when the larvae enter hibernation, although they will still emerge to feed during winter months when the weather is warm. In common with their blue lycaenid

Underside, male and female similar

Aberration *caeruleopunctata*

cousins, "It is thought that pupae are tended by ants among leaf litter but little is known about the pupal stage in the wild" (Asher *et al.*, 2001: 138). The pupal stage lasts for three to four weeks.

With its relatively short life cycle, the Small Copper usually produces three broods in Cornwall, and perhaps even a fourth in favourable summers, provided that drought has not withered the larval foodplants. Consequently, it can be seen in the County in almost any month from late March to mid-November, the later broods overlapping. In general, however, adults do not appear in good numbers until mid-April, and there is usually a gap in mid-June when few can be seen. Numbers are usually at their peak in July and August.

Analysis of trends

Populations fluctuate according to the weather. Cool, wet summers cause colonies in shadier, woodland areas to slump, whilst hot, dry summers can affect the larval foodplants, with a subsequent decline in later broods. In Cornwall

Egg

Flight curve based on number of records in Cornwall 1999–2018; and UK life cycle

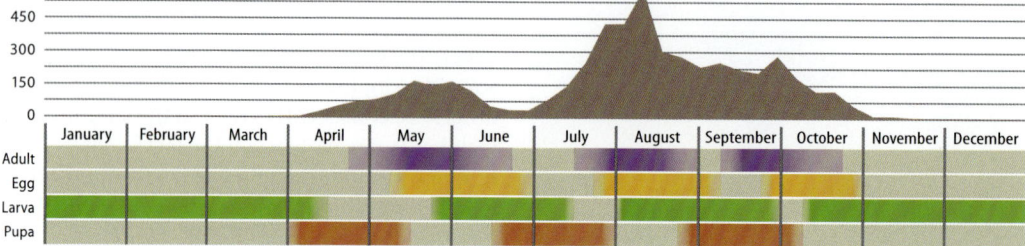

Coast near Pordenack Point, West Penwith

there was a slight decline in abundance over the period 2009–2018, based on the ratio of number of individuals seen per record. However, the unusually fine summer of 2018 proved to be the best year in recent times for both number of records and number of individuals, and this is also borne out by butterfly transect results. The national picture is more worrying. Butterfly Conservation was reported in the media as stating that 2015 was the worst year on record for Small Copper, and the overall national decline in abundance since 1976 is significant.

Conservation
No specific conservation recommendations for the Small Copper are necessary, although the species will obviously benefit from good conservation management, such as the establishment of wildlife areas on farms, clearance of scrub and rank vegetation, and the reduction of pesticide use.

Best places to see
Almost anywhere in Cornwall, particularly along the coast.

Upperside, male

Purple Hairstreak
Favonius quercus

Sponsored by
Sarah Board, Tegen Higgs
and Jenna Wearne

This small, elusive butterfly is always a treat to observe but is not easy to see, so it is reported infrequently in Cornwall. Success in pursuit of the Purple Hairstreak depends on knowing where and at what time of the day to look. The upperside of the male's wings have a bluish-purple sheen all over, whilst the female boasts purple patches near the body on each forewing. The underside of both sexes is similar, being pale grey with a fine white line running parallel to the outer margins of both forewings and hindwings, and an orange eyespot near the short tail. When flying in numbers during the evening, they resemble silver coins tossed in the air.

Distribution in Cornwall
The Purple Hairstreak is one of the few butterflies in Cornwall whose distribution has substantially reduced in recent years despite an increase in recorders. This trend is difficult to interpret, but under-recording probably plays a large part, as the butterfly is rarely noted by inexperienced recorders. Observations in recent years back this up, the butterfly being recorded in several new areas where it had never been recorded before. However, loss of woodland and hedgerows containing suitable oak *Quercus*

Resident in Cornwall
(not on the Isles of Scilly)

CONSERVATION STATUS
■ Butterfly Conservation priority: Low

TRENDS AND DISTRIBUTION
UK rate of change in abundance
↓ 56% 1976–2018
South-west England rate of change in abundance
↑ 95% 2009–2018
↓ 95% 1976–2018
% of 1km squares occupied in Cornwall
1% 2009–2018 (39% 10km squares)
4% 1976–2008 (52% 10km squares)

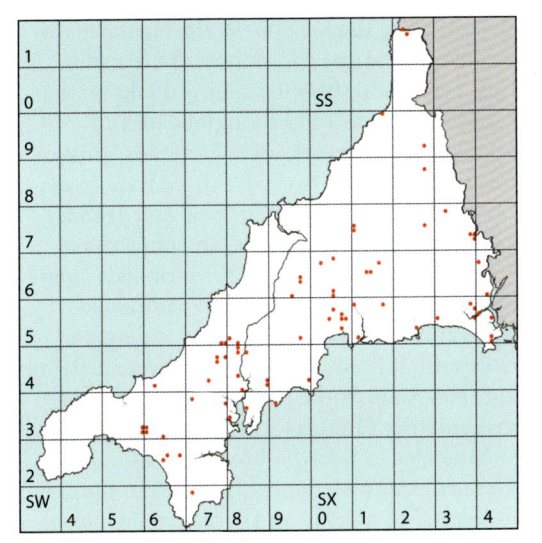

Distribution 2009–2018

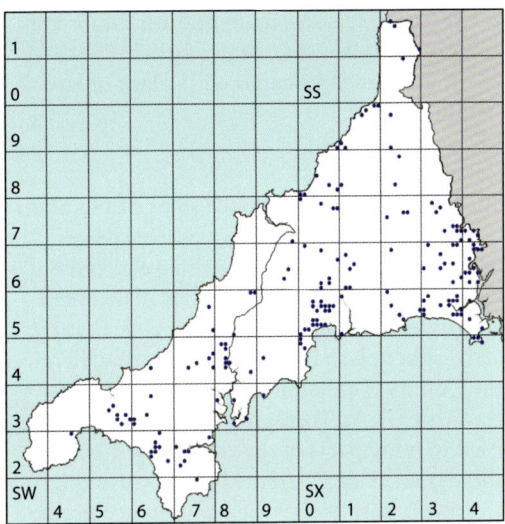

Distribution 1976–2008

spp. trees may also be a contributory factor to the decline of the species.

One of the best places to see Purple Hairstreak on a warm, still summer's evening is in the woodland of the National Trust's Godolphin Estate in west Cornwall. At that time of day they are more likely to congregate at low level with their wings open, basking in the last rays of the setting sun. However, due to the low elevation of many of the oaks on this site, observations can sometimes be made at close quarters even earlier in the day. In July 1997, over 90 adults were seen flying simultaneously around a single tree. But, in common with many other species, the butterfly is subject to annual fluctuations. In the following year, only a few adults were seen on that same tree, even though it was observed on several occasions. Such large numbers have not been recorded since.

Underside, male and female similar

The species is scarce on Bodmin Moor and areas of West Penwith, probably because these are more open landscapes with a lack of larval foodplants.

Larval foodplants

The leaves of oak are the sole source of food for the larvae, primarily the two native species, Pedunculate Oak *Quercus robur* and Sessile Oak *Q. petraea*, which are widely distributed across most of Cornwall. Turkey Oak *Q. cerris* is thought to be a less favoured alternative in the County (Penhallurick, 1996). There are records of Holm (Evergreen) Oak *Q. ilex* being used in other parts of the country, although there is no evidence of the butterfly using it in Cornwall, despite the tree's abundance in the County (Eeles, 2019).

Habitat and ecology

The Purple Hairstreak primarily inhabits woodlands, provided that they contain oak, but the butterflies can also be found on solitary trees, such as those growing in hedges or in the centre of a field; lone trees are capable of supporting a colony. The adults can also frequent other trees that are not larval foodplants, such as Beech *Fagus sylvatica* and more particularly Ash *Fraxinus excelsior*, where they enjoy feeding on the honeydew provided by aphids. Likewise they have been seen clustering around a small Alder Buckthorn *Frangula alnus* situated among oaks at Godolphin and nectaring on its flowers; they also nectar on a variety of other flowers, including Bramble *Rubus fruticosus* agg. In years with hot weather or drought, honeydew may be in short supply and the butterfly is more likely to come down from the treetops to seek other nectar sources or minerals and moisture from the ground (Eeles, 2019). They seldom fly during the main part of the day and spend most of their adult lives perched in sheltered, sunny nooks on the canopy (Thomas and Lewington, 2014).

As is the case with other lycaenids, ants are involved in the life cycle of Purple Hairstreak. Both larvae and pupae can 'sing' and produce substances that attract ants, and pupae have been found in ants' nests. *Myrmica scabrinodis* and *M. ruginodis* are known to be suitable hosts and both are common and widespread in Cornwall. It is not yet known, however, whether ants play an essential part in the life cycle of this butterfly (Thomas and Lewington, 2014).

The species is single-brooded. The adult flies in Cornwall from the last week of June to the end of August, and occasionally into September, with a peak in the middle of this period. Mating usually takes place in the evening. The male perches on higher branches of an oak, on the lookout for a passing female, at the same time skirmishing with any rival. Courtship is perfunctory. Eggs, which resemble tiny sea urchins, are laid at the base of a leaf bud or on an adjoining twig on sheltered, south-facing, usually gnarled branches of older oak trees, which contain the biggest buds (Eeles, 2019). The larva is fully developed within the egg after about three weeks but will remain in that state for about eight months. It emerges the following spring as the leaf buds start to open, when it burrows into a developing bud, feeding on the young leaves as they unfurl. After the first moult the woodlouse-shaped larva spins itself a loose web, which catches debris and provides it with some protection when resting beneath it during the day. The larva moves out to feed at night. Its mottled colour provides effective camouflage on oak branches, helping to protect it from predators.

Flight curve based on number of records in Cornwall 1999–2018; and UK life cycle

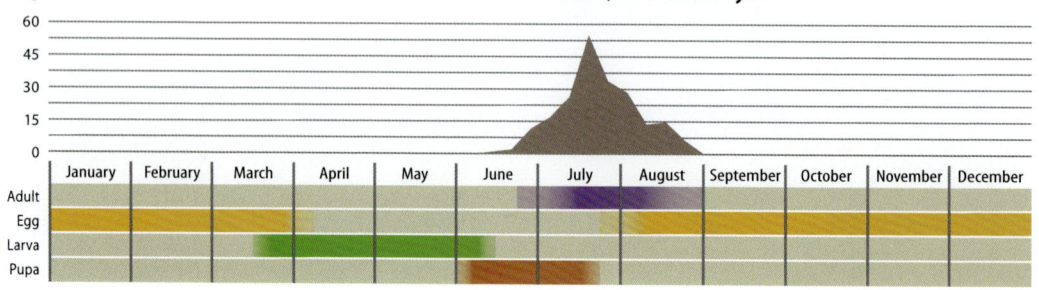

Breney Common, one of the best places to see Purple Hairstreak

It pupates at the end of May or the beginning of June, the mottled brown pupa forming on the ground, under soil or vegetation, or in an ants' nest. The adult emerges three to four weeks later and can be found on grass stems as it expands its wings beneath the oak canopy (Eeles, 2019).

Analysis of trends
Abundance trends from South-west England appear contradictory: the steep decline over four decades reversed with a spectacular recovery over the last 10 years, a trend that does not tally with the Cornish data for 2009–2018. The number and distribution of records in Cornwall have decreased. Unfortunately there are very few transects in Cornwall where the butterfly is present, and as transects are mostly walked in daytime the actual numbers are probably underestimated. Over 1,000 were reported from the Marsland Nature Reserve on a single day in 2013. Lower but reasonable numbers have been recorded at a few other locations, but most records come from just a few isolated colonies where numbers are generally small. It has already been mentioned that this butterfly is probably under-recorded, so it is important for visits to both known and likely sites to be made at the right time of day and in the right weather, as Purple Hairstreak are so well hidden in the oak canopy for most of the day. The best time to see this butterfly is around 7pm on a warm, still summer's day, when they are most active.

Conservation
The butterfly depends on the presence of oaks of varying ages, in woodland, to provide an undulating canopy, and with sufficient clearings to maintain warm, sunny lower branches. Loss of oak trees and fragmentation of wooded landscapes are the chief threats, although the latter is perhaps not quite so serious, given the butterfly's habit of colonising individual trees (Spalding with Bourn, 2000).

Best places to see
- Godolphin Woods SW600323
- Cabilla Wood SX134653
- Marsland Nature Reserve SS2316
- Breney Common SX055610
- St Allen SW824505

Underside, male and female similar

Green Hairstreak
Callophrys rubi

Sponsored by
*Poppy Besterman, Jen Bousfield
and in memory of Delcie Burgoyne*

With its unmistakable iridescent wing colour, the Green Hairstreak is the UK's only truly green butterfly. The upperside is a dull brown colour, seen only in flight, when it can be mistaken for a moth. It always rests or basks with its wings closed and tilted to optimise body temperature. The vivid green underside of the wings has a line of small white spots, although these can sometimes be absent or very lightly marked. Despite its startling colour, it is perfectly camouflaged on green vegetation, such as gorse *Ulex* spp., so it is probably one of our most under-recorded butterflies. Male and female are similar in appearance, but the female is very slightly larger.

Distribution in Cornwall

The Green Hairstreak is widespread both nationally and in Cornwall, and at one time or another it has been recorded in the majority of Cornwall's 10km squares. However, there

Resident in Cornwall
(not on the Isles of Scilly)

CONSERVATION STATUS
■ Butterfly Conservation priority: Medium

TRENDS AND DISTRIBUTION
UK rate of change in abundance
↓ 45% 1976–2018
South-west England rate of change in abundance
↓ 13% 2009–2018
↓ 73% 1976–2018
% of 1km squares occupied in Cornwall
3% 2009–2018 (56% 10km squares)
6% 1976–2008 (71% 10km squares)

is a decline in the number of locations where it has been spotted in recent years, compared with records dating back to 1976, despite the increase in recorders. This is in line with a 30% decline in occurrence nationally between 1976 and 2014 (Fox *et al.*, 2015). Although the butterfly is seen mainly in ones or twos, over 20

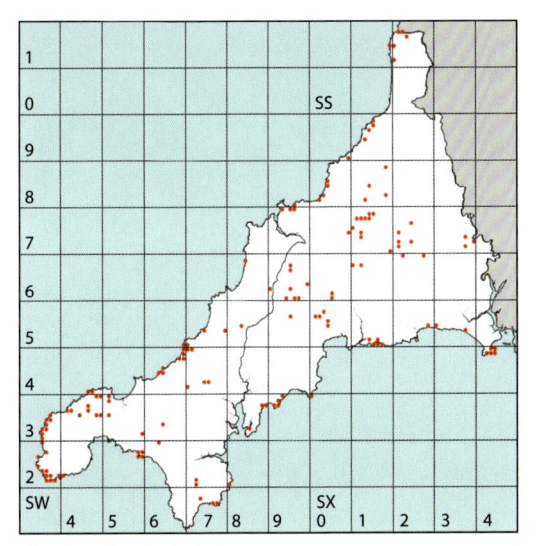

Green Hairstreak

Distribution 2009–2018

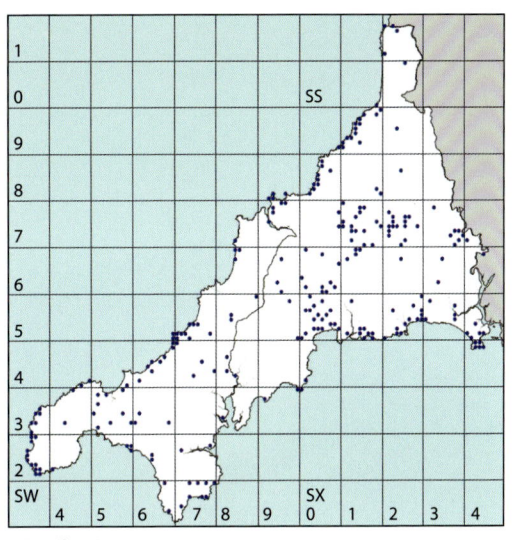

Distribution 1976–2008

at a time have been counted in a few locations on several occasions in the last 10 years. These include sections of the West Penwith coast, such as the Levant to Botallack old mining area and further west at Porthgwarra. On the north coast, the Trevaunance Cove to Chapel Porth coastal path near St Agnes is also a good place to spot relatively large numbers, and occasionally surprising numbers have been seen on Bodmin Moor.

Larval foodplants

Green Hairstreak caterpillars feed on the widest range of foodplants of any British butterfly. These include gorse, Common Bird's-foot-trefoil *Lotus corniculatus*, Bramble *Rubus fruticosus* agg., Ling/Heather *Calluna vulgaris*, Dyer's Greenweed *Genista tinctoria* and Petty Whin *G. anglica* (largely restricted to The Lizard in Cornwall), Broom *Cytisus scoparius*, Bilberry *Vaccinium myrtillus*, Dogwood

Mating pair

Resting with wings tilted to optimise body temperature

Cornus sanguinea, Cross-leaved Heath *Erica tetralix* and Alder Buckthorn *Frangula alnus*.

Habitat and ecology

This species inhabits unimproved land where there is usually some scrub cover, such as Blackthorn *Prunus spinosa*, gorse and Broom, willow *Salix* spp., Holly *Ilex aquifolium* and Hawthorn *Crataegus monogyna*. Consequently it is most likely to be seen on coastal grassland and heaths, in inland river valleys where steep ground has prevented cultivation, in woodland rides, or along old railway lines, footpaths and bridleways.

In Cornwall Green Hairstreak are usually first seen in April and fly into July, although they have rarely been recorded as late as September. There is only one brood. The male tends to be more conspicuous than the female and can sometimes be seen flying above the tops of bushes fending off rivals. The female spends her time resting or looking for suitable young and succulent shoots on which to lay her eggs. The larva is woodlouse-shaped, pale brown initially and becoming bright green by the third instar.

In common with other lycaenids, the Green Hairstreak enjoys a symbiotic relationship with ants. Secretions from the hairy cuticle of the pupa are highly attractive to ants, which are known to bury the pupa deep inside their nest during hibernation. In addition, the pupa emits sounds that are attractive to ants and are so loud that they are audible to the human ear (Eeles, 2019).

Analysis of trends

Although under-recording may play a part in the number of squares where this butterfly has been seen in Cornwall during recent years, it is possible that there may be a genuine loss of range. This reflects the UK picture, which, in addition to indicating a drop in occurrence, shows a significant decline in abundance of 45% between 1976 and 2018, with an even

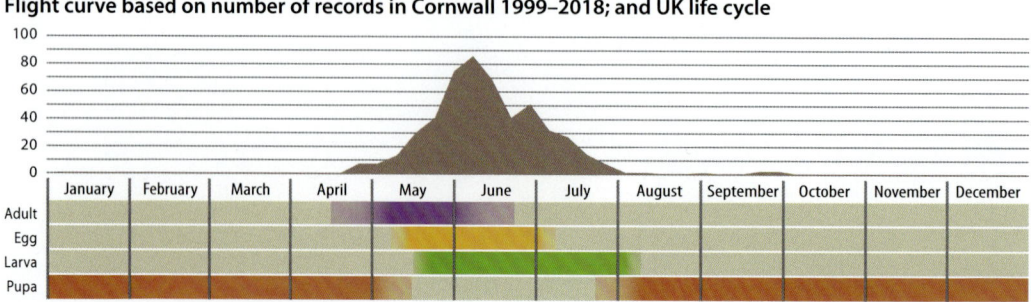

Flight curve based on number of records in Cornwall 1999–2018; and UK life cycle

Chapel Porth, one of the best places to see Green Hairstreak

greater decline in South-west England, although this has slowed in the last 10 years.

In Cornwall, an increase in the number of transects where the Green Hairstreak is present has contributed to boosting overall numbers although yearly fluctuations on these transects are considerable. South-west England shows a major long term downward trend which has continued, to a lesser extent, in the last 10 years. In Cornwall, however, the number of both records and individuals counted in the good summer of 2018 were the best in at least a decade, showing the important influence of weather and how difficult it is to draw conclusions about how this particular species is faring.

Nationally, it is thought that destruction of habitat is the primary reason contributing to the decline of the Green Hairstreak (Asher et al., 2001), although the Cornish populations may not have been so much affected, since most of its known habitats are unsuitable for agricultural 'improvement' or development.

An additional reason for localism (and perhaps decline) could be the butterfly's "need for an abundance of succulent foodplants and high densities of ants, possibly of particular species" (Thomas and Lewington, 2014: 84).

Conservation

Maintenance of habitat is the foremost consideration. The most serious threat to the Green Hairstreak is probably destruction caused by development for houses, caravan sites or industry. Moreover, the uncontrolled clearance of scrub in so-called tidying-up operations at former mine sites or along footpaths and bridleways is likely to hasten the decline of this butterfly in Cornwall.

Best places to see
- Levant Zawn SW365342
- Roskestal West Cliff (Porthgwarra) SW366220
- Trevaunance to Chapel Porth SW698498
- Breney Common SX055610

Butterflies of Cornwall — Lycaenidae

Underside, male and female similar

White-letter Hairstreak
Satyrium w-album

Sponsored by
Dr Richard Ashford, Jamie Burston
and Jo Poland

After an almost 30-year absence of records in Cornwall, this rare native species nearly didn't make it into this chapter. Nonetheless, and against all the odds, a handful of sightings in the last few years show that the White-letter Hairstreak is still present in Cornwall. The fate of this elusive little butterfly in the UK is inextricably linked to the demise of the elm *Ulmus* spp., stricken by Dutch elm disease since the 1970s.

Hairstreak species are generally under-recorded, and the White-letter Hairstreak is harder to spot than most because of its dun colour and its habit of flying high up in the tree canopy. The sooty-brown uppersides of the wings are usually only seen as small, spiralling silhouettes in the distance, when the butterfly is in flight. The undersides are also brown, and the hindwing is inscribed with a thin white zigzag line resembling the letter W, giving the species its name. There is also a wavy reddish-

Resident in Cornwall
(not on the Isles of Scilly)

CONSERVATION STATUS
- Schedule 5 Wildlife and Countryside Act 1981 (as amended) (for sale only)
- Section 41 Species of Principal Importance (NERC Act 2006)
- Red List of British Butterflies; IUCN criteria: Endangered (EN)
- Butterfly Conservation priority: High

TRENDS AND DISTRIBUTION
UK rate of change in abundance
↓ 92% 1979–2018
South-west England rate of change in abundance
↓ 38% 2009–2018
No data available 1976–2018
% of 1km squares occupied in Cornwall
<1% 2009–2018 (5% 10km squares)
<1% 1976–2008 (9% 10km squares)

White-letter Hairstreak

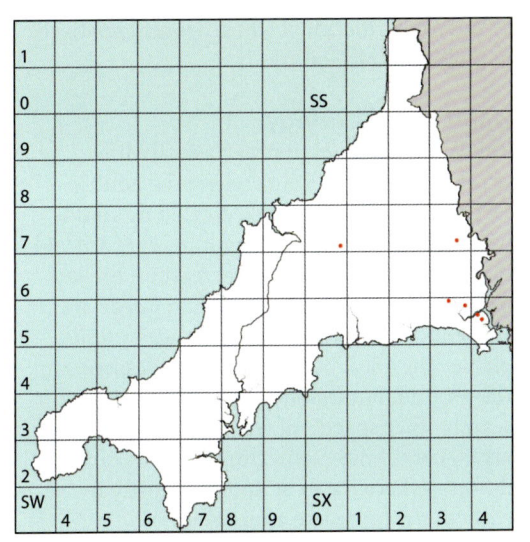

Distribution 2009–2018

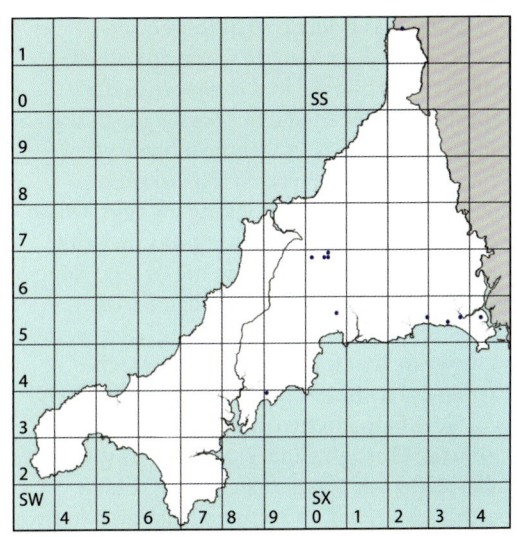

Distribution 1976–2008

orange band edged in black on the outer margin of the hindwing, which has a small 'tail'. If an observer is lucky enough to catch a rare glimpse of a White-letter Hairstreak nectaring on a flower away from the treetops, it is solely the underside that is seen, as this butterfly invariably perches with its wings closed, angling them towards the sun.

Although their respective undersides are quite different, there could be confusion with the Purple Hairstreak when seen from a distance, as the two butterflies are the same shape and not vastly different in size. But a glimpse of the unmistakable bluish-purple sheen of the Purple Hairstreak's open wings immediately excludes the White-letter Hairstreak. The trees on which the two species depend are also different: Purple Hairstreak need oak *Quercus* spp., and White-letter Hairstreak need elm. Male and female White-letter Hairstreak are similar, with only slight differences in size and colour.

Distribution in Cornwall

Frohawk (1934) stated that the White-letter Hairstreak was absent from Cornwall, and the first records were not authenticated until 1945, at Pendennis, Falmouth. In the years following the first sighting, it was recorded in several different areas of the County, notably at Torpoint, where 30 butterflies were reported on a Wych Elm *Ulmus glabra* in 1976, with only a further 19 records reported for the remainder of the century. A single butterfly was seen nectaring on Hemp-agrimony *Eupatorium cannabinum* at Ponts Mill, in 1985. It was recorded about 300m away from the nearest mature elm, which died soon afterwards. This report aroused great interest, as there had been no previous records from the Luxulyan Valley, and there have been none since. The last record at the Marsland Nature Reserve was in 1986, after five consecutive years of sightings; the species was not reported in Cornwall again in the twentieth century.

Although Dutch elm disease is responsible for the decline of the White-letter Hairstreak more widely in the UK, this species has always been scarce in Cornwall. The known colonies were confined to the Roseland peninsula, the Marsland Nature Reserve, the far south-east of the County, and the west side of Bodmin Moor. This last colony was assumed to be extinct, since no adults had been seen since the 1970s (Penhallurick, 1996; Smith, 1997). However, in 2016 a record was received from a Butterfly Conservation specialist, who reported an adult in Little Shell Wood, just a few miles away from Dunmere Wood, where the butterfly had last been seen 40 years previously. The lesson is not to give up on this butterfly, but to keep searching, particularly in places where it has been previously reported, regardless of the gap in time.

So, there is evidence that this elusive butterfly is still resident in Cornwall, albeit probably unobserved and therefore under-reported. A concerted effort was made to search for the species in 2010 and the following few years by Cornwall Butterfly Conservation volunteers, particularly looking for eggs, which are easier to find than the adults, as they can be spotted when the trees are not in leaf. None were found, despite a prize being offered! However, there were two important discoveries in the years that followed. During searches carried out by Butterfly Conservation in 2013, eggs were found in two places in south-east Cornwall (Torpoint and Tideford). This was followed by two separate sightings of adult butterflies in 2014, in a garden belonging to two longstanding Butterfly Conservation members. They had a single Wych Elm in Downgate, near Callington, which unfortunately died in 2020, but not before adults were seen twice again in their garden, in 2019. The only other adults seen in recent times were on the Antony House transect near Torpoint, where one was recorded in 2018 (having been recorded there in the past), and at Trevollard, near St Germans, in 2017.

These recent sightings show that colonies persist in secluded Cornish locations, albeit in very small numbers.

Larval foodplants

The White-letter Hairstreak was once thought to breed solely on mature flowering elm species, of which Wych Elm and English Elm *Ulmus procera* are the most favoured. Since the loss of so many elms to Dutch elm disease, the butterfly has adapted to this depleted environment and occasionally uses the regenerating suckers from the roots of stricken trees, and it has been seen in other parts of the country breeding on younger elms in hedges and copses. It also sometimes uses less favoured species, although in Cornwall the butterfly does not appear to have much liking for Cornish Elm *U. minor* subsp. *cornubiensis*, Dutch Elm *U.* × *hollandica* or other hybrids, all of which are still relatively common in the County.

Habitat and ecology

The White-letter Hairstreak's entire life revolves around the elm, where the adult feeds largely on honeydew created by aphids, mating takes place in the canopy, eggs are laid on the branches, caterpillars feast on the leaf buds, flower buds and leaves, and pupae are suspended from the leaves or twigs by a silk girdle. The adult butterfly seldom descends to the ground, but if honeydew is scarce it will search other species of tree or, in less wooded areas, may choose to nectar on a range of flowers, where it can be spotted mostly in the early morning or late afternoon.

The species is generally on the wing from late June to mid-August. The distinctive eggs, which are shaped like miniature flying saucers, are laid singly on the rough bark of twigs in sunny sheltered locations, usually beneath flower buds, where new growth joins the previous year's growth. The eggs start off green with a whitish dome but soon turn brown. Some of them are laid on lower boughs, and this makes them easier to find after the leaves have dropped. In the early stages, they are often infested by a parasitic wasp, but by mid-August the caterpillar is fully developed within the egg and is less vulnerable. The egg remains in this state until March, enabling searches for the species to be carried out during the winter.

The first instar larvae were known traditionally to feed exclusively on flower buds and developing seeds. As they matured, the larvae gravitated to leaf buds, and then to leaves, which are reduced to a characteristic diamond shape by their feeding pattern. As they develop, the larvae change colour to suit their microhabitat, the green fully grown larvae perfectly blending with unopened leaves, whilst the pupae resemble dead leaves. Recent evidence suggests that the newly hatched larvae now feed on leaf buds, and this is thought to

UK life cycle (too few Cornish records to include a flight curve)

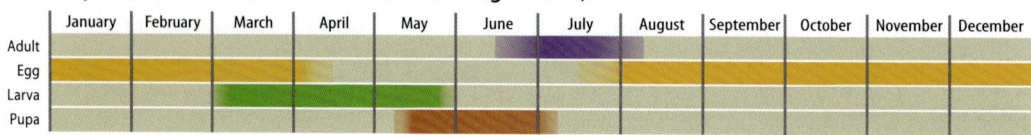

	January	February	March	April	May	June	July	August	September	October	November	December
Adult												
Egg												
Larva												
Pupa												

Antony House, Torpoint, where the White-letter Hairstreak was recorded in 2018

be an adaptive response to the lack of mature, flowering elms available. Elm suckers or young saplings that do not flower, but come into leaf earlier than mature trees, may be used if they synchronise with the butterfly's life cycle.

Analysis of trends

UK trends for this species show a 92% decline in abundance since 1976, and a 45% decrease in occurrence (1976–2014) (Fox *et al.*, 2015). In South-west England the butterfly has declined in abundance by 38% in the last 10 years.

On the available evidence, the White-letter Hairstreak was scarce in Cornwall even before the 1970s, when Dutch elm disease wiped out most of the trees on which it depended, because its favoured foodplant, Wych Elm, has always been rare, although widespread in the County. Nonetheless, the disease had an impact on the Cornish population of the butterfly, resulting in the complete absence of records over three decades. However, recent sightings of both eggs and adults prove that at least some remnants of the population survive, albeit undetected for a time and still vulnerable to the continuing ravages of Dutch elm disease. Recent records provide good reason to look for more colonies of White-letter Hairstreak in Cornwall.

Conservation

Limiting flail-cutting and the clearance of regenerating sucker growth, as well as young elm saplings, from hedgerows and elsewhere is particularly important for the future of this butterfly in Cornwall.

Research to find out whether White-letter Hairstreak may use Cornish Elm or Dutch Elm would also be beneficial.

Best places to see

Any mature Wych Elms are worth searching for, particularly where they are commonly found in the tributary valleys of the Helford, Camel and Tamar rivers (French, 2020).

Upperside, male

Holly Blue
Celastrina argiolus

Sponsored by
Robert Gunlack, Hannah Ben Matthams
and Leon Truscott

This small, bright blue butterfly is the first of the blues to emerge in the Cornish spring. It is a regular inhabitant of urban green spaces and is therefore seen in small numbers throughout most of England. This contrasts with other species of blue butterfly, which are more commonly found in rural areas. The Holly Blue is easily distinguished from the other blues found in Cornwall, exhibiting a peppering of black spots on the underside of its wings and no trace of the orange markings found on Common Blue *Polyommatus icarus* and Silver-studded Blue. It also tends to fly and perch around head height on bushes and trees, and this helps to distinguish it from the lower-flying blues, although Common Blue can also be found in gardens and are capable of quite powerful flight. The upperside of the wings of the Holly Blue is rarely seen when at rest, but the female's, unlike those of the male, have black borders, which tend to be broader in the second brood. This also distinguishes it from the Common Blue, which lacks a distinctive black border. Holly Blue in mainland Britain belong to the subspecies *britanna*.

Resident in Cornwall
(resident on the Isles of Scilly)

CONSERVATION STATUS
■ Butterfly Conservation priority: Low

TRENDS AND DISTRIBUTION
UK rate of change in abundance
↑ 40% 1976–2018
South-west England rate of change in abundance
↑ 135% 2009–2018
↑ 90% 1976–2018
% of 1km squares occupied in Cornwall
19% 2009–2018 (77% 10km squares)
18% 1976–2008 (75% 10km squares)

Distribution in Cornwall
The maps show that the geographical distribution and abundance of Holly Blue

Holly Blue

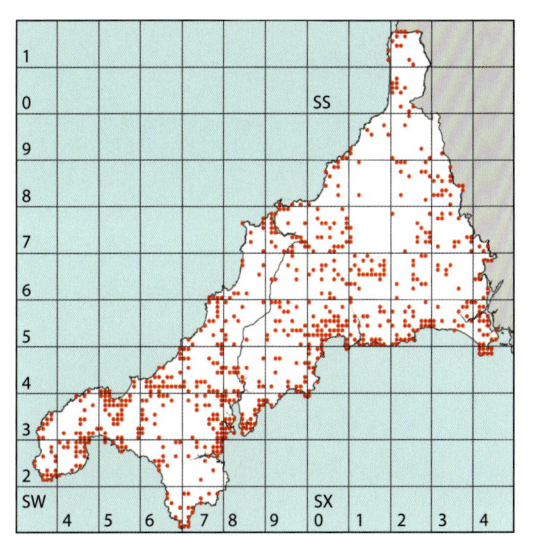

Distribution 2009–2018

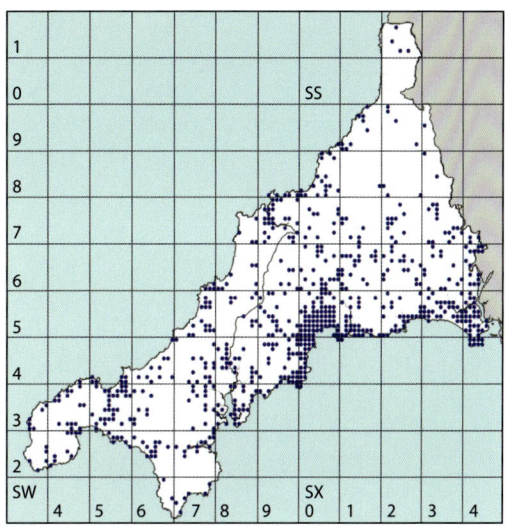

Distribution 1976–2008

have changed little when the last 10-year period is compared with the previous 32-year period. The species is widespread in both time periods, albeit that the distribution differs slightly. This can be explained, at least in part, by the participation of new recorders and the introduction of the annual Big Butterfly Count, when this butterfly is recorded regularly in gardens throughout Cornwall.

When Holly Blue are plentiful they can be found almost anywhere in the County, although there tend to be more records from coastal regions and the valleys leading inland from them. They are much scarcer on Bodmin Moor and on the high ground of the Carnmenellis area near Redruth.

Larval foodplants

The Holly Blue is very unusual among butterflies in preferring a different larval foodplant for its two main broods: Holly *Ilex aquifolium* in the spring, and Atlantic Ivy

Upperside, female

Undersides displayed during 'puddling'

Hedera hibernica in the summer (Atlantic Ivy is the common ivy species in Cornwall, whereas Common Ivy *H. helix*, which is found throughout the rest of England, occurs only as an introduced plant in the County.) The larvae feed on the buds, flowers and immature berries of these plants. The difference in flowering times of Holly and Ivy presumably accounts for the Holly Blue's varied seasonal menu. They can, however, use a wide range of other wild and cultivated plants, including Spindle *Euonymus europaeus*, heathers, dogwood *Cornus* spp., snowberry *Symphoricarpos* spp., cotoneaster *Cotoneaster* spp., gorse *Ulex* spp. and Bramble *Rubus fruticosus* agg. It is thought that on the coast, gorse can occasionally substitute for the larval foodplants normally preferred by the butterfly.

Habitat and ecology

Holly Blue are very mobile and are great wanderers, exploring sheltered sunny woodlands and gardens. They are also found in parks and churchyards and along lanes, tracks and footpaths. Their only requirements are ample trees and shrubs containing plenty of their larval foodplants, and consequently the butterfly is almost entirely absent from areas without these. The adults nectar greedily on Holly and ivy blossoms and the honeydew generated by aphids, as well as enjoying garden flowers. The Holly Blue, particularly the male, makes good a deficiency in certain nutrients by sucking these up from damp soil, mud and dung, a behaviour known as 'puddling'.

There are normally two broods, but in Cornwall a partial third brood occurs in some

Flight curve based on number of records in Cornwall 1999–2018; and UK life cycle

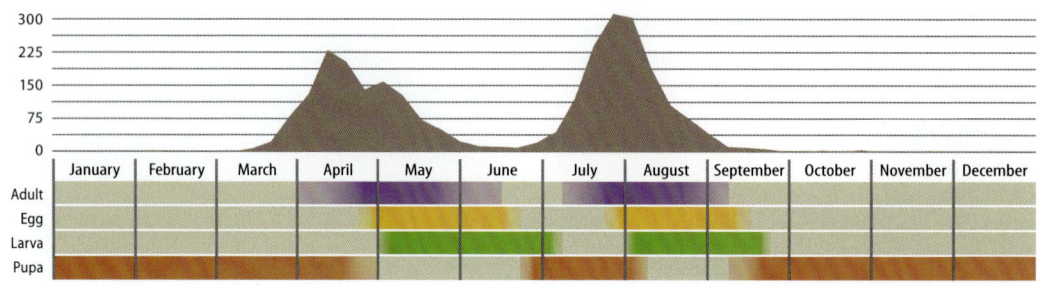

places in good years. The first brood emerges in March or April, although individuals have been recorded in the County as early as February, with Penhallurick noting their appearance as "the first real sign of spring" (Penhallurick, 1996: 64). The adults of the first generation fly until May or early June, although they are rarely seen in years when the butterfly is scarce. The second brood emerges in early July and continues flying until early September. There is a short gap between this and the partial third brood (when it occurs), which begins in late September. Adults of this brood have been seen as late as November, although they do not usually fly after late October.

The white eggs are laid singly, usually at the base of flower buds on the chosen foodplant, and hatch in around a week. The first instar caterpillar is honey-coloured, with long, backward-curving hairs. The second, third and fourth instars are a better-camouflaged green and are relatively hairless, but the later instars have various colour forms and can have rich pink markings. The larva pupates near the ground, and after several weeks the second (and third) brood butterflies emerge. The first brood of the year emerges from the overwintering pupa. Like other Lycaenidae the chrysalis and larva create secretions that are inviting to ants; both red ants *Myrmica* spp. and black ants *Lasius* spp. have been observed in attendance to the latter life stage (Asher et al., 2001).

Analysis of trends

Although both UK and South-west England abundance trends are very encouraging, the Holly Blue has the most widely fluctuating populations of almost any British butterfly species. This is attributed at least in part to its cyclical struggle with two parasitoid wasps, *Cotesia inducta* and *Listrodromus nycthemerus*, of which the latter is specific to the Holly Blue (Thomas and Lewington, 2014). *L. nycthemerus* injects a single egg into the caterpillar, eventually killing it during its chrysalis stage, from which the adult wasp emerges. When the butterfly host is plentiful, the parasitoid population rises, so that the increased parasitism causes the Holly Blue population to crash. The sparsity of the host results in a slump in the wasp population, allowing the butterfly population to recover. This 'boom-and-bust' cycle is repeated every four to six years (Eeles, 2019). Nationally this parasite is known to have a significant impact on Holly Blue populations, but until recently there had been no records of the wasp in Cornwall for over 100 years. However, an eagle-eyed Butterfly Conservation member recently identified one in his garden, so there are probably many more that have gone unnoticed.

Weather conditions also play an important part in annual fluctuations of the Holly Blue population, with numbers tending to be higher in hot summers that follow mild or dry winters, and lower after cold, wet winters. It is particularly important that the emergence of newly hatched caterpillars coincides with the availability of early, tender flower buds which they favour.

The highest number of Holly Blue recorded in Cornwall during the last 10 years was in 2015, supported by evidence from transect data and records from the Big Butterfly Count, when numbers increased by 151% on the previous year. Although the weather in summer 2015 was not particularly good, it was preceded by a warm and sunny 2014 followed by a mild winter, which probably increased survival rates.

Conservation

Nothing specific is required, due to the butterfly only needing a plentiful supply of Holly and ivy. Gardeners are advised not to prune back ivy until late autumn.

Best places to see

Anywhere in Cornwall where the foodplants are abundant. Holly Blue seems to do particularly well on the Isles of Scilly, although it was not recorded there until the 1960s.

Listrodromus nycthemerus, a wasp that parasitises the Holly Blue

Upperside, male

Silver-studded Blue
Plebejus argus

Sponsored by
*Pete Holdaway, Margaret Jamieson
and Pete Jamieson*

This conspicuous little butterfly is the smallest of the 'blues' resident in Cornwall. It is a flagship species for the County, and it indicates a well-functioning ecosystem, such as the dune systems at Penhale and Hayle. It can reliably be distinguished from other lycaenids once subtle but key differences have been identified. In both male and female Silver-studded Blue the black spots along the outer edge of the underside of the wings can have silvery-blue centres – hence the butterfly's name. These 'studs', however, are not always present or may be very small, especially in the female.

The male Silver-studded Blue is quite different from the female. The upperside of the male's wings is leaden blue with a metallic sheen, whilst the female upperside is brown, often with a powdering of blue scales, and has a fringe of orange crescents (lunules) of varying intensity around the outer margins of both fore- and hindwings. The markings on the underside

Resident in Cornwall
(not on the Isles of Scilly)

CONSERVATION STATUS
- Schedule 5 Wildlife and Countryside Act 1981 (as amended) (for sale only)
- Section 41 Species of Principal Importance (NERC Act 2006)
- Red List of British Butterflies; IUCN criteria: Vulnerable (VU)
- Butterfly Conservation priority: Medium

TRENDS AND DISTRIBUTION
UK rate of change in abundance
↑ 47% 1979–2018
South-west England rate of change in abundance
↑ 42% 2009–2018
↓ 28% 1979–2018
% of 1km squares occupied in Cornwall
3% 2009–2018 (38% 10km squares)
3% 1976–2008 (43% 10km squares)

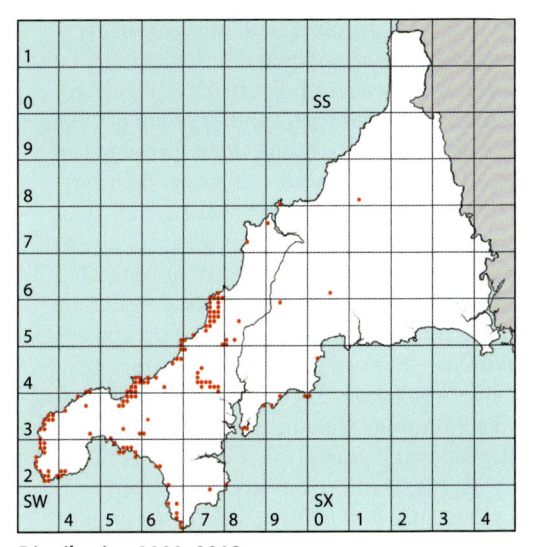

Distribution 2009–2018

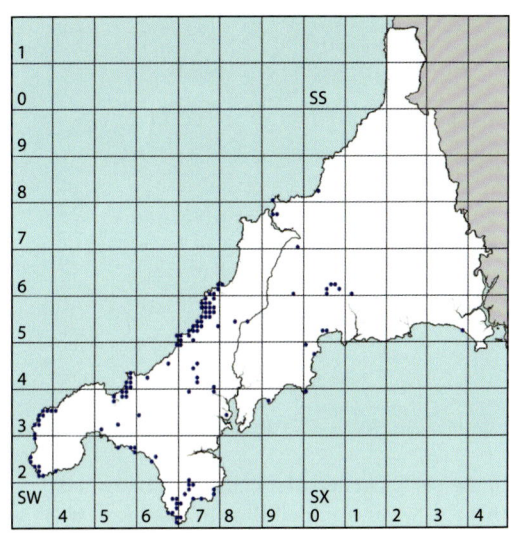

Distribution 1976–2008

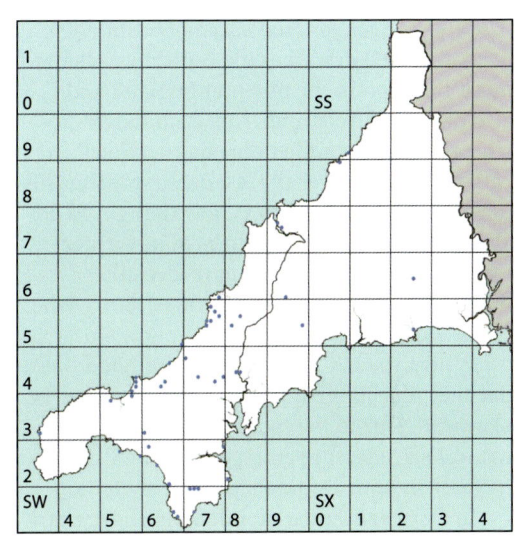

Distribution pre-1976

are similar in both sexes, but the background colour of the male is quite pale, suffused with blue towards the body, whereas the female is a darker beige, with little, if any, blue powdering.

Because of her brown colour and orange lunules, the female Silver-studded Blue can be mistaken for a female Common Blue, although the latter is usually noticeably larger, or a Brown Argus *Aricia agestis*. When freshly emerged, the male Silver-studded Blue, which can be confused with the male Common Blue, has a distinctive charcoal border on the upperside of both fore- and hindwings, a feature lacking in the male Common Blue. Such distinguishing features, however, are less reliable in faded, older individuals. The undersides are diagnostic when the silvery 'studs' are visible; in addition, the absence of a black spot on the underside of the forewing, close to the body, is a feature of Silver-studded Blue and Brown Argus, whereas the spot is present on Common Blue.

At present the British population is now represented mainly by the nominate subspecies *argus*, which is the subspecies found in Cornwall. (On the Great Orme and nearby areas in north Wales, it is replaced by the subspecies *caernensis*.)

The butterfly is mainly found in three habitats in the UK: lowland heath, calcicolous (lime-rich) grassland, and sand dunes. In Cornwall, as well as heathland populations, large colonies are found on the sand dunes of the north coast. The adults on these sand dunes appear to be larger and more brightly coloured than the heathland forms of the subspecies *argus*, and are similar to the extinct subspecies *cretaceus*. Rather than being regarded as separate subspecies, however, these forms are referred to as ecotypes or races and are not afforded special protection.

Distribution in Cornwall

Research by the late Professor John Wacher from towards the end of the twentieth century

Upperside, female

to the early twenty-first century established that the Silver-studded Blue in Cornwall was more widely distributed and more abundant than earlier butterfly publications indicated. *A Butterfly Book for the Pocket* by Edmund Sandars (1939) describes the butterfly as being more common in the north of England and also present in Scotland. This is no longer true: the Silver-studded Blue has declined catastrophically over the last century, primarily due to habitat loss, and it is now thought to be absent from around four-fifths of its former range. Mapped at 10km-square level, the butterfly has declined by 71% since 1800 (Asher *et al.*, 2001).

Whilst the UK as a whole has suffered these declines, the butterfly has fared much better in Cornwall. Due to the industriousness of early naturalists, a picture emerges of a genuine increase in both abundance and distribution.

In Cornwall, Silver-studded Blue have nearly always been restricted to the western half of the County, although the butterfly has historically been recorded in the east, in areas near Boscastle and Liskeard. In the west it is largely confined to the coastal regions, and some of the strongest colonies are found on the sand dunes at Penhale, and from Godrevy to Hayle, where they can be seen in their thousands. One very weak colony has been observed on Lelant Towans, but this may not survive, as the Hayle Estuary is likely to form a barrier across which reinforcing adults cannot fly.

There are also significant colonies inhabiting heathland, many of them on former mining sites. The butterfly is found in a succession of colonies on the coast of West Penwith. The stronghold of colonies originally ran northwards from Botallack to Pendeen Watch. Together with another group of colonies north of Gwennap Head, these spread out, so that by 2018 there were numerous sightings from Pen Enys Point west of St Ives, all along the coast of West Penwith to an area close to Penberth Cove in the south. West Penwith has consistently held resilient colonies of the butterfly, including some that report substantial numbers. For instance, at Enys Zawn, cliffs close to Pendeen, 214 butterflies were recorded in a 50-minute timed count in 2014, and 287 the following year.

On the south coast, stretching from east of Penzance to Porthleven, there is a large heathland colony east of Cudden Point, and good-sized colonies have been recorded from the Perranuthnoe area; however, colonies at Rinsey Cove and Trewavas Head are still small, with low counts of adults recorded. Colonies have also been reported on the dunes between Porthleven and Loe Bar. Many of these colonies appear to be isolated and may be vulnerable to local extinction.

On The Lizard, the butterfly has historically been recorded in a larger area, the most recent records show the species consistently reoccurring in the same locations but in smaller numbers. These colonies look to be small and fragmented but further investigation is needed in areas of suitable habitat where the species may be present.

Inland in Cornwall there are far fewer sites, many of which are small and remote. One such is Breney Common, where there have been records in the past but none since 2010, and the habitat has become increasingly unsuitable. There are also no records since 2010 from nearby Goss Moor. Most of the inland sites in Cornwall are heathland and the vast majority are inextricably linked to former mining sites, notably in the Carnon Valley between Redruth and Truro. The Lizard appears to be the exception. The presence of heavy metals in the mine spoil has arrested plant succession on these sites. Without mining and the resultant waste, much of the heathland would have disappeared to development or agriculture. Most of the agriculture has involved modifying soils in order to grow crops such as daffodils and brassicas, leaving little room for the butterfly's habitat. In addition, much heathland has succeeded to scrub and woodland. A further threat to Silver-studded Blue is the invasion of alien plants.

Larval foodplants

On calcareous habitats such as the sand dunes on the north coast, the caterpillar feeds on Common Bird's-foot-trefoil *Lotus corniculatus* and other members of the Fabaceae family such as Common Restharrow *Ononis repens* and possibly young gorse *Ulex* spp.

On heathland, the preferred foodplant is normally members of the Ericaceae family such as Bell Heather *Erica cinerea*, Cross-leaved Heath *E. tetralix*, Cornish Heath *E. vagans* and Ling/Heather *Calluna vulgaris*. However, within this habitat there may be members of the Fabaceae family present including gorse. Few observations are made of Silver-studded Blue larvae feeding, so further investigation is required to discover the species' specific foodplant preference within this habitat across the County's populations.

Habitat and ecology

The two main types of habitat in Cornwall for Silver-studded Blue are dunes and heaths. The dunes support a typical calcicole flora with a mosaic of bare ground, short turf often cropped by rabbits, and patches of longer grasses such as Marram *Ammophila arenaria* and scrub. This mosaic of habitat is essential to the butterfly's requirements for its complete life cycle. Scrub and Marram provide shelter, resting and roosting places. On heathland the required habitat includes a mixture of heathers, gorse, vetches and other plants. As in the dune habitat, areas for roosting are essential. Colonies are usually found on warm south-facing slopes or in other areas protected from strong winds. There is no limit to the areas where the butterfly will seek out shelter: old walls, sand-dune bowls, scrub, hedges, and even the bunkers of the old explosive works on Upton Towans.

Silver-studded Blue live in close-knit colonies of variable size, and were previously thought not to wander more than about 50m from them (Thomas and Lewington, 2014). More recently, though, this view has been modified and several observations have been made of the species outside known colonies. This revised view has been supported by observations in Cornwall, where movement between colonies has been recorded at Upton and Gwithian Towans, and also between the group of colonies north of Botallack (Foster and Dennis, 2020b).

The Silver-studded Blue usually produces only one brood of adult butterflies in Britain. However, John Wacher discovered evidence of a second brood in Cornwall (Wacher, 2002). This inspired others to investigate further, and in the warm summer of 2018 more supporting evidence for a second brood was found at Penhale by Cerin Poland when he discovered an early instar caterpillar on 1 August and, over three weeks later, a freshly emerged male adult butterfly in the same location.

An intimate association with ants is a characteristic shared by most, if not all, lycaenids, including the 'blues'. The Silver-studded Blue has a remarkable symbiotic

Mating pair, female above. Undersides showing silvery 'studs' near the outer margin.

Caterpillar attended by ant

relationship with black ants belonging to the genus *Lasius*. This amazing relationship never ceases to delight and inspire young and old when they learn of this fascinating intricate cycle of life.

The female lays her eggs only where she detects suitable ant pheromones, near nests of *Lasius* black ants. The eggs overwinter, hatching in late February or early March. Ants chaperone the larvae, which they defend against attacks by predators such as spiders, wasps and parasites. Although the interdependence between the larvae and ants is beyond doubt, there is still a lot to discover about this ecologically indispensable association. The ants probably pick up the larvae soon after they hatch and place them in ant chambers beneath plants, rocks or stones. The larval stage lasts for around three months and the larvae then pupate within or close to the ant nests, where each is tended by ants until the butterfly emerges, usually two or three weeks later. In return, the ants receive sustenance in the form of a sweet fluid produced from a gland on the back of the larva.

At present it is believed that five species of ant form a relationship with Silver-studded Blue in the UK, and two have been positively identified with the butterfly in Cornwall: *Lasius alienus* at Upton Towans and Godrevy Warren; and *L. niger* at St Gothian Sands Local Nature Reserve (LNR) and Godrevy Head.

Analysis of trends

In the UK, the long-term trend for this butterfly shows a significant decline in distribution and a moderate increase in abundance, with the trend over the last decade showing a moderate increase in distribution and a stable abundance.

Although a reasonable number of transects in Cornwall record Silver-studded Blue, very few have been walked consistently over the last 10 years. Of these, Upton Towans provides the most data on abundance over the long term. This transect, started in 2010, features regularly as one of the top Silver-studded Blue transects in the UK. The Upton Towans data indicates a stable population with a gentle upward trend. The Gwithian Green transect first recorded the butterfly in 2010, and in the last six years there has been a steady increase in numbers, indicating that a small breeding colony has been established.

On a site visit to Cudden Point in 1999, 500 Silver-studded Blue were recorded. During the early years of the twenty-first century, numbers decreased significantly. In the last few years, numbers appear to have been recovering, and timed count results indicate an increasing trend. This is very much in line with the trends indicated on the Cornwall transects.

Despite concern about a decline in numbers in Cornwall towards the end of the last century (Foster and Dennis, 2020b), distribution appears to have expanded in the last 10 years,

Flight curve based on number of records in Cornwall 1999–2018; and UK life cycle

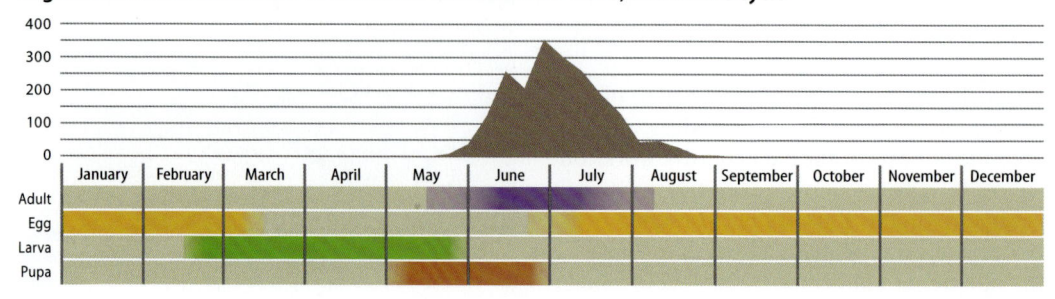

especially along the West Penwith coastline. The distribution of the species has also grown on inland mining areas, the core of which are spread from Wheal Busy to the bottom of the Carnon Valley (between Truro and Redruth). This is in part due to increased recording by the Mining for Butterflies project, in which Cornwall Butterfly Conservation volunteers searched over 37 mining sites for priority species of butterfly. (See chapter 3 for more information.)

Records also show an increase in numbers. The Silver-studded Blue has made a good recovery in the last decade, a trend also reflected in data for South-west England. In 2018, the butterfly underwent a healthy expansion in both range and numbers. This was possibly a 'dispersal' year, and future monitoring of new sites will establish whether new breeding colonies have been formed.

Understanding the reasons for the upturn in the fortunes of this rare butterfly over the last few years is open to debate. The relationship with ants is clearly important, and without the ant present on a site it is unlikely that the butterfly will form a new colony. In some locations conservation work has resulted in the creation of habitat more favourable to the butterfly. At Gwithian Green LNR, where the ant species is long-established but Silver-studded Blue had never been recorded, it is possible that the population at nearby St Gothian Sands LNR only spread onto Gwithian Green LNR when it had reached a critical level.

Previous under-recording is always a possibility, but many of the sites generating records, such as those along the north coast of West Penwith, have expanded; for these locations, under-recording cannot be the explanation for rising numbers, as they have been consistently and carefully searched in the past. One area that has probably suffered from under-recording is The Lizard.

Conservation
The maintenance of suitable habitats is of prime importance, particularly in and around the neighbourhood of existing colonies, and strategic partnerships with landowners and land managers should be fostered, based on an awareness of the butterfly's requirements (Ravenscroft and Warren, 1996). Landscape-scale conservation management is required in order to sustain metapopulations, and any local extinction can then be recolonised naturally. There are many factors to consider when undertaking management, and most sites will require a site-specific plan. The very particular requirements for the Silver-studded Blue's life cycle contribute to its highly restricted distribution and mean that it is threatened both by habitat loss from development, which can lead to the isolation of colonies, and by declining habitat quality through inappropriate site management. Godrevy Warren, which is currently in an agricultural stewardship scheme involving scrub removal and conservation grazing by native breeds of cattle, is an excellent example of best practice.

Silver-studded Blue are particularly reluctant to cross obstacles, so potential breeding areas that are separated by unsuitable habitat are effectively isolated. Dense Bracken *Pteridium aquilinum* and scrub seem to constitute insurmountable barriers on many heathland sites.

"Conserving ecologically distinct populations and maintaining genetic variation below the level of taxonomic species is critical to an effective conservation programme for the species, as the Cornish ecotypes are considered important and should be conserved" (pers. comm. Butterfly Conservation).

Best places to see
- Upton Towans SW576397 / Mexico Towans SW559388 / Godrevy SW582423
- Gear Sands (Penhale) SW769562 / Mount Field SW782570
- Cudden Point SW550277
- Porthgwarra SW370215
- Enys Zawn / Pendeen Watch SW378354
- Poldice Mine SW743429

Upperside, female

Brown Argus
Aricia agestis

Sponsored by
Perry Besterman, Ronald Crewes
and Cerin Poland

The range of the Brown Argus is restricted in Cornwall, concentrated mainly within sand dune habitats. It is not easy to identify this butterfly, as it is small and relatively inconspicuous, with a silvery appearance to the wings as it flies close to the ground, and it can resemble the females of the other blue butterfly species unless viewed closely, particularly since it often flies in similar locations and at similar times to the Common Blue and Silver-studded Blue. The lack of blue scales on the upperside of Brown Argus wings helps distinguish it from the other blues found in Cornwall, but the wings can take on a blue sheen under certain light conditions. The best way to be certain of identification is to view the underside of the hindwing, where two spots are situated next to each other below the top edge, towards the rear, creating a figure of eight; this feature is absent in the other blues found in the County.

Resident in Cornwall
(not on the Isles of Scilly)

CONSERVATION STATUS
■ Butterfly Conservation priority: Low

TRENDS AND DISTRIBUTION
UK rate of change in abundance
↓ 15% 1976–2018
South-west England rate of change in abundance
↑ 19% 2009–2018
↓ 54% 1976–2018
% of 1km squares occupied in Cornwall
1% 2009–2018 (22% 10km squares)
1% 1976–2008 (25% 10km squares)

There are also behavioural differences. Brown Argus males are fiercely territorial and fly up to investigate any butterfly that comes into view, a habit that helps to distinguish them from the more laid-back Common Blue females. The male and female Brown Argus are similar in appearance, but the female is slightly

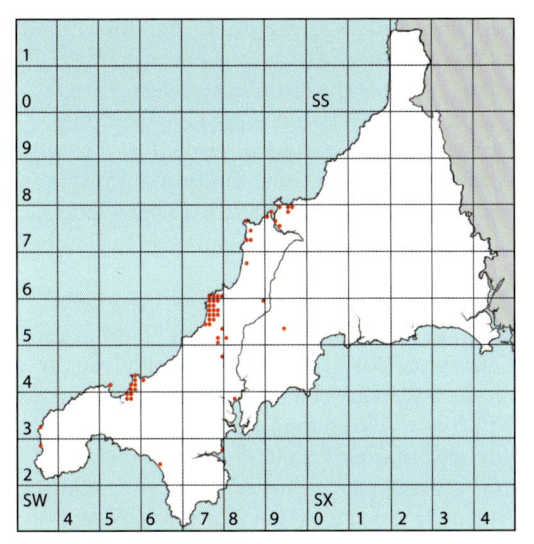

Distribution 2009–2018

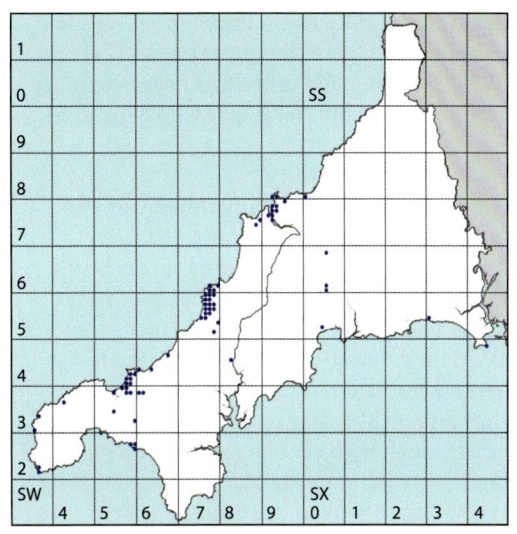

Distribution 1976–2008

larger and has a shorter and more rounded thorax, which tapers rapidly. On the upperside of the wings, the border of orange crescents in the female is more pronounced than in the male and extends further towards the tips of the forewings.

Distribution in Cornwall

The strongholds of the Brown Argus are the sand dunes along the north Cornwall coast, with a few scattered sightings inland and along the West Penwith coastline. The three main hotspots are Upton and Gwithian Towans, Gear and Penhale Sands, and the sand dunes around Rock and Padstow. From time to time, credible records are submitted from entirely new locations, of which an increasing number are private gardens, which might indicate that this butterfly is more widely distributed than previously thought. This may be due to under-recording and misidentification, as less experienced recorders might assume that they are observing the more widespread Common Blue female.

Despite this, the distribution maps from the two different time periods show a remarkably similar pattern, with the species living in discrete, localised colonies. Since the larvae are often attended by ants (Eeles, 2019), the long-term viability of these colonies may depend on the presence of ant species whose distribution is likely to have a close association with that of the butterfly.

Larval foodplants

Within calcareous grasslands, Common Rock-rose *Helianthemum nummularium* is the usual larval foodplant. However, it is a plant of chalk and limestone regions and therefore does not occur in Cornwall, where its place as the larval foodplant is taken by Common Stork's-bill *Erodium cicutarium* and Dove's-foot Crane's-bill *Geranium molle*. The former is especially common on the dunes of Cornwall, while the latter has a more general distribution in the central and western regions of the County, but

Underside, male and female similar

in east Cornwall Dove's-foot Crane's-bill is mainly confined to the coastal areas (French, 2020). Occasionally other plant species in the crane's-bill family Geraniaceae may be used.

Habitat and ecology

Adult Brown Argus usually inhabit small, loose-knit colonies containing just a few individuals, in areas where the turf is kept short by rabbits or other means and where larval foodplants grow in abundance, such as on the calcareous dunes. The Penhale Sands area (stretching from Gear Sands to the Penhale Ministry of Defence land) is exceptional, with very large numbers reported regularly. The butterfly can cope with taller vegetation so long as this is fairly sparse, enabling it to use habitats with a more varied vegetation structure, such as the margins of agricultural land.

There are normally two broods, the first mainly in May and June, although some have been seen in late April in Cornwall. The second appears in July and August or early September. Eeles refers to a partial third brood appearing "in southern England in good years, which is the norm in southern Europe" (Eeles, 2019: 367), so it is possible that some late emergences quoted by Penhallurick (1996) in late September belong to a partial third brood. In recent years there have been several reports of sightings of this butterfly in October, confirming this view.

As an adult, the Brown Argus has a very short life span of about four days (Thomas and Lewington, 2014). Eggs are laid singly, usually on the underside of leaves, and the green larvae, which have a line along each side, are usually attended in their fifth (and final) instar by ants attracted to the larvae's sugary secretions. Normally it is the red ant *Myrmica sabuleti*, which is common on south-facing, warm coastal slopes where the grass sward is short, or the black ant *Lasius alienus*, the most abundant ant of the sand dunes. The butterflies hibernate as larvae and pupate the following spring.

Analysis of trends

Nationally in the last century, there appeared to be a noticeable reduction in the number of colonies of Brown Argus, resulting mainly from habitat destruction through intensive agriculture, the abandonment of habitat management and scrub encroachment (Asher *et al.*, 2001). During that time, however, the butterfly remained reasonably stable in Cornwall, although Penhallurick (1996) mentions the disappearance of the species from some localities due to coastal development. The early 1990s saw a considerable expansion of its range in Britain, especially in central and eastern England; it has always been restricted to southern Britain (Asher *et al.*, 2001). There is evidence that the increases were related to

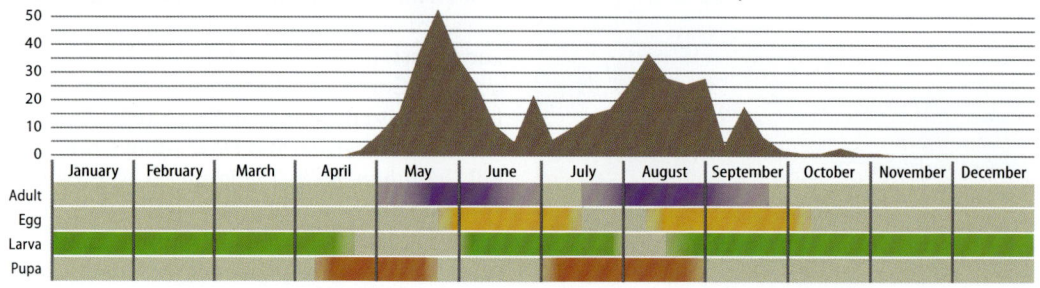

Caterpillar with ant

Flight curve based on number of records in Cornwall 1999–2018; and UK life cycle

Upton Towans, one of the best places to see Brown Argus

good summer temperatures between June and August, while wet, cool summers at the end of the 1990s had a negative effect and are thought to have arrested this expansion for a time. Since then it has been postulated that global warming has contributed to the northwards expansion of the butterfly's range in England over recent years, as warmer conditions have enabled it increasingly to use Dove's-foot Crane's-bill as its larval foodplant. This has led to a rise in available habitats, which include road verges and uncultivated field margins, and hence the consequent rapid range expansion (Pateman *et al.*, 2012).

In Cornwall, analysis of abundance trends is not easy. Increased monitoring of the Penhale Sands area in the 1990s by one particular recorder undoubtedly boosted the records. At the beginning of this last 10-year period there was serious concern about the reduced numbers of Brown Argus being reported. It may be that this was a genuine crash, as there were several poor summers in succession. It could also be that there was an absence of surveying. Since then there have been considerable fluctuations in the number of records and the number of individuals from year to year. Again, this seems to be partly weather-related but is also dependent on how often the Penhale Sands area is visited, since regular recording throughout this extensive sand dune complex will boost numbers considerably. Unfortunately, Penhale does not yet have a transect attached to it, so there is no consistency of recording on an annual basis. Brown Argus are seen regularly on only three transects in Cornwall, so it is difficult to draw conclusions from the small numbers recorded. Thomas and Lewington (2014) report significant local variation in population dynamics within the same geographical area, with one site having a good year, while numbers are down on nearby sites.

Conservation

Maintaining habitats to support the larval foodplants and associated ant species is important to conserving the Brown Argus. It is dependent on grazing and regular scrub management on many sites (Spalding with Bourn, 2000). The creation, retention and suitable management of field margins is also of importance to this species.

Best places to see

- Hayle to Godrevy sand dune system, especially Upton Towans SW576400
- Penhale area, including Gear Sands SW769562 and Mount/Pennans Field SW782568
- Coastal area from Porthcothan to Lundy Bay, including Rock Dunes SW928767 and Lundy Bay SW954797
- Cubert Common SW783597

Upperside, male

Common Blue
Polyommatus icarus

Sponsored by
*Francia Besterman, Nina May Besterman
and Anne Green*

The Common Blue, as its name suggests, is the most widespread of the UK's blue butterflies; this is also true for Cornwall. Male and female are quite different. The upperside of the male's wings is a vivid blue, whilst females are far more variable, their base colour varying from being almost completely brown to all blue. Females also have orange spots along the borders of the wings. The most commonly seen form of the female is brown, with a light shading of blue scales spreading out from the base of the wings. A good time to see a Common Blue with its wings open is early in the day when it warms itself in the sun. In the evening it can be found roosting upside down on long grass with its wings closed.

At a distance, male Common Blue can be confused with Holly Blue, although their flight behaviour is quite different. The Holly Blue flies high, alighting on the top of bushes, while the Common Blue usually flies closer to the ground in open, grassy areas. In addition, their markings are noticeably different when at rest with their wings closed. The Holly Blue has a scattering of black spots on the pale, silvery blue undersides of its wings, whereas the Common Blue has orange marks along the margins of the underside of the wings, and

Resident in Cornwall
(resident on the Isles of Scilly)

CONSERVATION STATUS
■ Butterfly Conservation priority: Low

TRENDS AND DISTRIBUTION
UK rate of change in abundance
↓ 16% 1976–2018
South-west England rate of change in abundance
↓ 17% 2009–2018
↓ 33% 1976–2018
% of 1km squares occupied in Cornwall
31% 2009–2018 (84% 10km squares)
23% 1976–2008 (81% 10km squares)

Common Blue

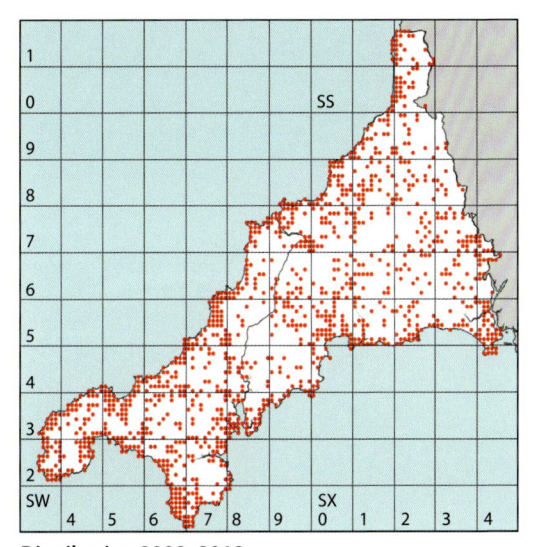

Distribution 2009–2018

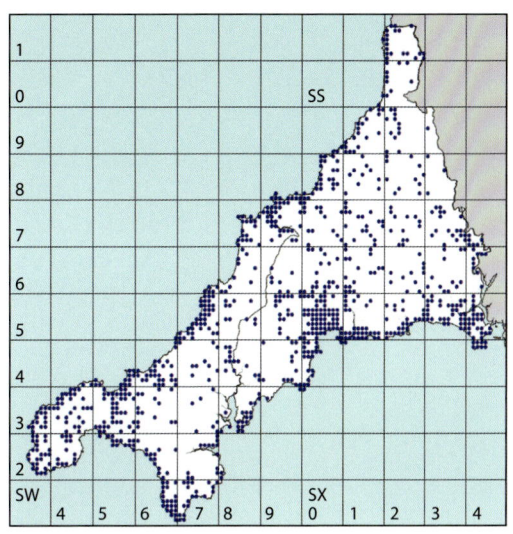

Distribution 1976–2008

larger black spots encircled by white on a dull beige background. In Cornwall the Holly Blue is the first blue butterfly to emerge in spring, flying from late March, whereas the Common Blue generally flies from May onwards.

It is not so easy to distinguish between Silver-studded Blue and Common Blue, particularly since they share the same habitat and have overlapping flight periods. The Common Blue tends to be a more vigorous flyer and, when seen close up, its key identifying features are the narrow white slash on the underside of its hindwings and the extra spot on the forewing close to its body, which are both absent in Silver-studded Blue and Brown Argus.

The brown female Common Blue can be mistaken for the Brown Argus. The latter is a richer chocolate brown with stronger orange markings on the upperside of fore- and hindwings. The Brown Argus never has blue scales on the upperside of its wings, although the wings can sometimes exhibit a deceiving blue sheen. To be certain of identification the undersides of the wings should be examined for the appropriate features.

Distribution in Cornwall

The distribution maps provide evidence of a slight increase in occurrence over the last 10 years, which might, however, be the result of increased recording rather than new colonisations. The butterfly is widespread throughout Cornwall, although there has been a drop in abundance in the more recent period.

Upperside, typical female

Upperside, more unusual female

Mating pair, female above

Fortunately Cornwall still has large tracts of unimproved grassland where the larval foodplants are to be found. Most of this is now confined to the coastal regions, but even the smallest areas inland often support colonies, although they are nowhere near as large as those on the coastal sites, where numbers can often be counted in hundreds. Penhallurick (1996) states that all writers concerned with butterflies from 1846 onwards have described this species as 'common' or 'abundant', with many going on to note 'especially near the sea'.

Larval foodplants
Most widely used in Cornwall is Common Bird's-foot-trefoil *Lotus corniculatus*, which is widespread in the County. The caterpillar also feeds on Greater Bird's-foot-trefoil *L. pedunculatus*, Black Medick *Medicago lupulina*, Lesser Trefoil *Trifolium dubium* and White Clover *T. repens*.

Habitat and ecology
The primary habitats of the Common Blue are sheltered situations with ample larval foodplant along the coastal grasslands and dunes, and on wastelands, heaths, former mining sites and any unimproved grassland containing bare ground and a varied sward structure below 15cm (Asher *et al.*, 2001). The butterfly is relatively mobile and can easily form new colonies. It can be seen almost anywhere in Cornwall, including urban

Flight curve based on number of records in Cornwall 1999–2018; and UK life cycle

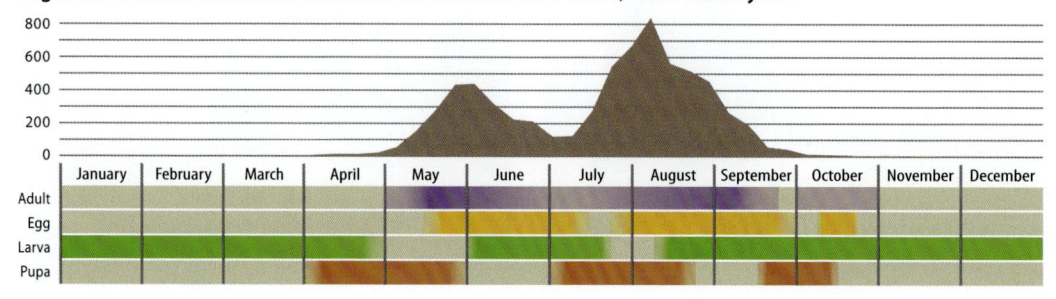

gardens, parkland and cemeteries, except on the highest moors, as it appears to seldom be found at heights of over 300m.

The larvae and pupae, as with so many 'blues', are attended by particular species of ant. In the case of the Common Blue, these are the Red Wood Ant *Formica rufa*, the red ant *Myrmica sabuleti* and *Lasius alienus* (Emmet and Heath, 1990). The relationship is not thought to be as significant as in the case of the Silver-studded Blue, but nevertheless both the larval and pupal stages produce a sugary cocktail that is appreciated by the ants, and both the larvae and the pupae emit singing noises that attract ants.

The butterfly normally has two broods, but in Cornwall a partial third brood sometimes occurs. The first brood emerges in early May, although individuals have been seen in late April, and this brood flies until the end of June or early July. The male is very territorial and is highly conspicuous as he perches and patrols in search of a female. She, on the other hand, spends most of her time resting or feeding. She lays her pale green eggs singly on the upperside of a leaflet of the foodplant after flying low over suitable areas, selecting her favoured plant by testing its quality with her feet. The eggs turn brilliant white when dry and are easy to spot. The larvae emerge after about 10 days and will feed on both flowers and leaves. They take about six weeks to develop. The pale, rather colourless first instar becomes greener and hairier with each moult. The fifth (and final) instar is a rich, well-camouflaged green, with a pale longitudinal stripe running up each side of its body, and a shiny black head. The pupa is formed on the ground, or it can be found just under the surface – which may be the work of ants in their efforts to conceal it. It spends two weeks in this stage.

The second brood starts to emerge in mid-July and flies until late September. A partial third brood can be seen flying from mid-September until October, with an overlap between this and the second brood. In many years it is thus possible to see Common Blue on the wing from early May and throughout the summer, with a few even spotted in late October. It is thought that not all the caterpillars produced by the first brood of butterflies will go on to produce a second brood, as some of the larvae will mature slowly and go into hibernation at the base of their foodplant when only partially developed, to be joined by caterpillars produced by broods later in the year. They become active again in early March.

Analysis of trends

The main cause for variation in annual numbers is probably the weather in both the preceding and the current year. Generally the warmer and drier it is, the higher the numbers, provided that there is no drought, which affects both nectar sources and foodplants, as it did particularly in 1976, causing a crash from which it took some years to recover.

Cornwall's records, arising from both transects and *ad hoc* records, show considerable fluctuations. In the last 10 years there has been a slight downward trend overall in terms of abundance, reflecting national and South-west England figures. However, the good summer of 2018 produced astonishing numbers of this butterfly, ameliorating this decline. Strangely, this was not echoed in the transect figures from St Mary's on the Isles of Scilly, which showed the worst count of Common Blue since the transect was set up in 2012.

Conservation

Livestock grazing can be an important tool used to create a varied sward structure that is of benefit to Common Blue. Potential threats are mostly the destruction of habitat by development and agriculture. However, the incorporation of herbal leys that include larval foodplants into farming systems may create new opportunities for the species.

Best places to see

- Gear Sands (Penhale) SW769562
- Penlee Battery SX438491
- Enys Zawn SW378354 to Levant Mine SW370346
- West of Porthgwarra SW367220
- Perranuthnoe SW540294 to Cudden Point SW551278 to Prussia Cove SW556280
- Mullion Cove SW666179 to Higher Predannack Cliff SW661164

CHAPTER 5
Occasional visitors and rarities

Swallowtail, a rare visitor to Cornwall, photographed in Spain

The butterfly species included in the previous chapter are all currently resident in Cornwall or are recorded regularly as migrant species. This chapter, by contrast, includes butterflies that are UK resident species not currently known to be breeding in Cornwall, and rare migrants, vagrants and other butterfly species that have been recorded only rarely in the County. All the records cited, including older historical records, are in the ERICA database. However, in the absence of supporting photographic evidence, the reliability of many of the reported sightings must be open to question.

UK resident species

Essex Skipper
Thymelicus lineola
The Essex Skipper is a resident species in many parts of England and is thought to have expanded its range in recent years, being found as far north as Yorkshire. Its status in Cornwall is uncertain. It is very similar to the Small Skipper *Thymelicus sylvestris* – see species account in chapter 4 – and uses the same grassland habitat. Consequently this could mean that the Essex Skipper has been overlooked in the past. However, none of the 30-plus sightings from various sources and recorders have been confirmed.

The distinguishing feature of the Essex Skipper is the sharply defined black underside of the antennae and the positioning and length of the male sex brands. Using the antenna tips for identification has been found to be unreliable as the sole method of distinguishing the two species. A female adult with black antenna tips collected in Devon had her genitalia examined and was determined to be a Small Skipper (Henwood, 2015). This has been backed up by sightings of male Small Skipper in Cornwall that had black antenna tips but could be clearly identified due to the distinctive sex brands. High-quality photographs of both features are the preferred method of ensuring correct identification in Cornwall.

There have been 14 records of Essex Skipper in various parts of the County this century, none of which have been confirmed. An attempt to introduce the species to the St Austell area in 1995 by an independent party was unsuccessful, so it is unlikely that those spotted in a variety of locations could be descendants. It has been suggested that eggs could arrive in imported hay, but the jury is still out on this matter.

The primary larval foodplants are Cock's-foot *Dactylis glomerata* and other grasses.

Wood White
Leptidea sinapis

The Wood White was first recorded in Cornwall in the nineteenth century but has not been reported in the County since 2006. In England the species favours sheltered and slightly shaded margins of sunny woodland rides with tall vegetation and ample larval foodplants and nectar sources. It has never been common in the County and historical records suggest it had a scattered but local distribution in east Cornwall. The coast in the far north-east and areas along the Devon–Cornwall border is where most substantiated sightings originated. Millook, south of Bude, is a particular hotspot where early naturalists such as F.W. Frohawk and W.A. Rollason documented the species (Penhallurick, 1996). There were several sightings in the late 1980s and early 1990s at Marsland near the Devon border, and a scattering of observations between St Austell and Lanhydrock from 1939 to 2004, although several of these, by competent recorders, are unconfirmed.

The butterfly has declined considerably in range across England and this is thought to have been caused by changes in woodland management, such as cessation of the coppicing that previously ensured the creation of marginal habitat and ride edges required by the species. It is unlikely that any substantial populations of Wood White remain in Cornwall, although it is possible that it has gone unrecorded, since populations still exist in Devon. Current policies to increase woodland cover in Cornwall may result in the return of Wood White to the County. The species can have two broods, and the best times to search for adults are May, June and early August.

The primary larval foodplants are Common Bird's-foot-trefoil *Lotus corniculatus*, Greater Bird's-foot-trefoil *Lotus pedunculatus*, Meadow Vetchling *Lathyrus pratensis*, Bitter-vetch *Lathyrus linifolius* and Tufted Vetch *Vicia cracca*.

High Brown Fritillary
Fabriciana adippe

This resident British butterfly was once widely distributed in England and Wales but has declined significantly since the 1950s and only remains on around 40 sites. It favours habitat covered with Bracken *Pteridium aquilinum*, on particularly warm south-facing slopes containing good amounts of the larval foodplants, violets *Viola* spp. This decline has been mirrored in Cornwall, where it previously had a scattered distribution across the County, extending as far as West Penwith. Since 2000, all records of the species have come from a single location, Marsland Nature Reserve in the far north of Cornwall, the last High Brown Fritillary being seen there in 2014. This butterfly may now be extinct in the County, although its similarity to the Dark Green Fritillary *Speyeria aglaja* could mean that its presence has been missed. However, this seems unlikely and it is possible that some historical sightings were in fact misidentified Dark Green Fritillary.

Lack of woodland management, the planting of coniferous woodland and reduction in grazing allowing scrub growth have all resulted in habitat loss, contributing to the decline of this species in the UK. Recent research from areas where the butterfly is found outside of Cornwall has also suggested that current habitat management practices may not be effective, due to influence from factors such as nitrogen deposition and climate change (Ellis *et al.*, 2019). Whilst populations are still present across the border in Devon, many of these are a considerable distance away from where the butterfly was historically recorded in Cornwall, so even if habitat restoration took place, natural recolonisation is unlikely. The best time to search for adults is July.

The primary larval foodplants are Common Dog-violet *Viola riviniana* and Hairy Violet *Viola hirta*.

White Admiral
Limenitis camilla
Despite its occurrence at numerous localities in neighbouring Devon, the White Admiral has rarely been reported in Cornwall, possibly due to the County's cooler climate and the lack of suitable woodland in some areas. There are historical records from several parts of the County, the furthest west sighting occurring at Godolphin in the nineteenth century. Marsland Nature Reserve accounts for the majority of records, with the butterfly being reported in low numbers from 1992 until 1994 and a single adult seen in 2011. However, this site appears to be a considerable distance from the localities where the butterfly has been found in Devon during recent years. The peak flight period is in July.

The primary larval foodplant is Honeysuckle *Lonicera periclymenum*.

Purple Emperor
Apatura iris
There are few records of this species in Cornwall, and early historical sightings lack any reliable confirmation. Several of the more recent sightings have been attributed to the release of captive-bred individuals (Penhallurick, 1996). The last known record in Cornwall was at an unspecified location along the River Lynher in 2006. The butterfly is now thought to be re-expanding its range in the UK after a period of decline in the twentieth century. Suitable habitat does exist in the County, and if the Purple Emperor continues to head south-west it may yet reach Cornwall (Oates, 2020: 346). Adults are elusive, but the best month to search for them is July.

The primary larval foodplants are Goat Willow *Salix caprea* and other willow species.

Brown Hairstreak
Thecla betulae
This elusive species has only ever been recorded reliably in the far east and north-east of the County, with the first sightings in the early twentieth century. Despite having a stronghold in Devon, where its historical distribution touched the Cornish border, the Brown Hairstreak has only rarely been reported in Cornwall. The butterfly uses a variety of habitats containing the principal larval foodplant, Blackthorn *Prunus spinosa*; these include hedgerows, scrubland and woodland edges. An important part of these habitats is the presence of an assembly tree, usually Ash *Fraxinus excelsior*, which is used by adults looking for a mate (Eeles, 2019). The main flight period is in August and September. Eggs can be found throughout the winter and this is usually the preferred stage to monitor, as the eggs are less elusive than adults when you know where to look.

Brown Hairstreak, Devon

There have been only four sightings of this species in Cornwall since 1938. One was of eggs found in the winter of 1988–1989 near Launceston on the banks of the River Tamar, which forms much of the border between Cornwall and Devon. Eggs were also recorded on the east side of Bodmin Moor at Bowithick in 1995. However, since then the only records in the County have been of single adults in 2014 and 2018 at the Marsland Nature Reserve. It is possible that populations of Brown Hairstreak are present in the far east of Cornwall and have gone unnoticed. Further searches for adults and eggs in areas of suitable habitat are required to determine the true status of this species.

The primary larval foodplant is Blackthorn.

Small Blue
Cupido minimus
Despite its larval foodplant Kidney Vetch *Anthyllis vulneraria* being found almost

continuously along Cornwall's coastline, the Small Blue has never been confirmed as a resident in the County. There has only been a handful of reliable records over the years, all in coastal locations. The most recent report was in 2015, just north of Bude; this was not, however, supported by a photograph, but was seen by a recorder who is used to identifying Small Blue in their home county. The species was unsuccessfully introduced to The Lizard and two locations in West Penwith in the 1980s by an independent body, but none were seen in subsequent years (Wacher *et al.*, 2003). This butterfly has two broods, and adults can be seen from May to early June and from late July to early August.

The primary larval foodplant is Kidney Vetch.

Large Blue
Phengaris arion

Cornwall was previously one of England's strongholds for the Large Blue, its core distribution spanning the coastal valleys east of Rock all the way north to the County border at Marsland. Records also exist from other parts of the County: notable sightings backed up by voucher specimens were made at Minack Head on the West Penwith peninsula in 1968 and in the St Austell area in 1906, suggesting that the species may have had a wider historical distribution in Cornwall. The butterfly experienced a sharp decline across the UK, becoming extinct in Cornwall in 1973 and nationally in 1979. This decline is thought to have resulted from loss of habitat caused by reduction of grazing on Large Blue sites due to changes in farming methods and the loss of rabbit populations to myxomatosis. Research by Jeremy Thomas in the 1970s discovered that the butterfly has a critically important relationship with the red ant *Myrmica sabuleti* (Thomas, 1980). He found that after feeding on Wild Thyme *Thymus drucei* in their early instars the larvae would fall to the ground and be picked up by *M. sabuleti* and taken into the ants' nests, where they would instead turn to feeding on the ant grubs. This revealed that the Large Blue is dependent on strong populations of *M. sabuleti* to survive; the ant species requires habitat with a short sward and warm temperatures to prosper (Eeles, 2019). In Cornwall the butterfly favoured well-grazed south-facing coastal slopes with occasional scattered scrub. However, large areas of this habitat are now dominated by species such as gorse *Ulex* spp. and Blackthorn, which out-compete the larval foodplant and create shaded conditions too cool for *M. sabuleti*. Adult Large Blue were historically recorded in late June and throughout July, slightly later than in other areas of the country.

As part of the national strategy to re-establish the Large Blue in the UK it was reintroduced to Dannonchapel on the north Cornwall coast in 2000, when 12 adults were released followed by 300 larvae. The butterfly was recorded on the site until 2007 but has not been seen since. This reintroduction was thought to be unsuccessful for several reasons. Primarily, the site required intensive management of targeted grazing and burning to suppress scrub regrowth. After key members of staff involved in the reintroduction project left or moved into different positions, the land managers felt they were unable to maintain the habitat for the butterfly. The site's thin soil also made the area vulnerable to drought, resulting in the inability of Dannonchapel to support far lower densities of the Large Blue compared to other locations (pers. comm. Thomas, 2021). However, due to successes in other areas, the Large Blue is now established on a number of sites elsewhere in South-west England and has been documented naturally expanding its range from these locations into suitably managed habitat. At the time of writing there are no immediate plans by Butterfly Conservation to attempt another reintroduction of the Large Blue in Cornwall, although this has not been ruled out for the future.

The primary larval foodplant is Wild Thyme.

Rare migrants, vagrants and others

Swallowtail
Papilio machaon gorganus

Any naturally occurring Swallowtail butterfly seen in Cornwall is likely to be the subspecies *gorganus*, which will have migrated from continental Europe. However, the butterfly is also resident in the UK, where it is restricted to the Norfolk Broads. There the subspecies is *britannicus*, which is slightly darker and smaller than its continental cousin. There have been few sightings of Swallowtail in Cornwall. Two separate adults were seen near Marazion in 2001. A single larva was found in a Hayle garden in October 2014 and a further two larvae were found in a garden in Looe during August 2018; on both occasions they were observed feeding on Fennel *Foeniculum vulgare*. The origin of these larvae is unknown.

Photos of so-called Swallowtail caterpillars are sometimes sent to the County Recorder and turn out to be larvae of the Mullein moth *Cucullia verbasci*. The two caterpillars have similar markings, but the Swallowtail has orange and black patterning on a pale green background, whereas the moth has yellow and black on a white background. A distinguishing feature of the Swallowtail larva is the *osmeterium*, a fleshy forked orange organ protruding from its first segment, raised when the caterpillar is threatened, and giving off a repellent smell.

The primary larval foodplants in Europe are Apiaceae. Milk-parsley *Thysselinum palustre* is used in Norfolk but is not present in Cornwall (French, 2020).

Bath White
Pontia daplidice

Although the Bath White is a rare migrant to the UK, it can appear in large numbers. In 1945, known as the '*daplidice* year', high numbers were seen in the southern counties of Britain, and Cornwall accounted for the most sightings, of over 450 adults. The butterfly is similar in appearance to the female Orange-tip *Anthocharis cardamines*, but the two species can be distinguished with relative ease in the field if the observer can get a good view. The majority of previous records of Bath White are coastal and it was last recorded in Cornwall in 2002. Most sightings are from July, although it has also been seen in August, September and October.

The primary larval foodplants are the cabbage family Brassicaceae.

Pale Clouded Yellow
Colias hyale
and Berger's Clouded Yellow
Colias alfacariensis

Both these butterflies are rare migrants to the UK and are difficult to distinguish from each other and from the commoner migrant Clouded Yellow *Colias croceus*. (See chapter 4.)

There are numerous historical records of Pale Clouded Yellow in Cornwall, but the species has not been recorded in the County since 2006. There is only one record of the even rarer Berger's Clouded Yellow, which was seen at Polruan in 1960.

The primary larval foodplants are all in the legume family. Pale Clouded Yellow uses clovers *Trifolium* spp. and Lucerne *Medicago sativa* subsp. *sativa*. Berger's Clouded Yellow uses Horseshoe Vetch *Hippocrepis comosa*, not recorded in Cornwall (French, 2020).

Monarch
Danaus plexippus

Monarch, photographed in Spain

There have been many sightings in Cornwall of this large, spectacular, orange and black migrant from North America, which at times has proved to be a frequent visitor to the coasts of West Penwith and The Lizard, notably in 1995, 1999 and 2001. In 1999, 70 of these butterflies were recorded in Cornwall. This contrasts with seven in 2014, which was the best year in the last decade. Two were seen in 2019 after a gap of several years. The majority of sightings are in September and October, and most records correlate with major migrations of Monarch on the east coast of the United States and with strong easterly flows of warm air across the Atlantic, which can bring an influx to our shores (Asher et al., 2001).

The primary larval foodplants are milkweeds *Asclepias* spp., which are not native to the British Isles.

Queen of Spain Fritillary
Issoria lathonia

This rare migrant butterfly has been recorded in Cornwall several times, including three records since 2000, the most recent of which was in 2008, near Tintagel. The characteristic wing shape and the prominent silver markings on the underside of the hindwings distinguish this butterfly from other fritillaries. In September 1945, around 25 fresh adults were observed near Portreath, and these were thought to be the offspring of an earlier migrant. However, the butterfly is believed to be unable to survive the British winter.

The primary larval foodplants are Field Pansy *Viola arvensis* and Wild Pansy *Viola tricolor*.

American Painted Lady
Vanessa virginiensis

This butterfly is similar in appearance to the Painted Lady *Vanessa cardui*, a common migrant to our shores, but is far rarer. Only three sightings are known for Cornwall: single adults were seen in 1826, 2012 and 2013. The butterfly could originate from North America, the Canary Isles, Madeira or Portugal.

There are several features that distinguish it from the Painted Lady. On the upperside of the wings there is an extra tiny white spot on the orange cell below the black wing tip, and faint eyespots near the border of the hindwing. On the underside there are just two large eyespots on the hindwing of the American Painted Lady, whereas the Painted Lady has four smaller ones.

The primary larval foodplants are thistles *Carduus* spp. and *Cirsium* spp.

Camberwell Beauty
Nymphalis antiopa

The Camberwell Beauty is a rare migrant from Scandinavia, so it is more commonly seen on the east side of Britain. Nonetheless, it has been recorded across Cornwall on various occasions and there have been five sightings since 2000, the most recent of which was in 2010. Its distinctive appearance – deep maroon/dark brown with a white/pale yellow fringe bordering a row of iridescent blue dots on both hind- and forewings – should avoid confusion with any other butterfly seen in Cornwall. The species tends to be recorded in August and September and then hibernates as an adult; however, it is thought to be unable to survive Cornwall's mild, wet winters.

The primary larval foodplants are birches *Betula* spp., elms *Ulmus* spp., poplars *Populus* spp. and willows *Salix* spp.

Large Tortoiseshell
Nymphalis polychloros

Once quite widespread across Britain, this species declined to extinction in the UK in the 1960s. It may never have been a permanent resident in Cornwall, but there have been occasional records from probable migrants since 1849. The butterfly is similar to the resident Small Tortoiseshell *Aglais urticae*, but they are possible to tell apart when examined closely, although a photograph is always valuable for verification.

There have been 16 records this century, three of them in 2019 and a further three in 2020. It is possible that climate change is bringing about an increase in migrants, which has been noted in other southern counties in Britain, although the possibility that this butterfly has been captive bred and released cannot be discounted. It is quite capable of surviving the winter in Cornwall, hibernating as an adult.

The primary larval foodplants are elms.

Long-tailed Blue
Lampides boeticus

This diminutive migrant butterfly visits the UK occasionally from the Mediterranean. Butterfly Conservation suggests that an increase in frequency over recent years could be the result of climate change. The species is so named because of the single, wispy tail on each hindwing, which has adjacent black-centred eyespots. Males are bright blue, whilst females are a duller mixture of brown and blue. The undersides of both are beige with many wavy white lines.

There have been more sightings in Cornwall of this butterfly in the present century than at any time previously. There were only five accepted records prior to 2001, whereas in the first decade of the twenty-first century there were four records, and in the second decade there have been seven. Most have been seen on the south coast, although there was a sighting in 2015 at Nanstallon, to the west of Bodmin. It may well be that many more have been visiting the County but have been overlooked. Most sightings have been in July or August, although in October 2020 two pristine males were recorded in separate parts of the County. Successful breeding and immature life cycle stages have been documented in other southern counties, such as Sussex, so it is possible that the fresh specimens seen in Cornwall were the locally hatched progeny of earlier migrants. The larvae feed on leguminous plants, and both eggs and caterpillars are worth searching for on foodplants in areas where adults have been observed.

The primary larval foodplants are Broad-leaved Everlasting-pea *Lathyrus latifolius*, Broom *Cytisus scoparius* and Bladder-senna *Colutea arborescens* (which is not native in the British Isles).

Long-tailed Blue, Roseland peninsula, October 2020

Geranium Bronze
Cacyreus marshalli

There was a single record of this South African species in Cornwall from a garden in Gwithian in 2002. It is thought to have been accidentally imported as an immature stage on plants from southern Spain.

The primary larval foodplants are *Pelargonium* spp. and *Geranium* spp.

Short-tailed Blue
Cupido argiades

This butterfly is a very rare migrant to Britain and has been recorded in Cornwall on only four occasions. All the records are from coastal locations; two were seen in July 1945 and another in July 1952, and the most recent sighting was in August 2003.

On the wing it can be mistaken for other 'blues', but the small 'tail' at the rear of each hindwing is distinctive – much shorter than those of the Long-tailed Blue – and the undersides of the hindwings, which otherwise resemble those of the Holly Blue *Celastrina argiolus*, have a pair of distinctive orange spots.

The primary larval foodplants are Common Bird's-foot-trefoil and Gorse *Ulex europaeus*.

Historical records

The following list includes species of butterfly that were sighted or captured in Cornwall more than 50 years ago and for which there is no more recent evidence of their presence in the County. Supporting evidence for some of these records can be patchy or absent altogether.

UK resident species

Lulworth Skipper
Thymelicus acteon

There are two historical records of the butterfly in Cornwall from the late nineteenth century, but the reliability of these is doubted and they are not backed up by specimens (Penhallurick, 1996). In 1979 a Lulworth Skipper was captured near Polperro and is now in a museum in Amsterdam. Polperro is not near the known historical locations, so the butterfly is thought to have been either a migrant or an escapee

(Bourn and Warren, 1997). The primary larval foodplant of this species, Tor-grass *Brachypodium rupestre*, is not found in Cornwall (French, 2020).

Duke of Burgundy
Hamearis lucina

This species may once have been a resident in Cornwall, although there is little evidence to support this, with only three known records from the County. The most recent sighting was from near Bodinnick in 1956. It seems highly unlikely that the butterfly is now present in Cornwall, and the nearest known populations are a considerable distance away in Somerset and Dorset.

Chalk Hill Blue
Polyommatus coridon

There are several historical records of Chalk Hill Blue in the County, the most recent of which is from 1947. As the larval foodplant (Horseshoe Vetch) does not occur in Cornwall it is unlikely that the butterfly was ever a resident here and any sightings were possibly of vagrants.

Rare migrants, vagrants and others

Apollo
Parnassius apollo

A single Apollo was recorded in Cornwall at Carclew, just north of Falmouth, in 1826, although it was thought to have been imported on plants from abroad.

Scarce Swallowtail
Iphiclides podalirius

The Scarce Swallowtail was recorded in Cornwall twice in the 1970s. The butterfly is thought to be a rare migrant, but human intervention cannot be ruled out.

Mediterranean Fritillary
Argynnis pandora

This butterfly is a very rare migrant and has only been recorded in the UK a handful of times. There was a single sighting in Cornwall in August 1911, when an individual was collected from a wooded valley near Tintagel.

CHAPTER 6
Butterflies of the Isles of Scilly

View across the islands from Teän

Introduction

The Isles of Scilly form an archipelago of over 200 islands, islets and rocks, rarely rising more than 30m above sea level and situated 45km (28 miles) west of Land's End. Rising sea level around 1,400 years ago resulted in the current island configuration, the remnants of a drowned landscape (Charman *et al.*, 2016). There are five inhabited islands: St Mary's, St Martin's, Tresco, Bryher and St Agnes (connected to Gugh at lower tides). The climate in Scilly is mild, with relatively humid conditions persisting all the year round. Any substantial frosts are rare, and snow rarer still. Even in the drought year of 2018, the temperature in Scilly reached a maximum of only 23°C on three dates in July (pers. comm. Headon, J., 2020). The current land area of the islands is about 1,600 hectares, so the scale compared with Cornwall is very different, the largest island, St Mary's, being just 680 hectares (including infrastructure, homes and shoreline).

However, there are 26 Sites of Special Scientific Interest (SSSI), most of which are managed by the Isles of Scilly Wildlife Trust.

The Isles of Scilly, with largest islands named

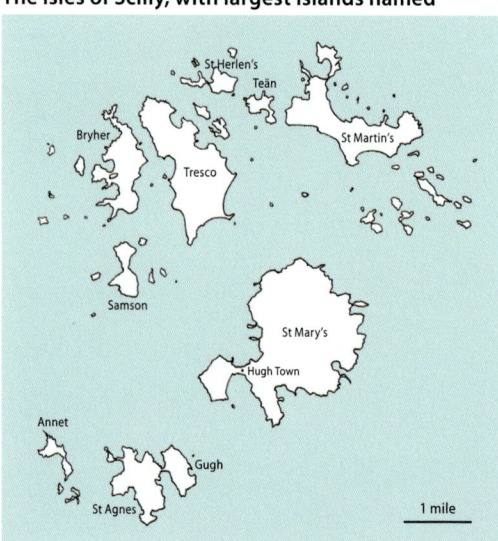

214

Island biodiversity: an overview

Colonisation and extinction are common themes for island biodiversity. Even small and seemingly sedentary butterfly species can make sea crossings, and a surprising amount of chance migration and vagrancy has been recorded in Scilly. The prevailing westerly and south-westerly winds may hamper natural dispersal from mainland Britain, for example of ballooning spiders, and may also hinder the likelihood that butterflies will reach the islands. However, the extent of drift migration of moths from the mainland is a striking feature in late spring and early summer during spells of more favourable weather, so why not butterflies too? Some may instead reach the islands through accidental introduction. Greater detail concerning island biogeography and the dynamic equilibrium between colonisation and extinction can be found elsewhere, for example in *British and Irish Butterflies: An Island Perspective* (Dennis and Hardy, 2018). These authors cite Edmund Brisco Ford's studies of Meadow Brown *Maniola jurtina cassiteridum* and Common Blue *Polyommatus icarus*, and "several interwoven strands of geographical variation" as a distinctive feature of British and Irish offshore islands (Dennis and Hardy, 2018: 141). In addition, they provide lists of species recorded by island and indicate the islands that have no butterfly records.

No local extinctions in Scilly have been recorded in recent years, in contrast to bumblebees *Bombus* spp., which have declined from eight species to four on the islands. However, during the 1950s and 1960s there were unexpected records for Small Heath *Coenonympha pamphilus* and Wall *Lasiommata megera* that, if not misidentifications, suggest the possibility that these species were at least temporarily resident then. Similarly, we will never know whether Large Tortoiseshell *Nymphalis polychloros* was ever established on the islands (Beavis, 2004).

Without a doubt, islands and their butterflies are fascinating subjects. The resident species may have local forms or demonstrate different phenology or population variation from their mainland counterparts. Expectations can and should be turned on their head – you are far more likely to see a Monarch *Danaus plexippus* on the islands than a Brimstone *Gonepteryx rhamni*, Gatekeeper *Pyronia tithonus* or fritillary!

For those requiring a more thorough overview of the butterflies of the islands, the 2004 paper by Ian Beavis is highly recommended reading, although we now have a greater understanding of the status of Comma *Polygonia c-album* and Peacock *Aglais io* on the islands. The article also covers the curious case of the 'Teän Blue' and the history and likely origin of the Speckled Wood *Pararge aegeria insula* in Scilly.

Comma

Scilly: an evolving environment for butterflies

Scilly looked very different about 11,700 years ago, at the end of the Last Glacial Period. It was then a single, large island, covered by a closed forest of birch *Betula* spp., oak *Quercus* spp. and Hazel *Corylus avellana*, with open areas in the low-lying wetlands. The early butterfly fauna is unknown, but perhaps there were woodland species that disappeared as open ground was created gradually by human activity and rising sea levels. The islands have been largely open for about 3,000 years (Charman *et al.*, 2016), with heathland gradually increasing. There are several small-scale dune systems.

The only common broadleaved trees today are willow *Salix* spp. and elm *Ulmus* spp., notably English Elm *Ulmus procera*, the islands having escaped Dutch Elm disease (Parslow, 2007). However, Hawthorn

Crataegus monogyna is reasonably widespread on St Mary's, and small numbers of other broadleaved species have been planted. The most significant recent vegetation change is the introduction of non-native species for gardens and amenity planting, coastline stabilisation (e.g. Hottentot-fig *Carpobrotus edulis*), hedges (particularly evergreen shrubs to protect flower crops) and tree cover (including conifer shelter belts) from the mid-nineteenth century onwards.

On Scilly, the lack of many butterfly species recorded in Cornwall can be explained by the absence of their larval foodplants. For instance, Alder Buckthorn *Frangula alnus*, larval foodplant of the Brimstone, and Devil's-bit Scabious *Succisa pratensis*, larval foodplant of the Marsh Fritillary *Euphydryas aurinia*, do not grow on the islands. Neither does Garlic Mustard *Alliaria petiolata*, larval foodplant of the Orange-tip *Anthocharis cardamines*, although the species is known to use a variety of crucifers, including Cuckooflower *Cardamine pratensis*, which, though present, is scarce on Scilly. On the other hand, despite an abundance of Common Dog-violet *Viola riviniana* and Bracken *Pteridium aquilinum*, there are no resident fritillaries on the islands; nor are there any hairstreaks or skippers, for which there is, arguably, suitable habitat. Similarly, despite the presence of suitable habitat, no Wall, Gatekeeper, Grayling *Hipparchia semele*, Marbled White *Melanargia galathea*, Brown Argus *Aricia agestis* or Small Heath are found on Scilly. Although the black ant *Lasius alienus* is absent from Scilly and *L. niger* is widespread and common (Beavis, 2012), the only resident blue species are Common Blue and Holly Blue *Celastrina argiolus*. The relatively cool temperatures may prevent establishment of some species on the islands should they arrive, while the number of broods of some resident species, such as Small Copper *Lycaena phlaeas*, appear to be fewer than on the mainland.

Well-informed amateur naturalists and professional scientists have visited the islands over many years. However, the first systematic accounts of the islands' butterflies had to wait until the twentieth century (Blair, 1925; Richardson and Mere, 1958; and Agassiz, 1981). This century, the inaugural issue of the annual *Isles of Scilly Bird and Natural History Review* included an account of the islands' butterflies (Hicks, 2000). Following a number of annual summaries (for example, Scott *et al.*, 2006), a further update was published in 2015 (Dawson, 2015), prompted by an increased interest in butterflies and improved recording, as a result of having more resident birdwatchers and naturalists on the islands. Included in the 2015 review was the addition of two vagrant species, Long-tailed Blue *Lampides boeticus*

Bulb field sheltered by hedging, St Mary's

and Queen of Spain Fritillary *Issoria lathonia*. Dawson (2015) and the ERICA database have been important sources for this chapter of the book. Predictably, most butterfly records and the greatest diversity of species are from the inhabited islands, particularly St Mary's. The uninhabited islands support good numbers of breeding Common Blue, Meadow Brown and Small Copper, with records of Large White *Pieris brassicae*, Small White *P. rapae* and regular migrant species on the few occasions that these islands are visited by observers.

There are 28 species reported from the islands, of which 15 are resident or regular visitors and 13 are either vagrants (for example, species that commonly breed in mainland UK, but do not breed in Scilly) or rare migrants (from continental Europe, North Africa or North America). (This contrasts with the 37 species known to be resident or regular visitors to Cornwall described in chapter 4, and those recorded in Cornwall as occasional visitors and rarities in chapter 5.) The summary presented here is made on the basis of butterfly records received in good faith. However, authentication is sometimes in doubt, such as frequent reports of Gatekeeper, Marbled White and Small Heath (in unexpected numbers), for example from Big Butterfly Count submissions.

Resident and regular visiting butterflies on Scilly

The 15 butterfly species currently resident or regularly occurring on the islands are listed in Table 1. Three of these are locally common on one or two of the islands only: Green-veined White *Pieris napi*, Comma and Ringlet *Aphantopus hyperantus*. Of the commoner migrants, Red Admiral *Vanessa atalanta* breeds annually, Painted Lady *V. cardui* breeds occasionally, and breeding has been suspected for Clouded Yellow *Colias croceus*.

Of the resident butterflies, year-to-year variation in numbers is particularly pronounced in vanessids, such as Small Tortoiseshell *Aglais urticae* and Peacock, in a way that is often out of step with fluctuations of the same species on the mainland. A possible explanation for these discrepancies is the impact of parasitic wasps in what is a more or less isolated, closed system on the islands. Variation for other species is less evident, and is often referred to as a 'good' or a 'poor' year by local observers (such as for Comma, Common Blue, Holly Blue and Small Copper).

Two species of butterfly found on the islands have distinctive island forms, Speckled Wood *Pararge aegeria insula* and Meadow Brown *Maniola jurtina cassiteridum*. The island form of Speckled Wood, with its cream and orange spots on a warmer ground colour than its mainland cousin, can give rise to confusion with Wall (a rare vagrant on Scilly). Its origin is something of a mystery. There are early twentieth-century records, followed by a gap until David Hunt, who arrived in 1964, recorded one adult on St Agnes in 1967. Within a short time, colonies had been discovered

Small Copper aberration schmidtii, St Agnes, 2015

Speckled Wood island form subspecies insula

on St Mary's and the islands' population of this subspecies, together with those from the Channel Islands, were described as a distinct form. Ian Beavis discusses two possible explanations: undetected residency from time immemorial (since there are old records), or, the favoured theory, colonisation from the Channel Islands, with their identical form (Beavis, 2004). This dappled-shade-loving species is now widespread and abundant across the islands.

The island form of Meadow Brown, subspecies *cassiteridum* can be confused with Gatekeeper owing to its brighter and more

Meadow Brown island form subspecies *cassiteridum*

Table 1 Resident and common migrant butterfly species of the Isles of Scilly

Species	Status	Notes
Large White *Pieris brassicae*	Common resident	Very common. Supplemented by migrants.
Small White *Pieris rapae*	Common resident	Very common. Supplemented by migrants.
Green-veined White *Pieris napi*	Locally common resident	Common only on St Mary's, where it seems to utilise a range of crucifers as larval foodplants and is often found on or around agricultural land.
Clouded Yellow *Colias croceus*	Common migrant	Recorded in most years. Assumed to be a rare breeding species based on observations of mating pairs and early spring records of likely local emergence.
Speckled Wood *Pararge aegeria insula*	Common resident	Distinctive island form subspecies *insula*. Possible accidental import. The second generation is much the larger and extends well into October.
Ringlet *Aphantopus hyperantus*	Locally common resident	Discovered on St Martin's in 1995 and St Mary's in 1996 and slowly extending its range. Records confined to June and July except one in early August 2020. Single records from Gugh, St Agnes and Bryher; none from Tresco.
Meadow Brown *Maniola jurtina cassiteridum*	Common resident	Distinctive island form subspecies *cassiteridum*. Commonest butterfly in summer.
Red Admiral *Vanessa atalanta*	Common migrant	Annual breeder and has been recorded in all months of the year, including all months in one year (2019).
Painted Lady *Vanessa cardui*	Common migrant	Annual migrant and occasional breeder, but does not overwinter.
Peacock *Aglais io*	Common resident	Two broods in Scilly, the second brood usually the more abundant.
Small Tortoiseshell *Aglais urticae*	Common resident	A boom year in 2018, but numbers crashed in 2019 and showed little improvement in 2020.
Comma *Polygonia c-album*	Locally common resident	Now resident on St Mary's and Tresco, probably also Bryher and St Martin's. May colonise St Agnes soon. Possible third generation on St Mary's in 2018.
Small Copper *Lycaena phlaeas*	Common resident	Typically three broods in Scilly. One example of the aberration *schmidtii* recorded (St Agnes, 28 September – 13 October 2015).
Holly Blue *Celastrina argiolus*	Common resident	First recorded in 1966 and presumed to be an accidental import. Some evidence for a partial third brood.
Common Blue *Polyommatus icarus*	Common resident	Two broods, the second typically being the larger. No evidence of a partial third brood, although worn second-brood individuals may be seen in October.

extensive orange patches, and a proportion – particularly females – exhibit twin white dots in the 'eye' of the forewing.

By contrast, Peacock has long been present on the islands, although it became established on St Agnes only from the late 1990s. Perhaps it had been temporarily resident on this most remote of the inhabited islands at some point in the past, but for such a distinctive species there is no evidence from islanders that this is the case.

An early colonist (and likely accidental introduction) was Holly Blue, which was first recorded in 1966 by G.M. Spooner (pers. comm. Beavis, I., 2020) and has since spread to all the inhabited islands, where it is common. There is little Holly *Ilex aquifolium* on the islands, and egg laying has been observed on Greater Bird's-foot-trefoil *Lotus pedunculatus*, Yarrow *Achillea millefolium* and Broom *Cytisus scoparius* in spring; ivy *Hedera* spp., the foodplant favoured by the summer brood, is abundant, however.

The most recent colonists (from the late 1990s) are Comma and Ringlet. The earliest records of Comma are from 1971 and 1982, but there is then a gap in records until the late 1990s. This species is locally common only on St Mary's and Tresco, with the first reports of egg laying and larvae as recently as 2016. Comma is thought to be on the verge of colonising St Agnes; suspected breeding there in 2015 is thought to have resulted from a mated female arriving from St Mary's that spring.

Individual islands differ remarkably in their resident butterflies and their distribution. For example, Green-veined White is common and widespread on St Mary's but generally scarce elsewhere in Scilly. It is an exceptionally scarce vagrant to St Agnes and Gugh (three records in 60 years); presumably those that are seen are drift migrants from St Mary's during favourable conditions. The drought conditions of 2018 appeared to prompt widespread dispersal of Ringlet on St Mary's, and the first record for the species on St Agnes occurred that year. Since then, the dispersal of Ringlet seems to have become consolidated, with increasing records away from its central and northern stronghold on St Mary's. Grasses such as Cock's-foot *Dactylis glomerata* and False Brome *Brachypodium sylvaticum*, typical foodplants for this species, are widespread on the islands.

There is no recent history of live butterfly releases on the islands, although there was a butterfly 'farm' at Longstone, St Mary's, for a number of years, which closed in 1986. An experimental release in 1949 of Gatekeeper to the uninhabited island of Teän failed.

Occasional chance arrivals on Scilly

Common mainland species that are extremely unusual on the islands or may be previously unrecorded on Scilly include Brimstone, Silver-washed Fritillary *Argynnis paphia*, Wall and Gatekeeper. To be accepted, a record of such species would need to be supported by photographic evidence. This would also apply to a number of species that are occasional or infrequent visitors to the islands. In some cases, of course, there is no chance for a photograph, for example of a 'fritillary' flying fast across the downs, or the tantalising glimpse of a 'swallowtail' flying in from the sea, never to be seen again. Monarch, however, is unlikely to be confused with anything else, and its chance arrival on the islands from North America frequently coincides with the arrival of large numbers of birdwatchers for the autumn bird migration.

Table 2 is a collated list of reports received over the past 20 years for species that we would not expect to see on the islands. Documentation, including either a detailed description or photographs, is more likely for rare migrants, with considerable interest shown among local and visiting naturalists that can result in a sudden surge of enthusiasm, and there is the incentive to take notes or

Queen of Spain Fritillary

many photographs of an unfamiliar species. By contrast, vagrants that a visitor might see commonly at home are not subject to the same level of scrutiny and are much less likely to have a photograph taken.

Ian Beavis has been recording butterflies on the islands nearly every year since 1975. He regards the twentieth-century records of Brimstone as examples of potential colonisation (Beavis, 2004), despite the complete absence of its foodplant, Alder Buckthorn, from the islands. Orange-tip is perhaps a more plausible candidate for becoming established as a number of its

Table 2 Other butterfly species recorded in the past 20 years on the Isles of Scilly
These include species that arrive as vagrants (breeding in mainland Britain and Ireland but not thought of as migratory), as well as rare migrants coming from further afield. This is a compilation from a number of Cornwall and Isles of Scilly sources as well as the internet, and accuracy cannot be guaranteed in all cases (for example, the 4 June 2006 record for the Silver-washed Fritillary is unusually early).

Species	Status	Notes
Swallowtail *Papilio* spp.	Rare migrant	One unconfirmed report on St Mary's, 27 September 2020.
Orange-tip *Anthocharis cardamines*	Vagrant	One male on St Martin's (estimated 2016) with a previous male on St Mary's in 1986. A possible accidental import, as the species overwinters as a pupa.
Brimstone *Gonepteryx rhamni*	Vagrant	1911, 1977, 1984, unlocalised (Wacher *et al.*, 2003). May 2004, 6 May 2016, St Mary's. 27 September 2013, St Martin's. Older reports inadequately documented.
Monarch *Danaus plexippus*	Rare migrant	Spectacular arrival in 1999 with 178 records (Tunmore, 2000). Recorded in 12 of the past 20 years, with most in 2006 (nine individuals estimated), and last seen in 2019 (photographed by visitors at Tresco Abbey Gardens).
Wall *Lasiommata megera*	Vagrant	September 1962, September 1963, October 1972, 20 May 1992, 26 August 2006, St Mary's. 1982 and 1985, St Martin's. 11 May 2008, Bryher.
Gatekeeper *Pyronia tithonus*	Vagrant	The only record bearing scrutiny appears to be from St Agnes, 1978, the observer being familiar with the islands' butterfly fauna. As noted before, any suspected sighting of Gatekeeper should be considered with care and preferably supported by a photograph.
Silver-washed Fritillary *Argynnis paphia*	Vagrant	4 June 2006, Bryher – no details.
Dark Green Fritillary *Speyeria aglaja*	Vagrant	1 August 2008, St Mary's – closely seen by two experienced observers.
Fritillary (unidentified)	Vagrant	3 June 2006, St Martin's – no details. 21 October 2018, St Mary's – no details.
Queen of Spain Fritillary *Issoria lathonia*	Rare migrant	Two females recorded two months apart at Trenoweth, St Mary's (August and October 2006). The abundance of the larval foodplant (Field Pansy *Viola arvensis*) raised the question of local breeding.
American Painted Lady *Vanessa virginiensis*	Rare migrant	10 September 1998, St Agnes. 18 October 2003, 12–13 October 2006, St Mary's.
Camberwell Beauty *Nymphalis antiopa*	Rare migrant	October 1983, 29 September 2006, St Mary's. 16 August 1984, Samson. 9–10 October 2002, St Martin's. 10 September 2004, St Agnes.
Large Tortoiseshell *Nymphalis polychloros*	Rare migrant	13 July 2011, 10 March 2012, 1 April 2012, St Mary's. Late 1980s, early/mid July 2011, 2 July 2019, St Agnes. Three individuals, seen singly, July and August 1934, Tresco.
Long-tailed Blue *Lampides boeticus*	Rare migrant	Earliest record 1996 (Wacher *et al.*, 2003). Mainly recorded in 2006 and 2011, but also 2007, 2013 and 2018. In 2006 at least six were reported 9–24 October, mostly from St Mary's. The October 2011 influx saw individuals on St Mary's and St Agnes on a number of dates, with egg laying on Gorse *Ulex europaeus* on St Agnes on 12 October.

alternative larval foodplants are found on Scilly, albeit in low numbers.

An unsubstantiated record of Silver-studded Blue *Plebejus argus* was undoubtedly a misidentification (Beavis, 2004), so is omitted from the list. The following four species have a more enigmatic history and are worthy of inclusion:

Bath White *Pontia daplidice*
Recorded 17 October 1977, St Mary's. Many of these immigrants from central and southern Europe must have arrived in the so-called invasion year of 1945, but no one in Scilly recorded them. There is a possibility that UK records of Bath White (as with recent records in the Netherlands) may refer to Eastern Bath White *Pontia edusa*.

Small Heath *Coenonympha pamphilus*
Smith (1997) cites an intriguing series of records in the 1950s and 1960s in Scilly that may have arisen from accidental introduction with temporary colonisation. Although speculative, it is possible that post-war changes in patterns of imports and exports may have contributed to this, such as an increase in agricultural and horticultural activity on the islands and an accompanying gradual introduction of containerised freight (Chudleigh, 1992).

White Admiral *Limenitis camilla*
Recorded 29 July 1999, St Mary's (Smith, 2002); 23 August 2018, Tresco. The 1999 observers were convinced by what they saw, nectaring on Buddleja *Buddleja davidii* (Smith, 2002).

Pale Clouded Yellow *Colias hyale*
Recorded 1900, 'many'; 22 September 1968, St Mary's (Smith, 1997). This rare immigrant to the British Isles is widespread in the Palaearctic, and is a good candidate to arrive in Scilly, but there is potential confusion with the *helice* form of Clouded Yellow.

Penninis, St Mary's

CHAPTER 7
Key sites for butterflies

Cliff top view of Mullion Cove, The Lizard

Introduction

Cornwall is richly diverse in its habitats and butterflies. In this chapter, we highlight 20 sites that could be considered as butterfly 'hotspots', with none having fewer than 20 of our 37 resident and common migrant species. The County database, ERICA, was used to generate a map showing the number of butterfly species per 1km grid square using all records made from 2008 to date. The list of 20 prime butterfly sites was drawn up using this hotspot map as the starting point, the selection then adjusted to ensure a good spread of sites across the County. The list includes inland sites whose important habitats can be overlooked compared with better-known coastal locations. That said, beyond the coastal sections described below, many other stretches of the South West Coast Path will reward the watchful walker with plenty of butterflies at the right time of year and in the right weather.

The 20 selected sites are listed in alphabetical order in the table opposite, and their locations shown on the accompanying map. Many of these sites are also mentioned in chapter 4 for each butterfly species, under **Best places to see** for that particular species.

Planning your visit

The access point for each site is given as both a six-figure Ordnance Survey grid reference and a postcode, determined as accurately as possible (using the website https://gridreferencefinder.com). Please beware, however, that when programmed with a postcode in Cornwall's rural areas, satnavs do not always lead you to where you want to go. The relevant 1:25,000 Ordnance Survey map is more reliable. It is also useful to familiarise yourself with the site before setting out, by consulting both Ordnance Survey maps and aerial/satellite imagery freely available on various websites. In this way, routes can be planned using footpath and habitat information.

All the sites have public rights of access, with many being 'open access land' on which the responsible walker can exercise their 'freedom to roam'. Wherever possible, places to park are

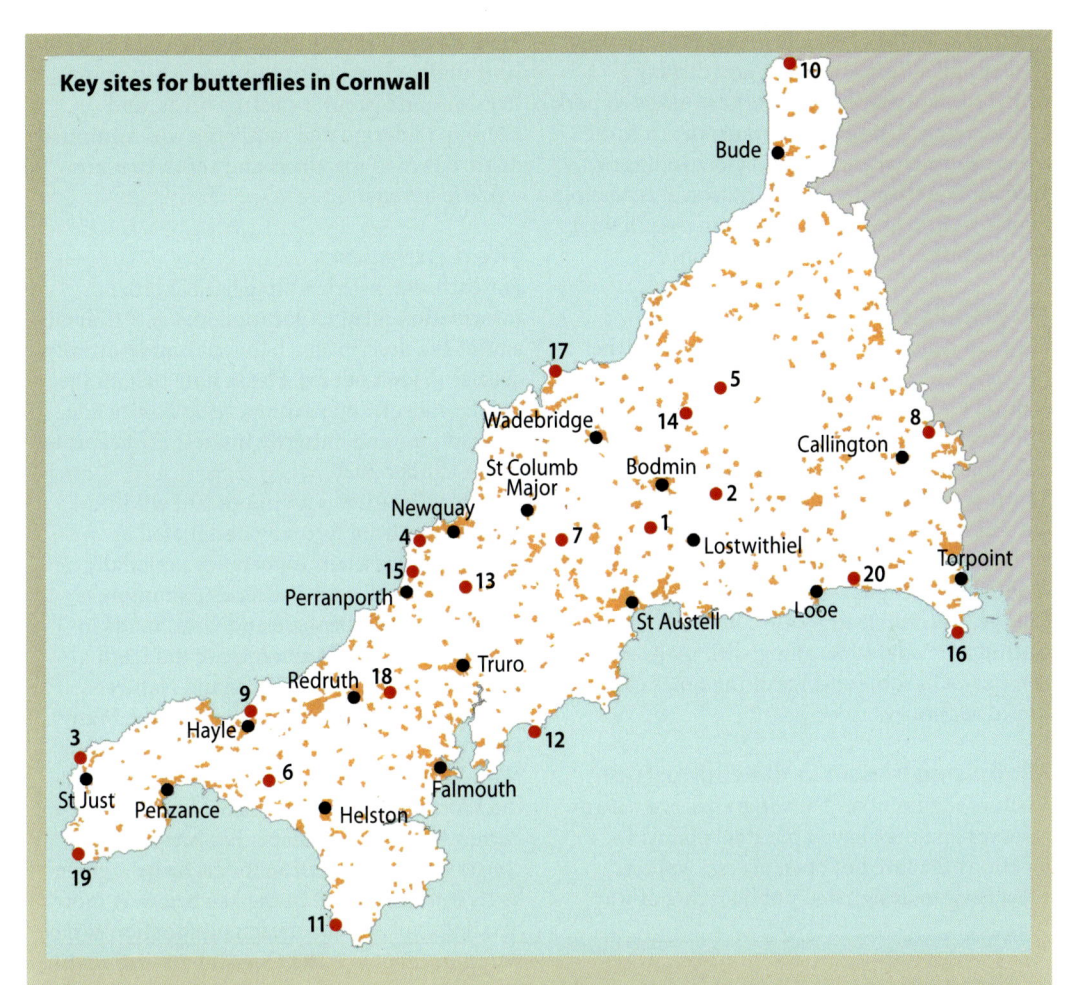

Key sites for butterflies   *Butterflies of Cornwall*

Key sites for butterflies in Cornwall

Site number and name	Nearest town/city	Number of butterfly species
1 Breney Common	Lostwithiel	31
2 Cabilla and Redrice Woods	Bodmin	24
3 Cape Cornwall, Botallack and Pendeen	St Just	27
4 Cubert Common and West Pentire	Newquay	26
5 Garrow Tor	Bodmin	20
6 Godolphin and Godolphin Hill	Helston	29
7 Goss Moor, including St Dennis Junction	St Columb Major	29
8 Greenscoombe Wood	Callington	30
9 Hayle to Gwithian Towans, including Gwithian Green	Hayle	29
10 Marsland Nature Reserve	Bude	30
11 Mullion Cove to Lizard Point	Helston	28
12 Nare Head	Truro	30
13 Newlyn Downs	Newquay	28
14 Pendrift Bottom	Bodmin	23
15 Penhale and Gear Sands	Perranporth	28
16 Penlee Point and Rame Head	Torpoint	27
17 Pentire Point to Port Quin	Wadebridge	30
18 Poldice Valley	Redruth	26
19 Porthgwarra to Pordenack Point	St Just	28
20 Seaton Valley Countryside Park	Looe	31

223

given close to the main site access point; many are car parks (including 'pay and display'), whilst some are laybys. Visitors are asked to park considerately. Using public transport in rural areas can be challenging, and Cornwall is no exception. At the time of publication, it is possible to plan your trip using the traveline south west website https://www.travelinesw.com/.

Accessibility and safety

For each site, information is provided on the nature of the terrain, including unevenness of ground and topography, so it is the responsibility of visitors to assess and mitigate risk. It is worth noting that sections of the South West Coast Path are steep in places, with uneven ground, and have high, unfenced cliffs. Many areas are remote and do not have a mobile phone signal. The weather can suddenly change, so sensible footwear and weatherproof clothing are advisable, along with maps and a compass. On some sites, walking poles can be a useful accessory.

Minimising impact in the countryside

Visitors should note that some sites are nature reserves and may have a 'no dog' policy. In locations that are not open access, walkers must keep to designated public rights of way.

The Countryside Code
1. **Respect everyone**
   - be considerate to those living in, working in and enjoying the countryside
   - leave gates and property as you find them
   - do not block access to gateways or driveways when parking
   - be nice, say hello, share the space
   - follow local signs and keep to marked paths unless wider access is available
2. **Protect the environment**
   - take your litter home – leave no trace of your visit
   - take care with BBQs and do not light fires
   - always keep dogs under control and in sight
   - dog poo – bag it and bin it – any public waste bin will do
   - care for nature – do not cause damage or disturbance
3. **Enjoy the outdoors**
   - check your route and local conditions
   - plan your adventure – know what to expect and what you can do
   - enjoy your visit, have fun, make a memory

Even on open access land, dogs should be kept under close control to avoid disturbing livestock and ground-nesting birds, and keeping to designated footpaths will minimise disturbance to meadows and scrub that are wildlife havens.

Site descriptions

For each site, listed in alphabetical order, information includes location and access details, a brief site description, habitat(s), and butterfly species to look out for. Please note that all the details concerning parking, accessibility, site descriptions and butterfly interest are accurate as of October 2020.

Every location is an important wildlife site and has been formally designated as such, some for their importance nationally or internationally, such as Breney Common, Goss Moor, and Penhale and Gear Sands in central Cornwall, Mullion Cove to Lizard Point in the south, and Marsland Nature Reserve in the far north of the County. Many are within the Cornwall Area of Outstanding Natural Beauty (AONB) and some lie within the World Heritage Site of Cornwall and West Devon Mining Landscape. For brevity, the conservation designation for each site is given by its initials, set out in the key below. A more detailed explanation of each designation can be found in chapter 2 **The Cornish environment**.

Initials	Site designation
AONB	Area of Outstanding Natural Beauty
CWS	County Wildlife Site
CWT Reserve	Cornwall Wildlife Trust Reserve
DWT Reserve	Devon Wildlife Trust Reserve
LNR	Local Nature Reserve
NNR	National Nature Reserve
SAC	Special Area of Conservation
SSSI	Site of Special Scientific Interest
WHS	World Heritage Site

Finally, have fun exploring these sites, as well as the numerous other beautiful places around Cornwall. Please don't forget to send any records of the butterflies you see to the County Recorder (records@cornwall-butterfly-conservation.org.uk). Records can also be submitted by using the iRecord Butterflies app or via a form on the Cornwall Butterfly Conservation website.

1. Breney Common, near Lostwithiel

Access point grid reference
SX053610

Nearest postcode
PL30 5DU

Site area 147 hectares

Total number of butterfly species 31

Site designation SAC, SSSI, CWT Reserve

Parking and accessibility
- Take minor roads off main A30 at Innis Downs roundabout towards Lowertown. Parking in layby opposite Gunwen Chapel (SX052612).
- Paths with some uneven rutted surfaces (often muddy) and sections of boardwalk.

Brief site description
Breney Common, part of Cornwall Wildlife Trust's Helman Tor Nature Reserve, is a mosaic of different types of habitat, including wet and dry heath, grassland, wet woodland and areas of wetland. Conservation work has aimed to reduce and control excess scrub and tree cover, supported by grazing cattle and ponies. Much of the site was mined for tin, and the wet pockets in the hollows that support rare plants were created by tin streaming.

Main habitats present
Heathland, fens, ponds/lakes, broadleaved/wet woodland, scrub, bracken, grassland, bare ground and river/stream.

Key butterfly species
Wall, Small Heath, Small Pearl-bordered Fritillary, Dark Green Fritillary, Marsh Fritillary.

Other points of interest
The site has many rare plants including sundews *Drosera* spp., Pillwort *Pilularia globulifera* and Lesser Bladderwort *Utricularia minor*, as well as reptiles. The ponds support numerous dragonflies and damselflies, along with amphibians. The adjacent Helman Tor is a Scheduled Ancient Monument and a County Geological Site. The Saints' Way (coast-to-coast footpath) passes close by.

2. Cabilla and Redrice Woods, near Bodmin

Access point grid reference
SX129652

Nearest postcode
PL30 4BE

Site area 77 hectares

Total number of butterfly species 24

Site designation CWS, CWT Reserve

Parking and accessibility
- Turn off the A38 at White Lodge, towards Cardinham. Take first track on right, and park opposite sawmill. Entrance to reserve is further along this track.
- There is a broad, fairly level track through the reserve (sometimes muddy), with other steeper tracks and paths.

Brief site description
Cabilla and Redrice Woods are working woodlands owned and managed by Cornwall Wildlife Trust. This area of mixed woodland, including coppiced Hazel *Corylus avellana*, is within a steep-sided valley, bordered by the River Fowey. There are also woodland glades and cattle-grazed fields. A walk along the valley bottom leads to a path up a side valley (just before a bridge) with an insect-rich open hillside.

Main habitats present
Broadleaved woodland including ancient woodland, coniferous woodland, river and streams, pond, bracken, scrub, grassland and bare ground.

Key butterfly species
Small Pearl-bordered Fritillary, Silver-washed Fritillary, Purple Hairstreak.

Other points of interest
Biologically diverse site with a variety of insects including the Blue Ground Beetle *Carabus intricatus*, dragonflies, damselflies and moths. There are several bat species, including Greater Horseshoe Bat *Rhinolophus ferrumequinum* and Lesser Horseshoe Bat *R. hipposideros* (hibernating in the mine adits), as well as the Hazel Dormouse *Muscardinus avellanarius*. Woodland birds abound, including the Pied Flycatcher *Ficedula hypoleuca*. Mosses and ferns adorn both the trees and the woodland floor.

3. Cape Cornwall, Botallack and Pendeen, near St Just

Access point grid reference
SW364332 (Botallack)

Nearest postcode
TR19 7QQ (Botallack)

Site area 705 hectares

Total number of butterfly species 27

Site designation SSSI, AONB, WHS

Parking and accessibility
- Use B3306 between St Ives and Sennen Cove. Single-track road access signposted to car parks at Botallack, Cape Cornwall, Levant and Pendeen.
- South West Coast Path and other footpaths cross uneven terrain.

Brief site description
This five-kilometre stretch of coast from Cape Cornwall to Pendeen is noted for both biological and geological interest. The dramatic rugged coastline here is dominated by vertical sea cliffs, with exposure to salt spray and the prevailing south-westerly winds, resulting in dwarfed heath vegetation. Tin mining was intensive here and the coastline is scarred with the remains of industrial buildings, mines and spoil heaps. Kenidjack Valley is a steep incision north of Cape Cornwall.

Main habitats present
Maritime cliff, heathland, coastal grassland, bracken, scrub, bare ground, mire, broadleaved woodland, Cornish hedges, rivers/streams and rock outcrops.

Key butterfly species
Wall, Small Heath, Grayling, Small Pearl-bordered Fritillary, Dark Green Fritillary, Green Hairstreak, Silver-studded Blue.

Other points of interest
Breeding birds include Fulmar *Fulmarus glacialis*, Shag *Phalacrocorax aristotelis*, Kittiwake *Rissa tridactyla*, Peregrine Falcon *Falco peregrinus*, Raven *Corvus corax* and Chough *Pyrrhocorax pyrrhocorax*, plus several warblers such as Whitethroat *Curruca communis* and Grasshopper Warbler *Locustella naevia*. A diverse and interesting flora with many rare and localised plants.

Important geological features of contact metamorphism can be seen along the coast. Major mining archaeology including numerous former mine sites, museums and mine tours. There is a working beam engine at Levant.

4. Cubert Common and West Pentire, near Newquay

Access point grid reference
SW776599

Nearest postcode
TR8 5SE

Site area 83 hectares

Total number of butterfly species 26

Site designation SSSI, CWS

Parking and accessibility
- Drive through Crantock village and turn left onto road to Treago Farm. Follow track through farm and on to National Trust car park at Polly/Porth Joke – uneven track surface with gates that will need to be opened and closed. There is another car park at West Pentire.
- Tracks and footpaths cross the site with a circular walk to the poppy fields at West Pentire and the South West Coast Path, which is steep in places. Cubert Common is open access land.

Brief site description
Cubert Common, a gently undulating area of calcareous grassland, which has developed over wind-blown sand, is one of the few enclosed commons in England. There are areas of short, rabbit-grazed grassland with a variety of flowers, and other more tussocky areas. Cattle and sheep also graze the common.

The arable fields on West Pentire are owned and managed by the National Trust specifically as a nature reserve for the flora and fauna associated with arable cultivation and are not commercially farmed. They are renowned for their riot of colour in June when the red poppies *Papaver* spp. and yellow Corn Marigold *Glebionis segetum* burst into flower.

Main habitats present
Grassland, heathland, maritime cliff, arable fields and river/stream with boggy areas.

Key butterfly species
Wall, Small Heath, Dark Green Fritillary, Silver-studded Blue, Brown Argus.

Other points of interest
A diverse flora, particularly on Cubert Common and the coastal grassland, supports a variety of day-flying moths, dragonflies, damselflies and the Great Green Bush-cricket *Tettigonia viridissima*.

5. Garrow Tor, near Bodmin

Access point grid reference
SX131767

Nearest postcode
PL30 4NN

Site area 420 hectares

Total number of butterfly species 20

Site designation SSSI, AONB

Parking and accessibility
- Parking on moorland edge at De Lank waterworks.
- Easy walking across open moorland, which can be boggy at times, with uneven tracks to the tor.

Brief site description
Garrow Tor lies within the northern half of Bodmin Moor and encompasses Emblance Downs, King Arthur's Downs and Garrow Tor itself. The open access moorland, grazed by sheep, ponies and cattle, has extensive views, including over to Rough Tor and Brown Willy. On the way to Garrow Tor an area of coniferous woodland in the valley acts as a natural amphitheatre for birdsong and the sound of the stream. The slopes of the tor are rocky, uneven and have areas of heathland, grassland, scrub and bracken. A visit to King Arthur's Hall will add interest, and is a haven for the Green Hairstreak. King Arthur's Downs, to the west, is a boggy area where Marsh Fritillary have been recorded.

Main habitats present
Moorland, heathland, grassland, scrub, bracken, coniferous woodland, river/stream and ponds.

Key butterfly species
Small Heath, Small Pearl-bordered Fritillary, Marsh Fritillary, Green Hairstreak.

Other points of interest
Bodmin Moor is of major importance for both nesting and wintering birds. Of the breeding species, the site is of county importance for Lapwing *Vanellus vanellus*, Snipe *Gallinago gallinago*, Redstart *Phoenicurus phoenicurus*, Stonechat *Saxicola torquata* and Black-headed Gull *Chroicocephalus ridibundus*. In winter Bodmin Moor supports a number of rare species of bird in small numbers, including the Hen Harrier *Circus cyaneus*, Merlin *Falco columbarius* and Peregrine Falcon *F. peregrinus*.

There are many archaeological features, including stone circles, King Arthur's Hall and a Bronze Age settlement on Garrow Tor.

6. Godolphin and Godolphin Hill, near Helston

Access point grid reference
SW600324 (River Hayle)
SW594307 (Godolphin Hill)

Nearest postcode
TR27 6AR (River Hayle)
TR20 9RZ (Godolphin Hill)

Site area 65 hectares

Total number of butterfly species 29

Site designation SSSI, CWS, WHS

Parking and accessibility
- River Hayle: Small car park down lane from sharp bend off minor road close to Godolphin House.
- Godolphin Hill: Small car park near Great Works Mine on minor road from Godolphin Cross.
- Numerous tracks and footpaths lead across open access land. Public footpaths link the two sites via Godolphin, the National Trust estate.
- Some uneven ground along valley bottom (River Hayle).

Brief site description
This site comprises two separate but linked areas, the first adjacent to the River Hayle and the second Godolphin Hill itself. Both have similar butterfly species, with the Purple Hairstreak easiest to find in the oak woodland adjacent to the River Hayle.

A path along the River Hayle leads both east and west from the car park. Lying to the east is mixed woodland, with many low-growing oak trees and areas of heath. To the west lies more open, disturbed ground, evidence of past mining activity, as well as scrub.

Godolphin Hill, with its bracken-covered slopes, heath and grassland, is cattle-grazed and gives fine views across west Cornwall – St Michael's Mount on the south coast and St Ives Bay on the north.

Main habitats present
Broadleaved and coniferous woodlands, scrub, heathland, grassland, bracken, bare ground, ponds and river.

Key butterfly species
Silver-washed Fritillary, Purple Hairstreak.

Other points of interest
Beneath the gorse and heather on Godolphin Hill lie the remains of ancient field systems, tinners' pits, artificial rabbit warrens and Bronze Age settlements.

The area along the river is designated as part of the wider West Cornwall Bryophytes SSSI and is noted for its population of rare bryophytes (mosses and liverworts), which are adapted to growing on the copper-rich substrates of the formerly mined land. The river also supports good numbers of dragonflies and damselflies.

7. Goss Moor, including St Dennis Junction, near St Columb Major

Access point grid reference
SW932599 (St Dennis Junction)
SW942587 (Goss Moor Trail)

Nearest postcode
TR9 6HW (St Dennis Junction)
PL26 8BY (Goss Moor Trail)

Site area 577 hectares

Total number of butterfly species 29

Site designation SAC, SSSI, NNR

Parking and accessibility
- Parking for St Dennis Junction near railway bridge on Moorland Road, Indian Queens. Path on south side.
- Goss Moor Trail car park is down a track to the east of electricity sub-station along B3279 Indian Queens to St Dennis road.
- Numerous footpaths and cycle tracks cross the moor (sometimes boggy and uneven). A small section by St Dennis Junction is used by motorcyclists.

Brief site description
Goss Moor is part of the Mid Cornwall Moors, occupying a mostly flat valley basin that is the source of the River Fal. It is a National Nature Reserve managed by Natural England, which is currently restoring the wetland habitats to benefit numerous species including the Marsh Fritillary and Narrow-bordered Bee Hawk-moth *Hemaris tityus*.

The site comprises a large area of mixed habitat, dominated at the present time by willow carr with areas of heathland. Rare flora recorded here include Cornish Moneywort *Sibthorpia europaea* and Yellow Centaury *Cicendia filiformis*. Tracks and trails make the site accessible to both walkers and cyclists. St Dennis Junction lies at the far north-western end of the site and has grassland, bare ground and scrub, along with ponds and limited heathland.

Main habitats present
Willow carr, heathland, scrub, bracken, grassland, bare ground, river/stream and ponds/lakes, fen, marsh, mire and swamp.

Key butterfly species
Dingy Skipper, Small Heath, Small Pearl-bordered Fritillary, Marsh Fritillary.

Other points of interest
Many of the tracks are the remains of old railway lines built to serve the china clay industry within the St Austell area. Previous tin streaming, gravel and sand extraction have taken place here.

Due to the variety of wetland habitats, the site has an abundance of dragonflies and damselflies, as well as being good for amphibians and reptiles. The diversity of moth species is also notable, along with birds and rare plants.

8. Greenscoombe Wood, Luckett, near Callington

Access point grid reference
SX390732

Nearest postcode
PL17 8NJ

Site area 29 hectares

Total number of butterfly species 30

Site designation SSSI, AONB, WHS

Parking and accessibility
- Car park in the hamlet of Luckett. Walk south-east about half a mile along road into the wood.
- Numerous tracks and paths through the wood.

Brief site description
Greenscoombe Wood, on the Cornwall side of the River Tamar, incorporates mixed heath and woodland, mainly coniferous plantation. It has open glades bordered by broadleaved woodland, which provide the habitat and main larval foodplants for Cornwall's sole breeding colony of Heath Fritillary, re-introduced to the site by Butterfly Conservation and partners in 2006 (having become locally extinct in 2002). The best places to see the Heath Fritillary are in the small glades at the top of the wood.

Main habitats present
Coniferous woodland, broadleaved woodland, heathland, grassland and river/streams.

Key butterfly species
Marbled White, Silver-washed Fritillary, Heath Fritillary, Purple Hairstreak.

Other points of interest
This site was extensively mined for copper, tin, tungsten and arsenic. It has been a deer park and has been used for forestry. The site is also of botanical interest, with both Lesser Butterfly-orchid *Platanthera bifolia* and Greater Butterfly-orchid *P. chlorantha* present.

9. Hayle to Gwithian Towans, including Gwithian Green, near Hayle

Access point grid reference
SW579413 (Gwithian)
SW579396 (Upton Towans)
SW586413 (Gwithian Green)

Nearest postcode
TR27 5BT (Gwithian)
TR27 5BL (Upton Towans)
TR27 5BX (Gwithian Green)

Site area 390 hectares

Total number of butterfly species 29

Site designation SSSI, LNR, CWT reserve

Parking and accessibility
- Several car parks giving access to different parts of this undulating dune landscape with well-used tracks and footpaths. Charges apply at Gwithian car park.

Brief site description
A large coastal dune system, stretching from Hayle to Gwithian Towans, which has a variety of habitats and is a stronghold for the Silver-studded Blue. The short, rabbit-grazed, grassy areas are home to rare plants. Gwithian Green is a small Local Nature Reserve of approximately 7 hectares, formerly registered as a common. It supports a wide variety of plants and invertebrates.

Main habitats present
Dunes and dune slacks, with associated grassland, scrub, broadleaved woodland, ponds and rivers.

Key butterfly species
Wall, Small Heath, Small Pearl-bordered Fritillary, Dark Green Fritillary, Silver-studded Blue, Brown Argus.

Other points of interest
There is an extensive history of industrial use, particularly the National Explosives Works, which have left a mark on the dunescape. The dunes were flattened and small enclosures made to contain the blast should an explosion occur.

A good place to see the Glow-worm *Lampyris noctiluca* and the Cinnabar *Tyria jacobaeae* moth in Cornwall.

10. Marsland Nature Reserve, near Bude

Access point grid reference
SS231171

Nearest postcode
EX23 9PQ

Site area 212 hectares

Total number of butterfly species 29

Site designation SAC, SSSI, CWS, CWT and DWT Reserve, AONB

Parking and accessibility
- Park in the village of Gooseham SS230164 (postcode EX23 9PF) and walk down Darracott Hill (very steep). Limited parking at SS217169 (postcode EX23 9SU), where there are public footpaths down the steep-sided valley.
- All access is on foot only, on steep and uneven paths. Part of the reserve is accessible only by prior arrangement with the warden.

Brief site description
Marsland Nature Reserve, on the Cornwall/Devon border, is predominantly steep-sided oak *Quercus* spp. woodland, with Ash *Fraxinus excelsior*, Holly *Ilex aquifolium*, Rowan *Sorbus aucuparia*, Beech *Fagus sylvatica*, Hazel *Corylus avellana* and Sycamore *Acer pseudoplatanus*. The valley bottom contains wet flushes with Alder *Alnus glutinosa* and willow *Salix* spp. Maritime grassland, heathland and scrub can be found along the coastal section.

The reserve is under an active management regime of coppicing, grazing, traditional haymaking and the cutting of rides and glades. It is managed by Devon Wildlife Trust.

Main habitats present
Coppiced broadleaved woodland, scrub, bracken, grassland, heathland, maritime cliff and river.

Key butterfly species
Marbled White, Pearl-bordered Fritillary, Small Pearl-Bordered Fritillary, Silver-washed Fritillary, Dark Green Fritillary, Purple Hairstreak.

Other points of interest
This area has been managed for wildlife for over 30 years. Its remoteness means that low levels of light pollution make it ideal for nocturnal visitors. Bats and moths make their homes here, as do European Otter *Lutra lutra*, many types of bird, dragonflies and hoverflies.

11. Mullion Cove to Lizard Point, near Helston

Access point grid reference
SW668179 (Mullion Cove)
SW701115 (Lizard Point)

Nearest postcode
TR12 7EG (Mullion Cove)
TR12 7NU (Lizard Point)

Site area 800 hectares

Total number of butterfly species 28

Site designation SAC, SSSI, NNR, CWS, CWT Reserve

Parking and accessibility
- Car parks at Mullion Cove, Lizard Point, Predannack Wollas, Windmill Farm and Kynance Cove.
- Access along the South West Coast Path, steep in places. Paths, including a signed nature walk at Windmill Farm. Much of the land is open access.

Brief site description
Stretch of coast on the west side of The Lizard peninsula, taking in Mullion Cove, Higher Predannack Cliff, Gew-graze, Kynance Cove, Lower Predannack Downs, Lizard Downs, Windmill Farm and Lizard Point. Maritime cliff and slope, grassland and heathland are found along the coastal strip, with purple moor-grass and rush pasture, heathland and ponds at Windmill Farm and Lower Predannack Downs, and heathland across Lizard Downs.

Main habitats present
Maritime cliff, heathland, grassland, bare ground, beaches and small streams.

Key butterfly species
Wall, Small Heath, Grayling, Small Pearl-bordered Fritillary, Dark Green Fritillary, Marsh Fritillary, Silver-studded Blue.

Other points of interest
Rare plant species of the Lizard trackways can be found, as well as rare clovers including Upright Clover *Trifolium strictum*, Twin-headed Clover *T. bocconei*, Rough Clover *T. scabrum* and Long-headed Clover *T. incarnatum* subsp. *molinerii*, as well as Cornish Heath *Erica vagans*. The 'Cornish' Chough *Pyrrhocorax pyrrhocorax* has returned and can be seen swooping and tumbling over the cliffs, or fossicking for invertebrates in the short grassy sward. There are good populations of dragonflies and damselflies, particularly at Windmill Farm.

The geology of the area is unique with its ancient metamorphic and igneous rocks, which are exposed in numerous quarries as well as the cliffs themselves. (The geology of The Lizard is discussed further in chapter 2.)

12. Nare Head, near Truro

Access point grid reference
SW921380

Nearest postcode
TR2 5PH

Site area 42 hectares

Total number of butterfly species 30

Site designation SAC, SSSI, AONB

Parking and accessibility
- Narrow lanes from Veryan to National Trust car park at Kiberick Cove, with good track and footpath to Nare Head.

Brief site description
The track to Nare Head passes arable fields and pasture with grazing cattle. The Cornish hedges support a variety of flowers and provide shelter for butterflies and other invertebrates. The headland is an open access area of short maritime grassland, scrub and Bracken *Pteridium aquilinum*. There are small pockets of woodland in the valleys along the South West Coast Path leading to Carne Beach.

Main habitats present
Maritime cliff, grassland, bracken, scrub, Cornish hedges and small streams.

Key butterfly species
Wall, Small Heath, Silver-washed Fritillary, Dark Green Fritillary, Green Hairstreak, Silver-studded Blue.

Other points of interest
Coastal site with sea views and coastal and sea birds. It was a second world war decoy site and has a cold war bunker.

13. Newlyn Downs, near Newquay

Access point grid reference
SW836551

Nearest postcode
TR8 5AU

Site area 115 hectares

Total number of butterfly species 28

Site designation SAC, SSSI

Parking and accessibility
- Along a minor road from Mitchell to St Newlyn East near East Wheal Rose Farm. Parking area near back entrance to Lappa Valley Railway.
- Broad tracks and footpaths across open access site, some uneven and muddy.

Brief site description
Newlyn Downs is a large area of wet and dry heath with four heather species, including the nationally rare Dorset Heath *Erica ciliaris*. It has previously been mined and quarried, resulting in bare ground and areas of gravel.

Main habitats present
Heathland, bare ground, scrub, bracken, marsh and mire, with some broadleaved woodland, ponds and streams.

Key butterfly species
Wall, Small Heath, Grayling, Silver-washed Fritillary.

Other points of interest
Heathland plants and insects including dragonflies and damselflies.

14. Pendrift Bottom, near Bodmin

Access point grid reference
SX099752 (near quarry)
SX086751 (Wenfordbridge)

Nearest postcode
PL30 4LH (near quarry)
PL30 3PN (Wenfordbridge)

Site area 32 hectares

Total number of butterfly species 23

Site designation SSSI

Parking and accessibility
- Small area on right at quarry entrance – please note that De Lank Quarry is a private working quarry so do not enter through main quarry entrance. Steep, uneven footpath down to the valley bottom.
- Car park at the end of the Camel Trail at Wenfordbridge (approximately 1½-mile walk to site).

Brief site description
Pendrift Bottom, lying along the De Lank River, comprises grassland, heathland and Bracken *Pteridium aquilinum*, with broadleaved woodland around the edges. The broad valley bottom is boggy in places. A lot of conservation work has been carried out around De Lank Quarry by Cornwall Butterfly Conservation to improve the environment for fritillaries. Local farmers and commoners are helping to manage this area with targeted conservation grazing using cattle, sheep and ponies.

Main habitats present
Grassland, heathland, bracken, scrub, broadleaved woodland, river and mire.

Key butterfly species
Brimstone, Pearl-bordered Fritillary, Small Pearl-bordered Fritillary, Silver-washed Fritillary, Marsh Fritillary.

Other points of interest
Granite from the adjacent De Lank Quarry was used to build the Eddystone Lighthouse and Tower Bridge.

Pendrift Downs and a visit to the Jubilee Rock could be incorporated into a longer walk.

15. Penhale and Gear Sands, near Perranporth

Access point grid reference
SW774553

Nearest postcode
TR4 9PP

Site area 650 hectares

Total number of butterfly species 28

Site designation SSSI, SAC, CWS

Parking and accessibility
- Laybys along road from Perran Sands Holiday Park to Mount; site can be accessed from any of these laybys. Undulating sandy paths through and around dune system.
- Penhale military training area (at northern end of site) is still active and not accessible.

Brief site description
Penhale Dunes are the highest dunes in Britain, rising to 90 metres, with a depth of 50 metres of sand before rock is reached. This is a large area of dunescape and short grassland adjacent to the north Cornish coast.

Main habitats present
Dunes and dune slacks, grassland, scrub and streams.

Key butterfly species
Dingy Skipper, Grizzled Skipper, Wall, Small Heath, Grayling, Dark Green Fritillary, Silver-studded Blue, Brown Argus.

Other points of interest
The dunes are nationally recognised as a significant bryophyte and lichen site, as well as being important for rare plants. There are sand dune specialist invertebrates, along with the Glow-worm *Lampyris noctiluca*, dragonflies and damselflies, and reptiles.

The site includes the remains of St Piran's Oratory, a chapel thought to have been built in the seventh century AD. It is one of the earliest surviving Christian sites in Britain. It remained in use until the 11th or 12th century. The remains of a later church can be seen a few hundred yards to the north. This met the same fate, abandoned when it was engulfed by the shifting sands of the dunes.

16. Penlee Point and Rame Head, near Torpoint

Access point grid reference
SX436491 (Penlee Point)
SX420487 (Rame Head)

Nearest postcode
PL10 1LG (Penlee Point)
PL10 1LG (Rame Head)

Site area 75 hectares

Total number of butterfly species 27

Site designation SSSI, CWS, CWT Reserve, AONB

Parking and accessibility
- From Millbrook head for Rame. Turn left before church and follow for ¾ mile to car park at Penlee Point or continue straight on to car park at Rame Head.
- Some steep sections of path, which can be slippery when wet.

Brief site description
With its meadows and coastal grassland, Penlee Point is the best site in Cornwall for the Marbled White, as well as having stunning panoramic views of Plymouth Sound. Walking westwards along the coast path brings you to Rame Head, with its prominent medieval chapel.

Main habitats present
Grassland, scrub, bracken, maritime cliff and broadleaved woodland.

Key butterfly species
Wall, Small Heath, Marbled White, Small Pearl-bordered Fritillary, Dark Green Fritillary.

Other points of interest
The Penlee Battery Cornwall Wildlife Trust Reserve is on the site of a former defensive and strategically placed battery used throughout both world wars. The site is good for birds including Tawny Owl *Strix aluco*, as well as the Six-belted Clearwing *Bembecia ichneumoniformis* moth.

Rame Head, an Iron Age cliff castle, with its old chapel, is a haunt of the Dartford Warbler *Sylvia undata*.

17. Pentire Point to Port Quin, near Wadebridge

Access point grid reference
SW941800

Nearest postcode
PL27 6UA

Site area 192 hectares

Total number of butterfly species 29

Site designation SSSI, CWS, AONB

Parking and accessibility
- Several National Trust car parks lead to this stretch of the South West Coast Path; charges may apply. Easily accessible, but busy in the summer season.
- Coast path is uneven and ranges from easy to strenuous.

Brief site description
This section of the coast path from Pentire Point to Port Quin has spectacular views overlooking the Camel estuary, The Rumps and the north coast. There are steep cliffs and wooded valleys with a network of paths leading to different habitats, giving this site broad appeal. The land is sympathetically managed for nature by the National Trust, including both conservation-grazed grasslands and arable fields, maintaining a diversity of species throughout.

Main habitats present
Maritime cliff, grassland, bare ground, scrub, arable and streams.

Key butterfly species
Wall, Small Heath, Grayling, Small Pearl-bordered Fritillary, Dark Green Fritillary.

Other points of interest
This area is of geological interest, with its mining history and slate quarries. The Rumps is an Iron Age defensive site with a series of ditches and ramparts and is a good place to 'seawatch' for dolphins, porpoises and the Basking Shark *Cetorhinus maximus*, along with resident and passage seabirds. Bird interest includes Peregrine Falcon *Falco peregrinus*, Fulmar *Fulmarus glacialis*, Linnet *Linaria cannabina*, Skylark *Alauda arvensis* and the ubiquitous Stonechat *Saxicola rupicola*.

18. Poldice Valley, near Redruth

Access point grid reference
SW737430 (layby)
SW737429 (car park)

Nearest postcode
TR16 5PT (layby)
TR16 5QA (car park)

Site area 26 hectares

Total number of butterfly species 25

Site designation SSSI, WHS

Parking and accessibility
- Layby opposite Truro Auction Centre; small car park further south.
- Many tracks throughout the site, some steep and uneven.
- The area is used by cyclists and horse riders.

Brief site description
Poldice Valley is a former industrial site with many ruined buildings in this area of past heavy mining activity. As a result of this and the associated pollution by arsenic and heavy metals, a landscape dominated by bare ground, grassland and heathland with mineral-tolerant plants has evolved, suiting a variety of Lepidoptera and other insects. The site is criss-crossed by numerous tracks.

Main habitats present
Heathland, bare ground, grassland, bracken with scrub, broadleaved woodland, ponds and a river.

Key butterfly species
Small Heath, Grayling, Silver-studded Blue.

Other interests onsite
The site is important for industrial archaeology as well as invertebrates such as dragonflies and damselflies.

The Mineral Tramways Coast to Coast Trail (Portreath to Devoran) runs through the site and links with the Cornwall Wildlife Trust Bissoe Valley Nature Reserve, which is also good for butterflies, dragonflies and damselflies.

19. Porthgwarra to Pordenack Point, near St Just

Access point grid reference
SW370217

Nearest postcode
TR19 6JP

Site area 158 hectares

Total number of butterfly species 28

Site designation SSSI, AONB

Parking and accessibility
- B3283 road from Penzance and single-track road from Polgigga. Pay and display car park at Porthgwarra.
- South West Coast Path and other footpaths within open access land across uneven terrain.

Brief site description
Coastal section southwards from Porthgwarra village to the Coastwatch Lookout Station at Gwennap Head and then westwards to Carn Barra, Carn Lês Boel and Pordenack Point beyond Nanjizal Bay. This site comprises a number of small bays and an extensive stretch of high granite cliffs and maritime heathland on the western edge of the Penwith peninsula. The most extensive habitat type on these cliff tops is 'waved' maritime heathland, dominated by heathers and gorse *Ulex* spp.

Main habitats present
Maritime cliff, coastal grassland, heathland, scrub, bracken, bare ground and granite outcrops.

Key butterfly species
Wall, Small Heath, Grayling, Small Pearl-bordered Fritillary, Green Hairstreak, Silver-studded Blue.

Other interests onsite
Birds include breeding Chough *Pyrrhocorax pyrrhocorax*. One of Cornwall's principal sites for passage migrants and seabird watching. In addition to expanses of heath and coastal grass there is a valley with willow scrub and a seasonal pond with good numbers of dragonfly species.

20. Seaton Valley Countryside Park, near Looe

Access point grid reference
SX304545

Nearest postcode
PL11 3JD

Site area 54 hectares

Total number of butterfly species 30

Site designation CWS, LNR

Parking and accessibility
- Cornwall Council car parks – charges apply. This area is very busy in the summer season.
- Level track up the valley, with seats and picnic tables.

Brief site description
The entrance to the steep-sided, wooded Seaton Valley is through a well-used meadow. The meandering, accessible path reveals well-managed, scalloped open areas of flower-rich diverse habitat, suitable for a wide range of invertebrates.

Main habitats present
Broadleaved woodland, grassland, scrub, ponds and river.

Key butterfly species
Dingy Skipper, Wall, Marbled White, Silver-washed Fritillary.

Other points of interest
The river corridor, including the ponds, supports species including Kingfisher *Alcedo atthis*, European Otter *Lutra lutra*, dragonflies and damselflies. There is an adjacent sandy beach at the river mouth.

CHAPTER 8
Further reading

Introduction

Experienced students of the natural world will have their own well-thumbed reference books. This list of publications is intended to provide a helping hand to someone new to the field (in every sense). There is a huge choice of books on natural history, which can seem daunting at first. The list below is a starting point and includes publications the authors of *Butterflies of Cornwall* find indispensable.

As an individual's interest develops, along with their knowledge, they will add titles of their own. Many places rich in butterflies have other insects that catch the eye, such as dragonflies, bees, wasps and hoverflies. The sight and sound of birds accompanies every field trip, and many butterfly enthusiasts are also birders. These other species are excluded from the list below, but some of the publishers listed might suit the reader who wishes to extend their knowledge of natural history with other titles in the series.

We include publications on moths because there are many day-flying moths and some of the commonest and most striking caterpillars encountered are moth larvae.

Because the life cycle of butterflies is so entirely dependent on plants, which feature large in *Butterflies of Cornwall*, some books on flora are included. Most people find that their interest in botany grows alongside their engagement with Lepidoptera.

Also included are some online resources that can help in identification. One of the best aids is a camera. A smartphone might suffice, but the zoom on digital cameras can improve an image for detail. Whether it is an insect or a flower, taking an image home can be indispensable to identification. Sometimes a salient detail is frustratingly absent from an image, such as a pattern on the underside of the wing, or hairs on a plant stem, and turns out to be crucial to identification. But that's all part of the fun of learning, and everyone, however expert, is learning all the time.

Modern books on natural history have superb illustrations that make identification both easier and a pleasure. Some books, such as the WILDGuides series, use photographs of insects taken in the wild. This has the advantage of realism: the reader knows this is how the insect looks. However, hand-painted illustrations often have the edge because they faithfully reproduce defining characteristics more clearly than a photograph can. The wildlife illustrator Richard Lewington is widely regarded as a master of his art, and any book illustrated by him is guaranteed to be of high quality.

When choosing a book, nothing beats trying it out in a bookshop to see if it suits your particular needs. Test it by looking up a butterfly you know. Purchasing from an independent bookshop, if there is one near you, also supports a local business.

The date of publication for each book listed below is of the most up-to-date edition at the time of publishing *Butterflies of Cornwall*.

In the field: butterflies

Slim, portable, rucksack-friendly aids to identification, of variable weather-resistance:

Bebbington, J. and Lewington, R. 2019. *Guide to the Butterflies of Britain and Ireland.* **Field Studies Council Occasional Publication, Vol. 184. Fold-out identification chart. FSC, Shrewsbury.**

Lewington, R. 2019. *Pocket Guide to the Butterflies of Great Britain and Ireland.* **Bloomsbury Publishing, London.**

Newland, D., Still, R., Swash, A. and Tomlinson, D. 2020. *Britain's Butterflies. A Field Guide to the Butterflies of Great Britain and Ireland.* **WILDGuides. Princeton University Press, Princeton.**

In the field: moths

Lewington, R. 2019. *Guide to the Day-flying Moths of Britain.* Field Studies Council Occasional Publication, Vol. 106. Fold-out identification chart. FSC, Shrewsbury.

Manley, C. 2021. *British Moths. A Photographic Guide to the Moths of Britain and Ireland.* Bloomsbury Publishing.

Newland, D., Still, R. and Swash, A. 2019. *Britain's Day-flying Moths. A Field Guide to the Day-flying Moths of Great Britain and Ireland.* WILDGuides. Princeton University Press, Princeton.

Sterling, P., Parsons, M., Lewington, R. 2012. *Field Guide to the Micro-Moths of Britain and Ireland.* Bloomsbury Publishing.

Waring, P., Townsend, M. and Lewington, R. 2018. *Field Guide to the Moths of Great Britain and Ireland.* Bloomsbury Publishing, London.

In the field: caterpillars of butterflies and moths

Henwood, B. and Sterling, P. 2020. *Field Guide to the Caterpillars of Great Britain and Ireland.* Bloomsbury Publishing, London.
Illustrated with more than 900 detailed colour artworks of the caterpillars of butterflies and moths by Richard Lewington, this book includes detailed species accounts covering field characters, species status and distribution (with maps), foodplants, habitats, pointers on separating similar species, and notes to help find each caterpillar in the field.

In the field: wild flowers

Parslow, R. 2020. *Discovering Isles of Scilly Wild Flowers.* Parslow Press.
A pocket guide to the common plants, aliens and Scilly 'specials'.

Rose, F. and O'Reilly, C. 2006. *The Wild Flower Key. How to identify Wild Flowers, Trees and Shrubs in Britain and Ireland.* Frederick Warne, London.
This is used by most botanists, amateur or professional, and is a good general field guide.

Books at home: butterflies in depth

Asher, J., Warren, M., Fox, R., Harding, P., Jeffcoate, G. and Jeffcoate, S. 2001. *The Millennium Atlas of Butterflies in Britain and Ireland.* Oxford University Press, Oxford.
This atlas is the result of five years of recording by thousands of volunteers across Britain and Ireland at the end of the 1990s, providing distribution maps for all species, along with commentary on longer-term trends in distribution derived from 200 years of recording, as well as looking at changes across Europe. Despite the fact that it was published 20 years ago, this is a useful and readable book that provides an assessment of each butterfly, the habitats in which they live, the threats they face, and how they may be conserved, along with the major changes that occurred since the previous national atlas, published in 1984.

Eeles, P. 2019. *Life Cycles of British & Irish Butterflies.* Pisces Publications, Newbury.
This book explains and illustrates every stage in the life cycles of all the butterflies found in Britain and Ireland. The colour photographs are excellent and include all the larval stages.

Thomas, J. and Lewington, R. 2014. *The Butterflies of Britain & Ireland.* Bloomsbury Publishing, London.
This informative book teams up two of the UK's leading naturalists. Jeremy Thomas is a renowned ecologist whose pioneering work on the Large Blue *Phengaris arion* led to its successful re-introduction in Britain. The book is packed with information from the latest research on all our native butterflies as well as rare visitors, with all life stages superbly illustrated by the inimitable Richard Lewington. *Butterfly* magazine was unequivocal in its praise: "Both authors are justly renowned for their work and this book remains a must-have classic for anyone interested in butterflies."

Books at home: moths in depth

Randle, Z., Evans-Hill, L., Parsons, M., Tyner, A., Bourn, N., Davis, T., Dennis, E., O'Donell, M., Prescott, T., Tordoff, G. and Fox, R. 2019. *Atlas of Britain and Ireland's Larger Moths*. Pisces Publications, Newbury.

Around 25 million moth records have been combined to produce accounts for 866 macro-moth species, each with a distribution map showing current and historical occurrences, trends, status, a phenology chart and colour image. Brief introductory chapters detail the long-standing tradition of moth recording and the development of the National Moth Recording Scheme, methods used to collect and analyse the data, an overview of trends since the 1970s and the environmental drivers of change in moth populations and distributions.

Books at home: larval foodplants

Crafer, T. 2005. *Foodplant List for the Caterpillars of Britain's Butterflies and Larger Moths*. Atropos Publishing.

This book lists more than 450 foodplants of the caterpillars of over 800 species of British butterflies and larger moths, helping to identify the species of caterpillar by the plant that it is eating.

Books at home: wild flowers

French, C.N. 2020. *A Flora of Cornwall*. Wheal Seton Press, Camborne.

This is a reference book for the plant enthusiast who might want to look up the distribution of every plant species recorded in Cornwall. An expert in his field, the author set up and manages the ERICA database of biological records for Cornwall, including butterfly records used in the analyses in *Butterflies of Cornwall*.

Streeter, D., Hart-Davies, C., Hardcastle, A., Cole, F., and Harper, L. 2016. *Collins Wild Flower Guide*. HarperCollins Publishers, London.

This comprehensive and well-illustrated book will assist the reader to identify the wild flowers of Britain and Ireland. It is too heavy to take into the field, but it packs a huge amount into a small paperback. It includes grasses, which are important for some butterfly species, and useful keys to aid identification of trickier groups.

Books at home: the wildlife of the Isles of Scilly

Parslow, R. 2007. *The Isles of Scilly*. Collins New Naturalist Library, Book 103. HarperCollins Publishers, London.

This book covers the many aspects that make the Isles of Scilly and their flora and fauna so special, including their geography, geology, climate, habitats, people and the way they used the land, and its present-day management. There is a chapter covering insects and other terrestrial invertebrates found on the islands.

Parslow, R. and Bennallick, I. 2017. *The New Flora of the Isles of Scilly*. Parslow Press.

Winner of the 2018 BSBI/WFS Presidents' Award, this account brings the record up to date almost 50 years after Lousley's 1971 *Flora of the Isles of Scilly*. It covers nearly 1,000 species, both native and introduced, with distribution maps and many photographs.

Other resources

Butterfly Conservation (BC)
Members receive the magazine *Butterfly* four times a year. The BC website is freely available to anyone and has useful information on every butterfly species found in the UK. https://butterfly-conservation.org/butterflies/identify-a-butterfly On this page, every British species of butterfly is listed in alphabetical order by common name with an image. Click on the common name to access further information.

Cornwall Butterfly Conservation (CBC)
As this is a branch of Butterfly Conservation, members receive *Butterfly* as well as the twice-yearly *Cornwall Butterfly Observer*, which has articles on butterflies and moths relevant to the County: back issues are freely available to download as pdf documents on the CBC website (click on the 'Resources' tab). The CBC website has information on events, news and Cornish butterflies. http://www.cornwall-butterfly-conservation.org.uk/

UK Butterflies
This website is freely available to anyone, with every British butterfly species listed on the home page. https://www.ukbutterflies.co.uk/index.php Grouped in taxonomic order by family, within each family, butterflies are listed in alphabetical order by common name: just click on the species you want. The site is owned and managed by Peter Eeles, the author of *Life Cycles of British & Irish Butterflies*, listed above.

Atropos
The UK journal for butterfly, moth and dragonfly enthusiasts, published four times a year including an annual Migration Review. A central theme is insect migration, but all subjects relevant to British Lepidoptera and Odonata are covered. Subscription rates and portal are at https://www.atroposbooks.co.uk/atropos-subscription

British Wildlife
Published since 2016 by NHBS, Totnes. "Since its launch in 1989, *British Wildlife* has established its position as the leading natural history magazine in the UK, providing essential reading for both enthusiast and professional naturalists and wildlife conservationists." The articles, which cover a wide range of natural history, are readable, interesting and well illustrated. Several articles have covered butterflies, including one on re-establishing the Large Blue in Britain (published in Vol. 31 No. 1 October 2019). The magazine includes a section entitled 'Wildlife reports', which provides recent sightings and commentary for various groups, including butterflies (compiled by Butterfly Conservation), along with recent publications and research. Subscription rates and portal are at https://www.britishwildlife.com/

UK Leps
This website contains over 12,000 images illustrating the butterflies and larger moths of Britain and Northern Europe, freely available at http://www.ukleps.org/

UK Moths
This website contains 7,288 images illustrating 2,261 species of macro- and micro-moths found in the UK, freely available at https://ukmoths.org.uk/

References cited in the text

Agassiz, D. 1981. *A Revised List of the Lepidoptera of the Isles of Scilly*. Isles of Scilly Museum Association, St Mary's, Isles of Scilly.

Agassiz, D.J.L, Beavan, S.D. and Heckford, R.J. 2020. *Third Update to Checklist of the Lepidoptera of the British Isles, 2013*. Royal Entomological Society, St Albans.

Asher, J., Warren, M., Fox, R., Harding, P., Jeffcoate, G. and Jeffcoate, S. 2001. *The Millennium Atlas of Butterflies in Britain and Ireland*. Oxford University Press, Oxford.

Beavis, I.C. 2004. Resident and regular migrant butterflies on the Isles of Scilly. *The Entomologist's Record* 116: 97–102.

Beavis, I.C. 2012. A revised list of the bees, wasps and ants of Scilly. *Isles of Scilly Bird and Natural History Review 2012*: 171–182. Isles of Scilly Bird Group.

Bebbington, J. and Lewington, R. 2019. *Guide to the Butterflies of Britain and Ireland*. Field Studies Council Occasional Publication, Vol. 184. Fold-out identification chart. FSC, Shrewsbury.

Blair, K.G. 1925. Lepidoptera of the Scillies. *The Entomologist* 58: 3–10.

Bodin, M. 2020. Coronaviruses often start in animals: here's how those diseases can jump to humans. *Discover* magazine. Available at: https://www.discovermagazine.com/health/coronaviruses-often-start-in-animals-heres-how-those-diseases-can-jump-to (accessed 21 November 2020).

Bourn, N.A.D. and Warren, M.S. 1997. *Species Action Plan Lulworth Skipper* Thymelicus acteon. *May 1997*. Butterfly Conservation, Wareham.

Bracken, C.W. 1936. Westward drift of the comma butterfly. *Transactions of the Devonshire Association* 68: 135–137.

Brereton, T. M. 1997. *Ecology and conservation of the butterfly* Pyrgus malvae *(Grizzled Skipper) in south-east England*. PhD thesis. University of East London. Unpublished.

Brereton, T.M., Botham, M.S., Middlebrook, I., Randle, Z., Noble D., Harris, S., Dennis, E.B., Robinson, A.E., Peck, K. and Roy, D.B. 2019. *United Kingdom Butterfly Monitoring Scheme Report for 2018*. Centre for Ecology & Hydrology, Butterfly Conservation, British Trust for Ornithology and Joint Nature Conservation Committee.

Brereton, T.M., Botham, M.S., Middlebrook, I., Randle, Z., Noble D., Harris, S., Dennis, E.B., Robinson, A., Peck, K. and Roy, D.B. 2020. *United Kingdom Butterfly Monitoring Scheme Annual Report 2019*. UK Centre for Ecology & Hydrology, Butterfly Conservation, British Trust for Ornithology and Joint Nature Conservation Committee.

BRIG 2011. *UK Biodiversity Action Plan: Priority Habitat Descriptions*. JNCC, Peterborough.

Buczacki, S. 2002. *Fauna Britannica*. Hamlyn, London.

Butterfly Conservation 2018. *South West England Regional Conservation Strategy 2025*. Available at: https://butterfly-conservation.org/our-work/our-conservation-strategies (accessed 3 November 2020).

Butterfly Conservation 2019. *Annual Review 2018/2019*. Available at: https://butterfly-conservation.org/sites/default/files/2019-10/Annual%20Review%20FINAL.pdf (accessed 25 November 2020).

Butterfly Conservation a. *All the Moor Butterflies*. Available at: https://butterfly-conservation.org/our-work/conservation-projects/england/all-the-moor-butterflies (accessed 22 November 2020).

Butterfly Conservation b. *Why Butterflies Matter*. Available at: https://butterfly-conservation.org/butterflies/why-butterflies-matter (accessed 19 November 2020).

Charman, D.J., Johns, C., Camidge, K., Marshall, P., Mills, S., Mulville, J., Roberts, H. and Stevens, T. 2016. *The Lyonesse Project: A Study of the Evolution of the Coastal and Marine Environment of the Isles of Scilly*. Cornwall Archaeological Unit, Cornwall Council.

Chudleigh, D. 1992. *Bridge over Lyonesse: Over 70 years of the Isles of Scilly Steamship Company*.

Clark, J. 1906. 'Lepidoptera' (Butterflies: 203 – 7) in Page, W. (ed.) *Victoria County History, Cornwall*, Vol. 1, London.

Cornwall Council 2013. *Economy and Culture Strategy 2013–2020.* Available at: https://www.cornwall.gov.uk/media/3624006/Economy-and-Culture-Strategy.pdf (accessed 3 December 2020).

Cornwall Council 2016. *Cornwall's Environmental Growth Strategy 2015–2065.* Available at: https://www.cornwall.gov.uk/media/24212257/environmental-growth-strategy_jan17_proof.pdf (accessed 18 January 2021).

Cornwall Council 2018. *Cornwall Planning for Biodiversity Guide.* Available at: https://www.cornwall.gov.uk/media/38341273/biodiversity-guide.pdf (accessed 19 November 2020).

Cornwall Council 2019a. *Draft Cornwall Council's Farms Strategy 2019–2039.* Available at: https://www.cornwall.gov.uk/media/38860030/draft-cornwall-council-farms-strategy.pdf (accessed 19 November 2020).

Cornwall Council 2019b. *Climate Change Plan: Creating the Conditions for Change through Direct Action and a New Form of Place-based Leadership for Cornwall to Become Net Carbon Neutral.* Available at: https://www.cornwall.gov.uk/media/40176082/climate-change-action-plan.pdf (accessed 19 November 2020).

Crafer, T. 2005. *Foodplant List for the Caterpillars of Britain's Butterflies and Larger Moths.* Atropos Publishing.

Dawson, R.J.G. 2015. Butterflies of Scilly: An Update for 2000–2015. *Isles of Scilly Bird and Natural History Review 2015*: 210–218. Isles of Scilly Bird Group.

Defra 2020a. *Butterflies in the United Kingdom: habitat specialists and species of the wider countryside, 1976 to 2019.* Available at: https://assets.publishing.service.gov.uk/government/uploads/system/uploads/attachment_data/file/924554/Butterflies_in_the_UK_1976_2019_accessible_rev_Oct_20.pdf (accessed 4 February 2021).

Defra 2020b. *Environmental Land Management Policy Discussion Document.* Available at: https://consult.defra.gov.uk/elm/elmpolicyconsultation/supporting_documents/ELM%20Policy%20Discussion%20Document%20230620.pdf (accessed 19 November 2020).

Defra 2020c. *Farming for the Future: Policy and Progress Update.* Available at: https://assets.publishing.service.gov.uk/government/uploads/system/uploads/attachment_data/file/868041/future-farming-policy-update1.pdf (accessed 19 November 2020).

Dennis, J. 2021. *Small Pearl-bordered Fritillary. An analysis of the first and second broods.* Available at: www.cornwall-butterfly-conservation.org.uk/resources.html (accessed 23 March 2021).

Dennis, R.L.H. and Hardy, P.B. 2018. *British and Irish Butterflies: An Island Perspective.* CABI.

Ecologic 2014. *Robert Swan: Stories from Beyond Two Poles.* Available at: https://www.ecologic.eu/11745#:~:text=On%20the%2015%20of%20September,both%20geographical%20poles%20on%20foot (accessed 22 November 2020).

Eeles, P. 2019. *Life Cycles of British & Irish Butterflies.* Pisces Publications, Newbury.

Ellis, S., Bourn, N.A.D. and Bulman, C.R. 2012. *Landscape-scale Conservation for Butterflies and Moths: Lessons from the UK.* Butterfly Conservation, Wareham.

Ellis, S., Wainwright, D., Dennis, E.B., Bourn, N.A.D., Bulman, C.R., Hobson, R., Jones, R., Middlebrook, I., Plackett, J., Smith, R.G., Wain, M. and Warren, M.S. 2019. Are habitat changes driving the decline of the UK's most threatened butterfly: the High Brown Fritillary *Argynnis adippe* (Lepidoptera: Nymphalidae)? *Journal of Insect Conservation* 23: 351–367.

Emmet, A.M. and Heath, J. 1990. *The Moths and Butterflies of Great Britain and Ireland, 7: Part 1 – The Butterflies.* Harley Books, Colchester.

Ford, E.B. 1990. *Butterflies.* Collins New Naturalist Series. Bloomsbury Books, London.

Foster, S. 2015. *Marsh Fritillary Larval Web Survey. Bodmin Moor Sites 2015.* Butterfly Conservation, Wareham. Available on application from Butterfly Conservation.

Foster, S. and Dennis, J. 2020a. *Gwithian Green Butterfly Monitoring Report 1997–2020.* Available at: www.cornwall-butterfly-conservation.org.uk/resources.html (accessed 27 February 2021).

Foster, S. and Dennis, J. 2020b. *A Review of the Silver-studded Blue in Cornwall 2020.* Available at: https://dynamicdunescapes.co.uk/species/silver-studded-blue-butterfly/ (accessed 27 February 2021).

Fox, R., Brereton, T.M., Asher, J., August, T.A., Botham, M.S., Bourn, N.A.D., Cruickshanks, K.L., Bulman, C.R., Ellis, S., Harrower, C.A., Middlebrook, I., Noble, D.G., Powney, G.D., Randle, Z., Warren, M.S. and Roy, D.B. 2015. *The State of the UK's Butterflies 2015.* Butterfly Conservation and the Centre for Ecology & Hydrology, Wareham, Dorset.

Fox, R, Warren, M.S. and Brereton, T.M. 2010. A new Red List of British butterflies. *Species Status* 12: 1–32. Joint Nature Conservation Committee, Peterborough.

French, C.N. 2020. *A Flora of Cornwall.* Wheal Seton Press, Camborne.

Frohawk, F.W. 1934. *The Complete Book of British Butterflies.* Ward, Lock & Co. Ltd, London.

Hayhow, D.B., Eaton, M.A., Stanbury, A.J., Burns, F., Kirby, W.B., Bailey, N., Beckmann, B., Bedford, J., Boersch-Supan, P.H., Coomber, F., Dennis, E.B., Dolman, S.J., Dunn, E., Hall, J., Harrower, C., Hatfield, J.H., Hawley, J., Haysom, K., Hughes, J., Johns, D.G., Mathews, F., McQuatters-Gollop, A., Noble, D.G., Outhwaite, C.L., Pearce-Higgins, J.W., Pescott, O.L., Powney, G.D. and Symes, N. 2019. *The State of Nature 2019.* The State of Nature Partnership. Available at: https://nbn.org.uk/wp-content/uploads/2019/09/State-of-Nature-2019-UK-full-report.pdf (accessed 19 November 2020).

Henwood, B. 2015. A Small Skipper *Thymelicus sylvestris* with Black Undersides of the Antennal Tips. *Atropos* 53: 30–31.

Henwood, B. and Sterling, P. 2020. *Field Guide to the Caterpillars of Great Britain and Ireland.* Bloomsbury Publishing, London.

Hicks, M. 2000. Butterflies on Scilly. *Isles of Scilly Bird and Natural History Review 2000*: 173–174. Isles of Scilly Bird Group.

HM Government 2018. *A Green Future: Our 25 Year Plan to Improve the Environment.* Available at: https://assets.publishing.service.gov.uk/government/uploads/system/uploads/attachment_data/file/693158/25-year-environment-plan.pdf (accessed 3 December 2020).

Hobson, R. and Budd, P. 2001. *The Marsh Fritillary in Cornwall: Site Dossier. Incorporating the Bodmin Moor Larval Survey 2000. Butterfly Conservation Report No. S01-32.* Butterfly Conservation, Wareham. Available at: www.cornwall-butterfly-conservation.org.uk/resources.html (accessed 27 February 2021).

Lane, R. 2001. Painted Lady: immigrations during the 2000/2001 winter. *The Butterfly Observer* 20: 12–14.

Lane, R. 2009. Winter breeding of the Red Admiral in south central Cornwall (2006/08). *The Butterfly Observer* 44: 4–8.

Lewington, R. 2019. *Guide to the Day-flying Moths of Britain.* Field Studies Council Occasional Publication, Vol. 106. Fold-out identification chart. FSC, Shrewsbury.

Lewington, R. 2019. *Pocket Guide to the Butterflies of Great Britain and Ireland.* Bloomsbury Publishing, London.

Lousley, J.E. 1971. *The Flora of the Isles of Scilly.* David and Charles, Newton Abbot.

Manley, C. 2021. *British Moths. A Photographic Guide to the Moths of Britain and Ireland.* Bloomsbury Publishing.

Marsden, P. 2014. *Rising Ground: A Search for the Spirit of Place.* Granta Books, London.

NASA, 2021. 2020 tied for warmest year on record, NASA analysis shows. *ScienceDaily* 15 January 2021. Available at: www.sciencedaily.com/releases/2021/01/210115103020.htm (accessed 18 January 2021).

Natural England 2013a. *National Character Area Profile: 153. Bodmin Moor.* Available at: http://publications.naturalengland.org.uk/publication/5032336 (accessed 6 February 2021).

Natural England 2013b. *National Character Area Profile: 154. Hensbarrow.* Available at: http://publications.naturalengland.org.uk/publication/6278141332946944 (accessed 6 February 2021).

Natural England 2013c. *National Character Area Profile: 156. West Penwith.* Available at: http://publications.naturalengland.org.uk/publication/3510328 (accessed 6 February 2021).

Natural England 2013d. *National Character Area Profile: 157. The Lizard.* Available at: http://publications.naturalengland.org.uk/publication/6949119 (accessed 6 February 2021).

Natural England 2013e. *National Character Area Profile: 158. Isles of Scilly.* Available at: http://publications.naturalengland.org.uk/publication/6566056445345792 (accessed 6 February 2021).

Natural England 2014a. *National Character Area Profile: 152. Cornish Killas.* Available at: http://publications.naturalengland.org.uk/publication/6654414139949056 (accessed 6 February 2021).

Natural England 2014b. *National Character Area Profile: 155. Carnmenellis.* Available at: http://publications.naturalengland.org.uk/publication/6254102417768448 (accessed 6 February 2021).

Natural England 2015. *National Character Area Profile: 149. The Culm.* Available at: http://publications.naturalengland.org.uk/publication/4292167 (accessed 6 February 2021).

Newland, D., Still, R. and Swash, A. 2019. *Britain's Day-flying Moths. A Field Guide to the Day-flying Moths of Great Britain and Ireland.* WILDGuides. Princeton University Press, Princeton.

Newland, D., Still, R., Swash, A. and Tomlinson, D. 2020. *Britain's Butterflies. A Field Guide to the Butterflies of Great Britain and Ireland.* WILDGuides. Princeton University Press, Princeton.

Nurse, P. 2020. *What Is Life? Understand Biology in Five Steps.* David Fickling Books, Oxford.

Nylin, S. 1989. Effects of changing photoperiods in the life cycle regulation of the comma butterfly, *Polygonia c-album* (Nymphalidae). *Ecological Entomology* Vol. 14, 2: 209–218. Available at: http://doi.org/10.1111/j.1365-2311.1989.tb00771.x (accessed 20 March 2021).

Oates, M. 2016. *In Pursuit of Butterflies. A Fifty-Year Affair.* Bloomsbury, London.

Oates, M. 2020. *His Imperial Majesty. A Natural History of the Purple Emperor.* Bloomsbury Wildlife, London.

Parslow, R. 2007. *The Isles of Scilly.* Collins New Naturalist Library, Book 103. HarperCollins Publishers, London.

Parslow, R. 2020. *Discovering Isles of Scilly Wild Flowers.* Parslow Press.

Parslow, R. and Bennallick, I. 2017. *The New Flora of the Isles of Scilly.* Parslow Press.

Pateman, R.M., Hill, J.K, Roy, D.B., Fox, R. and Thomas, C.D. 2012. Temperature-dependent alterations in host use drive rapid range expansion in a butterfly. *Science* 336: 1028–1030. Available at: http://doi.org/10.1126/science.1216980 (accessed on 22 March 2021).

Penhallurick, R.D. 1996. *The Butterflies of Cornwall and the Isles of Scilly.* Dyllansow Pengwella, Truro.

Phelps, S. 2019. *Marsh Fritillary on Bodmin Moor 2019 report.* Butterfly Conservation, Wareham.

Plackett, J. 2018. *Heath Fritillary in the Tamar Valley 2018 Status Report. Butterfly Conservation Report No. S18 – 07.* Butterfly Conservation, Wareham. Available at: www.cornwall-butterfly-conservation.org.uk/resources.html (accessed 27 February 2021).

Poland, C. 2020. *Grizzled Skipper* Pyrgus malvae *Species Report Cornwall 2018 & 2019.* Available at: www.cornwall-butterfly-conservation.org.uk/resources.html (accessed 27 February 2021).

Randle, Z., Evans-Hill, L., Parson, M., Tyner, A., Bourn, N., Davis, T., Dennis, E., O'Donell, M., Prescott, T., Tordoff, G. and Fox, R. 2019. *Atlas of Britain and Ireland's Larger Moths.* Pisces Publications, Newbury.

Ravenscroft, N.O.M. and Warren, M.S. 1996. *Species Action Plan: The Silver-studded Blue* Plebejus argus. Butterfly Conservation, Wareham.

Richardson, A. and Mere, R.M. 1958. Some preliminary observations on the lepidoptera of the Isles of Scilly with particular reference to Tresco. *Entomologist's Gazette* 9: 115–147.

Roer, H. 1968. Weitere Untersuchungen über die Auswirkungen der Witterung auf Richtung und Distanz der Flüge des Kleinen Fuchses (*Aglais urticae* L.) im Rheinland. *Decheniana* 120: 313–334.

Rose, F. and O'Reilly, C. 2006. *The Wild Flower Key. How to identify Wild Flowers, Trees and Shrubs in Britain and Ireland.* Frederick Warne, London.

Sandars, E. 1939. *A Butterfly Book for the Pocket.* Oxford University Press, Oxford.

Scott, M.A., Scott, W.J. and Scott, T.R. 2006. Longstone Heritage Centre, St Mary's, Isles of Scilly. *Isles of Scilly Bird and Natural History Review 2006*: 144–146. Isles of Scilly Bird Group.

Smith, F.H.N. 1997. *The Moths and Butterflies of Cornwall and the Isles of Scilly.* Gem Publishing Co., Wallingford.

Smith, F.H.N. 2002. *A Supplement to the Moths and Butterflies of Cornwall and the Isles of Scilly.* Gem Publishing Co., Wallingford.

Spalding, A. 1992. *Cornwall's Butterfly & Moth Heritage.* Twelveheads Press, Truro.

Spalding, A. with Bourn, N. 2000. *Regional Action Plan: South-west England.* Butterfly Conservation, Dedham.

Stace, C. 2019. *New Flora of the British Isles. Fourth Edition.* C&M Floristics, Suffolk.

Sterling, P., Parsons, M., Lewington, R. 2012. *Field Guide to the Micro-Moths of Britain and Ireland.* Bloomsbury Publishing.

Streeter, D., Hart-Davies, C., Hardcastle, A., Cole, F., and Harper, L. 2016. *Collins Wild Flower Guide.* HarperCollins Publishers, London.

Talavera, G., Bataille, C., Benyamini, D., Gascoigne-Pees, M., and Vila, R. 2018. Round-trip across the Sahara: Afrotropical Painted Lady butterflies recolonize the Mediterranean in early spring. *Biology Letters* 14(6): 274.

Thomas, C. 1985. *Exploration of a Drowned Landscape: Archaeology and History of the Isles of Scilly.*

Thomas, J. 1980. Why did the Large Blue Become Extinct in Britain? *Oryx*, 15(3), 243–247. Available at: http://doi.org/10.1017/S0030605300024625 (accessed 17 April 2021).

Thomas, J. and Lewington, R. 2014. *The Butterflies of Britain & Ireland.* Bloomsbury Publishing, London.

Tollefson, J. 2019. One million species face extinction. *Nature* 569: 171. Available at: https://media.nature.com/original/magazine-assets/d41586-019-01448-4/d41586-019-01448-4.pdf (accessed 19 November 2020).

Tunmore, M. 2000. The 1999 Monarch *Danaus plexippus* (Linn.) influx into the British Isles. *Atropos* 9: 4.

UKBMS. Butterflies as indicators. Available at https://www.ukbms.org/indicators#:~:text=Butterflies%20are%20increasingly%20being%20recognised,and%20responses%20of%20other%20wildlife (accessed 22 November 2020).

Van Dyck, H., Bonte, D., Puls, R., Gotthard, K. and Maes, D. 2014. The lost generation hypothesis: could climate change drive ectotherms into a developmental trap? *Oikos* 124 (1): 54–61.

Wacher, J. 1998. Successful overwintering of Painted Lady *Cynthia cardui* in the UK. *Atropos* 5: 19–20.

Wacher, J.S. 2002. A Second Brood Silver-Studded Blue *Plebejus argus. Atropos* 15.

Wacher, J., Worth, J. and Spalding, A. 2003. *A Cornwall Butterfly Atlas.* Pisces Publications, with Cornwall Butterfly Conservation. Newbury, Berkshire.

Waring, P., Townsend, M. and Lewington, R. 2018. *Field Guide to the Moths of Great Britain and Ireland.* Bloomsbury Publishing, London.

Warren, M.S. 1987. The Ecology and Conservation of the Heath Fritillary Butterfly, *Mellicta athalia*. III Population Dynamics and the Effect of Habitat Management. *Journal of Applied Ecology* 24 (2): 499–513. Available at: http://doi.org/10.2307/2403889 (accessed 22 March 2021).

Wedell, N. 1996. Mate Quality Affects Reproductive Effort in a Paternally Investing Species. *The American Naturalist* 148 (6): 1075–1088.

Wider Countryside Butterfly Survey Team 2019. *Wider Countryside Butterfly Survey Annual Newsletter 2019 Season* https://www.ukbms.org/Downloads/WCBS 2019 final.pdf (accessed 25 November 2020).

Locations mentioned in the text

Listed here, in alphabetical order, are all the locations named in this book, along with their nearest town or city, grid reference and postcode. The inclusion of a location on the list does not imply that it is publicly accessible. The six-figure Ordnance Survey national grid reference (NGR) and postcode for each location have been determined as accurately as possible (using the website https://gridreferencefinder.com) and lead to the specific site of interest, leaving it to the reader to determine the best access point and place to park. **Please be aware that satnavs can be unreliable in Cornwall's rural areas when programmed with a postcode**. Use of the relevant 1:25,000 Ordnance Survey maps is recommended.

Locations that are identified as key sites for a range of butterflies are in **bold type**; details of these 20 butterfly hotspots are given in chapter 7 **Key sites for butterflies**. Also included in the list below are the '**Best places to see**' locations given at the end of each species account in chapter 4. A map of Cornwall and the Isles of Scilly showing principal towns and roads can be found at the beginning of the book.

Location name	Nearest town or city	NGR	Postcode approximation
Aire Point	St Just	SW358281	TR19 7BD
Antony House	Torpoint	SX416563	PL11 2QA
Ardensawah Cliff	St Just	SW360227	TR19 6JW
Bissoe Valley Nature Reserve	Truro	SW771413	TR4 8QZ
Bodinnick	Fowey	SX129521	PL23 1LX
Boscastle	Camelford	SX097906	PL35 0BA
Botallack	St Just	SW368328	TR19 7QG
Botallack Mine	St Just	SW363332	TR19 7QQ
Bowithick	Camelford	SX181829	PL15 7SH
Breney Common	Lostwithiel	SX056612	PL30 5DU
Brown Willy	Camelford	SX158799	PL32 9QG
Bunny's Hill	Bodmin	SX117675	PL30 4EB
Cabilla and Redrice Woods	Bodmin	SX131653	PL30 4BE
Caerloggas Downs	St Austell	SX018566	PL26 8TE
Cape Cornwall, Botallack and Pendeen	St Just	SW364332	TR19 7QQ
Carbis	St Austell	SW999601	PL26 8JZ
Carclew	Falmouth	SW788380	TR11 5UW
Cardinham Woods	Bodmin	SX098667	PL30 4AL
Carkeet	Liskeard	SX219733	PL14 6SD
Carn Brea Castle	Redruth	SW686408	TR16 6SL
Carne Beach	Truro	SW904382	TR2 5PF
Carnmenellis	Redruth	SW695365	TR16 6NE
Carnon Valley	Truro	SW776412	TR4 8RA
Carrick Du	St Ives	SW512409	TR26 1EJ
Chapel Porth	Redruth	SW697495	TR5 0NS
Cheesewring	Liskeard	SX257724	PL14 5AT
Chynhalls Cliff	Helston	SW780169	TR12 6SB
Cotehele Woods	Callington	SX423683	PL12 6TA
Coverack	Helston	SW782183	TR12 6TE
Crackington Haven	Bude	SX143968	EX23 0JZ
Crantock	Newquay	SW789603	TR8 5SB
Criggan Moors	St Austell	SX016611	PL26 8QX
Crousa Downs	Helston	SW763189	TR12 6SG
Crowan Beacon	Camborne	SW664350	TR14 9JG
Cubert Common and West Pentire	Newquay	SW776599	TR8 5SE
Cudden Point	Marazion	SW550278	No suitable postcode
Dannonchapel	Camelford	SX037824	PL30 3LP
Deer Park Wood, Luckett	Callington	SX376734	PL17 8LQ
De Lank Quarry	Bodmin	SX099750	PL30 4LH

Locations mentioned in the text *Butterflies of Cornwall*

Location name	Nearest town or city	NGR	Postcode approximation
Devoran	Falmouth	SW794392	TR3 6PT
Dizzard (Wood)	Bude	SX167991	EX23 0NX
Downgate	Callington	SX368728	PL17 8JT
Dunmere Wood	Bodmin	SX045687	PL30 3AU
Eden Project	St Austell	SX050546	PL24 2SG
Emblance Downs	Camelford	SX128772	PL30 4NL
Enys Zawn	St Just	SW377354	TR19 7ED
Erisey Barton (Goonhilly Downs)	Helston	SW713190	TR12 7LJ
Fellover Brake	Camelford	SX091774	PL30 4PJ
Garrow Tor	Bodmin	SX145785	PL30 4NL
Gaverigan (Goss Moor)	St Columb Major	SW935589	TR9 6HE
Gear Sands (Penhale)	Perranporth	SW773557	TR4 9PP
Gew-graze	Helston	SW675144	TR12 7PJ
Godolphin	Helston	SW601318	TR13 9RE
Godolphin and Godolphin Hill	Helston	SW595308	TR20 9RY
Godolphin Warren	Helston	SW595308	TR20 9RY
Godolphin Woods	Helston	SW604321	TR13 9RE
Godrevy	Hayle	SW584424	TR27 5ED
Godrevy Head / Point	Hayle	SW580432	TR27 5ED
Godrevy Towans / Warren (no public access)	Hayle	SW586424	TR27 5ED
Golitha Falls	Liskeard	SX223686	PL14 6RU
Goonhilly Downs	Helston	SW720196	TR12 6NT
Goss Moor, including St Dennis Junction	St Columb Major	SW946598	PL26 8BY
Great Pool, Bryher, Isles of Scilly	Hugh Town, St Mary's, Isles of Scilly	SV874148	TR23 0PR
Greena Moor	Bude	SX234963	EX22 6UN
Greenscoombe Wood	Callington	SX392728	PL17 8NL
Gwennap Head	St Just	SW367216	TR19 6JW
Gwithian	Hayle	SW585411	TR27 5BS
Gwithian Green	Hayle	SW587413	TR27 5BX
Gwithian Towans	Hayle	SW578407	TR27 5BT
Harpur's Downs	Camelford	SX115797	PL32 9PZ
Hawkstor Marsh	Liskeard	SX260762	PL15 7NL
Hayle Kimbro	Helston	SW696169	TR12 7LJ
Hayle to Gwithian Towans, including Gwithian Green	Hayle	SW579413	TR27 5BT
Hayle Towans	Hayle	SW551383	TR27 5AS
Hell's Mouth	Camborne	SW603428	TR27 5EE
Helman Tor	Lostwithiel	SX061615	PL30 5HP
Hensbarrow Beacon	St Austell	SW996575	PL26 8HZ
Hensbarrow Downs	St Austell	SW996579	PL26 8HZ
Herodsfoot	Liskeard	SX213604	PL14 4QX
Higher Predannack Cliff	Helston	SW661168	TR12 7EZ
Holywell Bay	Newquay	SW766595	TR8 5PQ
Hudder Down	Camborne	SW607430	TR27 5EE
Kelsey Head	Newquay	SW769606	TR8 5SE
Kenidjack	St Just	SW355325	TR19 7QW
Kennack Sands	Helston	SW734165	TR12 7LT
Kerrow	Camelford	SX109744	PL30 4LG
Keveral Wood	Looe	SX295558	PL11 3JL
Kiberick Cove	Truro	SW924379	TR2 5PH
King Arthur's Downs	Camelford	SX128781	PL30 4NL
King Arthur's Hall	Camelford	SX129776	PL30 4NL
Kit Hill	Callington	SX375714	PL17 8HS
Kynance / Kynance Cove	Helston	SW683134	TR12 7PJ
Land's End	St Just	SW342251	TR19 7AA
Lanhydrock	Bodmin	SX086637	PL30 5AD
Lelant Towans	St Ives	SW545382	TR26 3DZ
Levant	St Just	SW368345	TR19 7SX
Levant Mine	St Just	SW368345	TR19 7SX

255

Location name	Nearest town or city	NGR	Postcode approximation
Levant Zawn	St Just	SW364343	TR19 7SX
Little Shell Wood	Bodmin	SX082718	PL30 4QF
Lizard Downs	Helston	SW694139	TR12 7PJ
Lizard Point	Helston	SW695116	TR12 7NU
Loe Bar	Porthleven	SW643241	TR13 9EP
Long Cove and Will's Rock, Porthcothan	Padstow	SW853724	PL28 8PW
Longstone, St Mary's, Isles of Scilly	Hugh Town, St Mary's, Isles of Scilly	SV919112	TR21 0NT
Lower Predannack Downs	Helston	SW688149	TR12 7LH
Lowertown Moor	Lostwithiel	SX051614	PL30 5DU
Lowland Point	Helston	SW802196	TR12 6NY
Luckett	Callington	SX388737	PL17 8NJ
Luckett Wood	Callington	SX392728	PL17 8NL
Lundy Bay	Wadebridge	SW956798	PL27 6QZ
Luxulyan	St Austell	SX049580	PL30 5EB
Maer Lake	Bude	SS207075	EX23 8NN
Main Dale	Helston	SW782198	TR12 6NT
Marsland Mouth	Bude	SS212174	EX39 6HQ
Marsland Nature Reserve	Bude	SS231171	EX23 9PQ
Mexico Towans	Hayle	SW559388	TR27 5AX
Millendreath	Looe	SX268541	PL13 1NY
Millook	Bude	SX184999	EX23 0DQ
Minack Head / Point	Penzance	SW387220	TR19 6JT
Molinnis Downs	St Austell	SX025593	PL26 8QS
Morvah	St Just	SW401354	TR20 8YT
Mount / Pennans Field (Penhale)	Perranporth	SW782570	TR4 9PP
Mullion Cliff	Helston	SW666176	TR12 7ET
Mullion Cove to Lizard Point	Helston	SW668179 / SW701115	TR12 7EG / TR12 7NU
Murrayton	Looe	SX284544	PL13 1NZ
Nanstallon	Bodmin	SX037669	PL30 5GD
Nare Head	Truro	SW916371	TR2 5PH
Newlyn Downs	Newquay	SW834543	TR8 5AU
North Corner	Helston	SW781188	TR12 6TQ
Penberth Cove	Penzance	SW402226	TR19 6HJ
Pendeen	St Just	SW382343	TR19 7DN
Pendeen Watch	St Just	SW378358	TR19 7ED
Pendennis	Falmouth	SW824318	TR11 4LP
Pendrift Bottom	Bodmin	SX096748	PL30 4LH
Pendrift Downs	Bodmin	SX102746	PL30 4JT
Pen Enys Point	St Ives	SW490410	TR26 3AE
Penhale and Gear Sands	Perranporth	SW767566	TR4 9PP
Penhale Dunes (Sands)	Perranporth	SW469569	TR4 9PP
Penkestle Moor	Liskeard	SX174703	PL14 6QA
Penlee Battery Nature Reserve	Torpoint	SX438491	PL10 1LG
Penlee Point and Rame Head	Torpoint	SX436491 / SX420487	PL10 1LG / PL10 1LG
Penrose Woods	Helston	SW640257	TR13 0RB
Pentire	Wadebridge	SW791614	TR7 1PJ
Pentire Point to Port Quin	Wadebridge	SW941800	PL27 6UA
Pentreath (Kynance)	Helston	SW692128	TR12 7PJ
Perranuthnoe	Marazion	SW538293	TR20 9NH
Phoenix Mine, Bodmin Moor	Liskeard	SX265723	PL14 5LH
Poldice Mine	Redruth	SW743429	TR16 5PT
Poldice Valley	Redruth	SW739430	TR16 5QA
Poldhu	Helston	SW665199	TR12 7JB
Polly Joke (Porth Joke)	Newquay	SW770605	TR8 5SE
Polperro	Looe	SX208509	PL13 2QX
Polruan	Fowey	SX126507	PL23 1PS
Poltesco	Helston	SW725157	TR12 7LR
Polyphant	Launceston	SX262820	PL15 7PS

Locations mentioned in the text

Location name	Nearest town or city	NGR	Postcode approximation
Ponts Mill	St Austell	SX073562	PL24 2RR
Pordenack Point	St Just	SW346242	TR19 7AF
Porkellis Moor	Helston	SW687325	TR13 0JT
Porthcothan	Padstow	SW859721	PL28 8LW
Porthgwarra to Pordenack Point	St Just	SW370217	TR19 6JP
Porth Joke	Newquay	SW773602	TR8 5SE
Portreath	Redruth	SW657453	TR16 4NR
Predannack Airfield (no public access)	Helston	SW685162	TR12 7LJ
Predannack Cliff	Helston	SW661164	TR12 7EZ
Predannack Head	Helston	SW660163	TR12 7EZ
Prussia Cove	Marazion	SW556280	TR20 9BA
Rame Head	Torpoint	SX418482	PL10 1LG
Red Moor	Lostwithiel	SX068615	PL30 5AN
Retire Common	Bodmin	SX003632	PL26 8LL
Rinsey Cove	Porthleven	SW594271	TR13 9TR
Roche Rock	St Austell	SW991596	PL26 8FJ
Rock	Padstow	SW942761	PL27 6NW
Rock Dunes	Padstow	SW927764	PL27 6LL
Rosenannon Downs	St Columb Major	SW954670	PL30 5PW
Roskestal and Ardensawah Cliff	St Just	SW361224	TR19 6JW
Rough Tor	Camelford	SX144807	PL32 9QG
Ruan Lanihorne	Truro	SW894419	TR2 5NZ
St Allen	Truro	SW824505	TR4 9QX
St Breward	Camelford	SW095764	PL30 4LL
St Dennis Junction (Goss Moor)	St Columb Major	SW936597	TR9 6HE
St Gothian Sands	Hayle	SW583418	TR27 5BX
St Keverne (Dean Quarries)	Helston	SW800205	TR12 6NY
St Michael's Mount	Marazion	SW514298	TR17 0EN
St Neot	Liskeard	SX184678	PL14 6NG
St Piran's Oratory	Perranporth	SW768563	TR4 9PP
Seaton	Looe	SX304546	PL11 3JF
Seaton Sea Wall	Looe	SX307542	PL11 3JB
Seaton Valley Countryside Park	Looe	SX303549	PL11 3JD
Stithians	Redruth	SW733368	TR3 7BH
Stithians Reservoir	Redruth	SW713363	TR3 7AS
Struddicks (near Murrayton)	Looe	SX289544	PL13 1PA
The Rumps	Padstow	SW933811	PL27 6UQ
Tideford	Saltash	SX347597	PL12 5HW
Tidna Valley	Bude	SS197148	EX23 9SR
Tintagel	Camelford	SX060884	PL34 0AD
Treburley	Callington	SX342772	PL15 9PN
Tregoss Moor	St Columb Major	SW971595	PL26 8NA
Treluggan Cliffs	Truro	SW887374	TR2 5EN
Trenoweth, St Mary's, Isles of Scilly	Hugh Town, St Mary's, Isles of Scilly	SV918123	TR21 0NT
Tresco Abbey Gardens, Tresco, Isles of Scilly	Hugh Town, St Mary's, Isles of Scilly	SV893142	TR24 0QQ
Treskilling Downs	St Austell	SX034577	PL26 8RN
Trevaunance to Chapel Porth	Perranporth	SW699500	TR5 0NS
Trevollard	Saltash	SX386583	PL12 4RX
Trewavas Head	Porthleven	SW596265	TR13 9TR
Trewint	Launceston	SX219805	PL15 7TG
Upton Towans	Hayle	SW576397	TR27 5BZ
Valency Valley	Camelford	SX103913	PL35 0HE
Watch House Field	Torpoint	SX438491	PL10 1LG
Wheal Busy	Redruth	SW740447	TR4 8NZ
Wheal Peevor	Redruth	SW708442	TR16 4AT
Windmill Farm	Helston	SW693152	TR12 7LH
Zelah	Perranporth	SW811518	TR4 9HS
Zennor	St Ives	SW454384	TR26 3BY

Larval foodplants found in Cornwall

In the complex interaction of factors on which the ecology of a butterfly depends, one of the most important is the plant on which the caterpillar feeds, the larval foodplant. Each species of butterfly requires its own larval foodplants, which are listed in the species accounts in chapter 4. The adult butterfly is an accomplished botanist, and is adept at laying her eggs on or near the right foodplant, so that the young caterpillars hatch close to a ready-made meal.

In the list below, you will find all the butterfly larval foodplants found in Cornwall, which are mentioned in chapter 4. (Many of these foodplants are rare or do not grow at all on the Isles of Scilly, as described in chapter 6.) The foodplants are listed by common name, in alphabetical order, followed by the scientific name of the plant and the common name of the butterfly species whose caterpillars feed on that plant. Some plants are used by only one species of butterfly, others by several.

Although the list includes plants on which the caterpillars of particular butterflies are known to feed in Cornwall, the list cannot claim to be exhaustive, insofar as alternative foodplants may yet be discovered in the County. For many butterflies there are one or more favoured larval foodplants, and these are indicated by the name of the butterfly species for those plants appearing in **bold** type.

Larval foodplants known to be used elsewhere in England, but which are absent from Cornwall, are omitted. The list includes wild, native species, as well as introduced and cultivated plants found in the County.

Agrimony	*Agrimonia eupatoria*	**Grizzled Skipper**
Avens, Wood	*Geum urbanum*	Grizzled Skipper
Bents	*Agrostis* spp.	**Small Heath**, Gatekeeper
Bent, Black	*Agrostis gigantea*	**Wall**
Bent, Bristle	*Agrostis curtisii*	Grayling
Bent, Common	*Agrostis capillaris*	**Wall**
Bent, Creeping	*Agrostis stolonifera*	Ringlet
Bilberry	*Vaccinium myrtillus*	**Green Hairstreak**
Bird's-foot-trefoil, Common	*Lotus corniculatus*	**Dingy Skipper**, Clouded Yellow, **Green Hairstreak**, **Silver-studded Blue**, **Common Blue**
Bird's-foot-trefoil, Greater	*Lotus pedunculatus*	Dingy Skipper, Common Blue
Bramble	*Rubus fruticosus* agg.	**Grizzled Skipper**, **Green Hairstreak**, Holly Blue
Brome, False	*Brachypodium sylvaticum*	**Wall**, **Speckled Wood**, **Ringlet**, **Meadow Brown**
Broom	*Cytisus scoparius*	**Green Hairstreak**
Buckthorn	*Rhamnus cathartica*	Brimstone
Buckthorn, Alder	*Frangula alnus*	**Brimstone**, **Green Hairstreak**
Burdocks	*Arctium* spp.	Painted Lady
Burnet, Salad	*Poterium sanguisorba* subsp. *sanguisorba*	**Grizzled Skipper**
Cabbages, cultivated	Brassicales	**Large White**, **Small White**
Cabbage, Wild	*Brassica oleracea* var. *oleracea*	**Small White**
Celandine, Lesser	*Ficaria verna*	Heath Fritillary
Charlock	*Sinapis arvensis*	**Small White**, **Green-veined White**
Cinquefoil, Creeping	*Potentilla reptans*	**Grizzled Skipper**
Clovers	*Trifolium* spp.	**Clouded Yellow**
Clover, White	*Trifolium repens*	Common Blue
Cock's-foot	*Dactylis glomerata*	**Large Skipper**, **Wall**, **Speckled Wood**, **Ringlet**, **Meadow Brown**, Marbled White
Cotoneasters	*Cotoneaster* spp.	Holly Blue
Couch, Common	*Elymus repens*	**Speckled Wood**, Gatekeeper
Cow-wheat, Common	*Melampyrum pratense*	Heath Fritillary
Crane's-bill, Dove's-foot	*Geranium molle*	**Brown Argus**
Cuckooflower	*Cardamine pratensis*	**Orange-tip**, **Green-veined White**
Currants	*Ribes* spp.	Comma

Larval foodplants found in Cornwall

Common name	Scientific name	Butterfly species
Dock, Broad-leaved	*Rumex obtusifolius*	Small Copper
Dog-rose	*Rosa canina* agg.	Grizzled Skipper
Dog-violet, Common	*Viola riviniana*	**Pearl-bordered Fritillary, Small Pearl-bordered Fritillary, Silver-washed Fritillary, Dark Green Fritillary**
Dogwoods	*Cornus* spp.	Holly Blue
Dogwood	*Cornus sanguinea*	**Green Hairstreak**
Elms	*Ulmus* spp.	Comma, White-letter Hairstreak
Elm, English	*Ulmus procera*	**White-letter Hairstreak**
Elm, Wych	*Ulmus glabra*	**White-letter Hairstreak**
Fescues	*Festuca* spp.	**Small Heath, Meadow Brown, Gatekeeper**
Fescue, Red	*Festuca rubra* agg.	**Gatekeeper, Marbled White, Grayling**
Foxglove	*Digitalis purpurea*	Heath Fritillary
Gorses	*Ulex* spp.	**Green Hairstreak**, Holly Blue, Silver-studded Blue
Greenweed, Dyer's	*Genista tinctoria*	**Green Hairstreak**
Hair-grass, Early	*Aira praecox*	Grayling
Hair-grass, Tufted	*Deschampsia cespitosa*	Ringlet, Grayling
Hair-grass, Wavy	*Avenella flexuosa*	**Wall**
Hazel	*Corylus avellana*	Comma
Heath, Cornish	*Erica vagans*	**Silver-studded Blue**
Heath, Cross-leaved	*Erica tetralix*	**Green Hairstreak, Silver-studded Blue**
Heather (Ling)	*Calluna vulgaris*	**Green Hairstreak, Silver-studded Blue**
Heather, Bell	*Erica cinerea*	**Silver-studded Blue**
Holly	*Ilex aquifolium*	**Holly Blue**
Honesty	*Lunaria annua*	**Orange-tip, Large White**
Hop	*Humulus lupulus*	Red Admiral, Peacock, Comma
Horse-radish	*Armoracia rusticana*	Large White
Ivy, Atlantic	*Hedera hibernica*	**Holly Blue**
Jack-by-the-Hedge	*Alliaria petiolata*	**Orange-tip, Small White, Green-veined White**
Ling (Heather)	*Calluna vulgaris*	**Green Hairstreak, Silver-studded Blue**
Lucerne	*Medicago sativa* subsp. *sativa*	Clouded Yellow
Mallows	*Malva* spp.	Painted Lady
Marram	*Ammophila arenaria*	Grayling
Meadow-grasses	*Poa* spp.	**Small Heath, Meadow Brown**, Gatekeeper
Meadow-grass, Annual	*Poa annua*	**Wall**
Medick, Black	*Medicago lupulina*	Common Blue
Mignonette, Wild	*Reseda lutea*	Large White
Mustard, Garlic	*Alliaria petiolata*	**Orange-tip, Small White, Green-veined White**
Mustard, Hedge	*Sisymbrium officinale*	**Orange-tip, Small White, Green-veined White**
Nasturtium	*Tropaeolum majus*	**Large White, Small White**
Nettles	*Urtica* spp.	Painted Lady
Nettle, Common (Stinging)	*Urtica dioica*	**Red Admiral, Peacock, Small Tortoiseshell, Comma**
Nettle, Small	*Urtica urens*	Red Admiral, Peacock, **Small Tortoiseshell**
Oak, Pedunculate	*Quercus robur*	**Purple Hairstreak**
Oak, Sessile	*Quercus petraea*	**Purple Hairstreak**
Oak, Turkey	*Quercus cerris*	Purple Hairstreak
Pellitory-of-the-wall	*Parietaria judaica*	Red Admiral
Plantain, Ribwort	*Plantago lanceolata*	**Heath Fritillary**
Restharrow, Common	*Ononis repens*	**Silver-studded Blue**
Rock-cress, Hairy	*Arabis hirsuta*	Green-veined White
Sage, Wood	*Teucrium scorodonia*	Heath Fritillary
Scabious, Devil's-bit	*Succisa pratensis*	**Marsh Fritillary**
Sea-kale	*Crambe maritima*	Large White
Sheep's-fescue	*Festuca ovina*	Small Heath, **Ringlet, Gatekeeper**, Marbled White, Grayling
Silverweed	*Potentilla anserina*	**Grizzled Skipper**
Smock, Lady's	*Cardamine pratensis*	**Orange-tip, Green-veined White**
Snowberries	*Symphoricarpos* spp.	Holly Blue
Sorrel, Common	*Rumex acetosa*	**Small Copper**
Sorrel, Sheep's	*Rumex acetosella*	**Small Copper**
Speedwell, Germander	*Veronica chamaedrys*	**Heath Fritillary**

Speedwell, Ivy-leaved	*Veronica hederifolia*	Heath Fritillary
Spindle	*Euonymus europaeus*	Holly Blue
Stock, Virginia	*Malcolmia maritima*	**Large White**
Stork's-bill, Common	*Erodium cicutarium*	**Brown Argus**
Strawberry, Barren	*Potentilla sterilis*	**Grizzled Skipper**
Strawberry, Wild	*Fragaria vesca*	**Grizzled Skipper**
Thistles		**Painted Lady**
Thistle, Creeping	*Cirsium arvense*	**Painted Lady**
Thistle, Marsh	*Cirsium palustre*	**Painted Lady**
Thistle, Spear	*Cirsium vulgare*	**Painted Lady**
Timothy	*Phleum pratense*	Marbled White
Tormentil	*Potentilla erecta*	Grizzled Skipper
Trefoil, Lesser	*Trifolium dubium*	Common Blue
Vetches	*Vicia* spp.	Clouded Yellow
Viper's-bugloss	*Echium vulgare*	Painted Lady
Violet, Marsh	*Viola palustris* subsp. *juressi*	**Small Pearl-bordered Fritillary**
Wallflower	*Erysimum cheiri*	**Large White**, **Small White**
Water-cress	*Nasturtium officinale*	**Green-veined White**
Whin, Petty	*Genista anglica*	**Green Hairstreak**
Willows	*Salix* spp.	Comma
Yarrow	*Achillea millefolium*	Heath Fritillary
Yorkshire-fog	*Holcus lanatus*	**Small Skipper**, **Wall**, **Speckled Wood**, Marbled White

Brimstone larva on Alder Buckthorn

Index of butterflies

This index lists in alphabetical order every butterfly mentioned in the book. Each butterfly has an entry under its common English name and its scientific name. Each entry includes all the pages on which the butterfly appears. The page number range in **bold** refers to the detailed species account in chapter 4.

Admiral, Red 29, **140–143**, 144, 145, 147, 152, 217, 218, 259
Admiral, White 208, 221
Aglais io 16, **148–151**, 215, 218
Aglais urticae 16, 144, **152–155**, 211, 217, 218
Anthocharis cardamines 29, **66–69**, 210, 216, 220
Apatura iris 35, 208
Aphantopus hyperantus 16, **102–105**, 217, 218
Apollo 213
Argus, Brown 5, 19, 26, 193, **198–201**, 203, 216, 228, 233, 239, 258, 260
Argynnis pandora 213
Argynnis paphia 20, **132–135**, 219, 220
Aricia agestis 19, 193, **198–201**, 216

Blue, Chalk Hill 213
Blue, Common 16, 19, 188, 193, 198, 199, **202–205**, 215, 216, 217, 218, 258, 259, 260
Blue, Holly 8, 16, 24, 26, 29, **188–191**, 202, 203, 212, 216, 217, 218, 219, 258, 259, 260
Blue, Large 7, 209, 246, 248
Blue, Long-tailed 211, 212, 216, 220
Blue, Short-tailed 212
Blue, Silver-studded 3, 4, 5, 7, 10, 16, 19, 23, 25, 26, 27, 28, 32, 35, 38, 39, 40, 42, 46, 188, **192–197**, 198, 203, 205, 221, 227, 228, 233, 235, 236, 239, 242, 243, 258, 259
Blue, Small 208
Boloria euphrosyne 16, 39, **122–127**
Boloria selene 16, 38, 122, **128–131**
Brimstone 1, 18, 21, 26, 29, 47, 82, **86–89**, 215, 216, 219, 220, 238, 258, 260
Brown, Hedge 112
Brown, Meadow 16, 29, 38, 98, 102, **106–109**, 110, 120, 215, 217, 218, 258, 259
Brown, Meadow subsp. *cassitertidum* 29, 108, 215, 217, 218

Cacyreus marshalli 212
Callophrys rubi 18, **180–183**
Camberwell Beauty 211, 220
Celastrina argiolus 16, **188–191**, 212, 216, 218
Celastrina argiolus britanna 188
Coenonympha pamphilus 16, 38, 47, **98–101**, 215, 221
Colias alfacariensis 83, 210
Colias croceus 29, **82–85**, 210, 217, 218
Colias hyale 83, 210, 221
Comma 16, 24, 29, 90, **156–159**, 215, 217, 218, 219, 258, 259, 260
Comma *hutchinsoni* form 156, 157, 159
Copper, Small 19, 24, 25, 28, 32, **172–175**, 216, 217, 218, 259
Copper, Small aberration *schmidtii* 217, 218
Copper, Small aberration *caeruleopunctata* 172, 174
Cupido argiades 212
Cupido minimus 208

Danaus plexippus 210, 215, 220
Duke of Burgundy 213

Erynnis tages 18, 39, **50–53**
Euphydryas aurinia 18, 31, 130, **160–167**, 216

Fabriciana adippe 39, 123, 207
Favonius quercus 18, **176–179**
Fritillary, Dark Green 18, 19, 23, 26, 28, 29, 132, **136–139**, 207, 220, 225, 227, 228, 233, 234, 235, 236, 239, 240, 241, 259
Fritillary, Heath 4, 5, 8, 19, 37, 39, 40, **168–171**, 232, 258, 259, 260
Fritillary, High Brown 39, 123, 136, 207
Fritillary, Marsh 4, 5, 6, 7, 8, 10, 18, 20, 21, 22, 23, 25, 27, 31, 39, 40, 41, 43, 130, **160–167**, 216, 225, 229, 231, 235, 238, 259
Fritillary, Mediterranean 213
Fritillary, Pearl-bordered 3, 4, 8, 16, 18, 21, 26, 39, 41, 43, 45, **122–127**, 128, 129, 130, 234, 238, 259
Fritillary, Queen of Spain 29, 211, 217, 219, 220
Fritillary, Silver-washed 20, 21, 26, 27, 29, **132–135**, 136, 137, 219, 220, 226, 230, 232, 234, 236, 237, 238, 244, 259
Fritillary, Silver-washed aberration *valezina* 132, 134
Fritillary, Small Pearl-bordered 4, 5, 8, 16, 19, 21, 23, 24, 25, 27, 28, 38, 39, 42, 43, 122, 123, 124, **128–131**, 137, 138, 225, 226, 227, 229, 231, 233, 234, 235, 238, 240, 241, 243, 259, 260

Gatekeeper 3, 16, 29, 38, 106, 107, **110–113**, 120, 215, 216, 217, 218, 219, 220, 258, 259
Geranium Bronze 212
Gonepteryx rhamni 18, 47, **86–89**, 215, 220
Grayling 4, 10, 18, 19, 21, 23, 25, 26, 27, 28, 32, 35, 39, 42, 46, 47, 98, **118–121**, 216, 227, 235, 237, 239, 241, 242, 243, 258, 259

Hairstreak, Brown 11, 208
Hairstreak, Green 18, 19, 23, 25, 28, 32, 36, 49, **180–183**, 227, 229, 236, 243, 258, 259, 260
Hairstreak, Purple 18, 19, 21, 27, 35, **176–179**, 185, 226, 230, 232, 234, 259
Hairstreak, White-letter 4, 19, **184–187**, 259
Hamearis lucina 213
Heath, Small 4, 16, 18, 19, 21, 23, 24, 25, 27, 38, 42, 47, **98–101**, 215, 216, 217, 221, 225, 227, 228, 229, 231, 233, 235, 236, 237, 239, 240, 241, 242, 243, 258, 259
Hipparchia semele 18, 32, 46, **118–121**, 216
Hipparchia semele semele 119

Iphiclides podalirius 213
Issoria lathonia 29, 211, 217, 220

Lady, American Painted 211, 220
Lady, Painted 141, **144–147**, 152, 211, 217, 218, 258, 259, 260
Lampides boeticus 211, 216, 220
Lasiommata megera 15, 38, **90–93**, 215, 220
Leptidea sinapis 207
Limenitis camilla 208, 221

261

Lycaena phlaeas 19, **172–175**, 216, 218

Maniola jurtina 16, 38, 98, **106–109**
Maniola jurtina cassiteridum 29, 108, 215, 217, 218
Maniola jurtina insularis 108
Melanargia galathea 18, **114–117**, 216
Melitaea athalia 19, 37, **168–171**
Monarch 11, 210, 215, 219, 220
Nymphalis antiopa 211, 220
Nymphalis polychloros 211, 215, 220

Ochlodes sylvanus 16, 38, 58, **62–65**
Orange-tip 2, 3, 29, **66–69**, 74, 78, 81, 210, 216, 220, 258

Papilio machaon britannicus 210
Papilio machaon gorganus 209
Pararge aegeria 16, **94–97**
Pararge aegeria insula 29, 94, 215, 217, 218
Pararge aegeria tircis 29, 94
Parnassius apollo 213
Peacock 16, 24, 29, 35, **148–151**, 215, 217, 218, 219, 259
Phengaris arion 209, 246
Pieris brassicae 16, **70–73**, 217, 218
Pieris napi 16, 74, **78–81**, 217, 218
Pieris napi sabellicae 79
Pieris rapae 16, 70, **74–77**, 217, 218
Plebejus argus 16, 38, 46, **192–197**, 221
Plebejus argus caernensis 193
Plebejus argus cretaceus 193
Polygonia c-album 16 ,90, **156–159**, 215, 218
Polyommatus coridon 213
Polyommatus icarus 16, 188, **202–205**, 215, 218
Pontia daplidice 210, 221
Pontia edusa 221
Purple Emperor 35, 208
Pyrgus malvae 19, 39, 46, **54–57**
Pyronia tithonus 16, 38, 106, **110–113**, 215, 220

Ringlet 8, 9, 16, 18, 24, 29, **102–105**, 107, 217, 218, 219, 258, 259

Satyrium w-album 19, **184–187**
Skipper, Dingy 4, 7, 18, 19, 23, 24, 26, 27, 39, 40, 42, 45, **50–53**, 54, 55, 56, 231, 239, 244, 258
Skipper, Essex 58, 59, 206

Skipper, Grizzled 4, 10, 19, 23, 39, 40, 41, 43, 44, 45, 46, 50, **54–57**, 239, 258, 259, 260
Skipper, Grizzled aberration *taras* 55
Skipper, Large 16, 23, 24, 38, 58, **62–65**, 258
Skipper, Lulworth 212
Skipper, Small 16, 23, 24, 38, **58–61**, 62, 206, 260
Speckled Wood 8, 16, 18, 21, 24, 27, 29, **94–97**, 215, 217, 218, 258, 260
Speckled Wood subsp. *insula* 29, 94, 215, 217, 218
Speckled Wood subsp. *turcis* 29, 94
Speyeria aglaja 18, 132, **136–139**, 207, 220
Swallowtail 11, 206, 209, 220
Swallowtail, Scarce 213

Thecla betulae 208
Thymelicus acteon 212
Thymelicus lineola 58, 206
Thymelicus sylvestris 16, 38, **58–61**, 206
Tortoiseshell, Large 211, 215, 220
Tortoiseshell, Small 3, 16, 24, 29, 35, 41, 144, 151, **152–155**, 156, 211, 217, 218, 259

Vanessa atalanta 29, **140–143**, 217, 218
Vanessa cardui 141, **144–147**, 211, 217, 218
Vanessa virginiensis 211, 220

Wall 4, 15, 16, 21, 24, 25, 28, 29, 38, 42, **90–93**, 215, 216, 217, 219, 220, 225, 227, 228, 233, 235, 236, 237, 239, 240, 241, 243, 244, 258, 259
White, Bath 210, 221
White, Eastern Bath 221
White, Green-veined 16, 29, 74, 75, 76, **78–81**, 217, 218, 219, 258, 259, 260
White, Large 16, 29, **70–73**, 74, 75, 76, 77, 80, 86, 217, 218, 258, 259, 260
White, Marbled 9, 18, 20, 21, **114–117**, 216, 217, 232, 234, 240, 244, 258, 259, 260
White, Small 16, 29, 70, 71, 72, **74–77**, 78, 80, 217, 218, 258, 259, 260
White, Wood 207

Yellow, Berger's Clouded 83, 210
Yellow, Clouded 5, 29, **82–85**, 86, 210, 217, 218, 258, 259, 260
Yellow, Clouded *helice* form 82, 83, 84, 221
Yellow, Pale Clouded 83, 210, 221